Handbook of Forest Nutrient Cycling

Handbook of Forest Nutrient Cycling

Edited by Austin Campbell

STATES
ACADEMIC PRESS
www.statesacademicpress.com

States Academic Press,
109 South 5th Street,
Brooklyn, NY 11249, USA

Visit us on the World Wide Web at:
www.statesacademicpress.com

ISBN: 978-1-63989-756-8

Cataloging-in-Publication Data

Handbook of forest nutrient cycling / edited by Austin Campbell.
p. cm.
Includes bibliographical references and index.
ISBN 978-1-63989-756-8
1. Forest plants--Nutrition. 2. Plant nutrients. 3. Biogeochemical cycles.
4. Plant nutrients. 5. Trees--Nutrition. I. Campbell, Austin.

QK867 .H36 2023
581.13--dc23

Table of Contents

Preface

I am honored to present to you this unique book which encompasses the most up-to-date data in the field. I was extremely pleased to get this opportunity of editing the work of experts from across the globe. I have also written papers in this field and researched the various aspects revolving around the progress of the discipline. I have tried to unify my knowledge along with that of stalwarts from every corner of the world, to produce a text which not only benefits the readers but also facilitates the growth of the field.

Nutrient cycling is a cyclic process that includes the use, transfer and recycling of nutrients in the ecosystem. It involves the transfer of nutrients from the physical environment to living organisms and back to the physical environment. Nutrient cycling involves both biotic and abiotic components. Abiotic components include air, water and soil, whereas trees, animals and plants are biotic components. Gaseous cycling and sedimentary cycling are the two types of nutrient cycling. Gaseous nutrients such as carbon, hydrogen, nitrogen and oxygen are recycled in water, air and soil. Contrarily, nutrients such as calcium, phosphorus, and potassium are recycled mostly in soil. Plants in forest ecosystems require these nutrients for their metabolism, development and reproduction. Nutrient cycle help ecosystems function and restore ecological equilibrium by sustaining species, and genetic and ecosystem variety. This book unravels the recent studies on nutrient cycling. It aims to serve as a resource guide for students and experts alike and contribute to the growth of research on this topic.

Finally, I would like to thank all the contributing authors for their valuable time and contributions. This book would not have been possible without their efforts. I would also like to thank my friends and family for their constant support.

Editor

1

Contribution of Harvest Residues to Nutrient Cycling in a Tropical *Acacia mangium* Willd. Plantation

Nguyen Van Bich [1,2,*], Alieta Eyles [1], Daniel Mendham [3], Tran Lam Dong [2], David Ratkowsky [1], Katherine J. Evans [1], Vo Dai Hai [2], Hoang Van Thanh [2], Nguyen Van Thinh [2] and Caroline Mohammed [1]

1 Tasmanian Institute of Agriculture (TIA), University of Tasmania, Private Bag 98, Hobart, Tasmania 7001, Australia; alieta.eyles@utas.edu.au (A.E.); d.ratkowsky@utas.edu.au (D.R.); katherine.evans@utas.edu.au (K.J.E.); Caroline.Mohammed@utas.edu.au (C.M.)
2 Silviculture Research Institute (SRI), Vietnamese Academy of Forest Sciences (VAFS), 46 Duc Thang, Bac Tu Liem, Hanoi 11910, Vietnam; dong.tran@vafs.gov.vn (T.L.D.); haivodai@gmail.com (V.D.H.); hoangvanthanhln@gmail.com (H.V.T.); nguyenthinhfsiv@gmail.com (N.V.T.)
3 CSIRO Land and Water, Private Bag 12, Hobart, Tasmania 7001, Australia; Daniel.Mendham@csiro.au
* Correspondence: NguyenvanBich@vafs.gov.vn

Abstract: Harvest residues can play a crucial role in conserving nutrients for recycling in forests, but little is known about the rates of decomposition and nutrient release from these residues following logging in tropical acacia plantations. In this study, we examined the biomass and nutrient content of harvest residue components (bark, leaves, and branches) using the litterbag technique for a 1.5-year-period following harvest of a seven-year-old *Acacia mangium* plantation in Northern Vietnam. At harvest, the total dry biomass of harvest residues was 18 t ha^{-1} comprising bark (8.9 t ha^{-1}), branches (6.6 t ha^{-1}), and leaves (2.5 t ha^{-1}). The retained bark on site conserved 51% N, 29% P, 32% K, 64% Ca, and 24% Mg content from harvest residues for recycling. Decomposition rate of the leaves was the most rapid (k = 1.47 $year^{-1}$; $t_{0.5}$ = 0.47 year), then branches (k = 0.54 $year^{-1}$; $t_{0.5}$ = 1.29 year), and bark (k = 0.22 $year^{-1}$; $t_{0.5}$ = 3.09 year). During decomposition, the loss of nutrients from harvest residues was K ≈ Ca > N > P > Mg. Decomposition of harvest residues and the associated rate of nutrient release can potentially supply a significant amount of nutrients required for stand development in the next rotation.

Keywords: nutrient loss; mass loss; nutrient release; nutrient dynamics; decay constant; residue half-life; nutrient cycling; tropical plantation

---

1. Introduction

In Vietnam, plantations are typically clear-felled at the end of the rotation. The stemwood that is down to 3 cm in diameter (over bark) is exported from the site as a harvest product [1,2]. Bark is often removed from the site with the commercial logs and sold for the production of charcoal, tannin, and garden compost [3–5], or is stripped after harvesting at the edge of the site or in a nearby wood yard, and not distributed over the logging area. The non-commercial logs and branches may be collected by locals for firewood, and the site subsequently burnt [6]. There are concerns that these practices may degrade the productive potential of the site over successive rotations [6–9]. While burning can initially improve soil fertility, it can also lead to large losses of N via volatilisation [10–13], and P, Ca, and K through leaching, as well as water and wind erosion [13–15]. Retention of harvest residues, especially bark, acts to conserve nutrients and leads to their controlled release in a way that minimizes losses from leaching and potentially supplies the amount of nutrients required for stand development in the next rotation [16–20]. Hence, understanding decomposition rates and nutrient release from harvest residues, including bark, is important for informing residue and nutrient management over successive rotations.

Management of residues after harvesting has been shown to improve site fertility [6,13,21–23]. For example, retention of harvest residues, including bark, branches, and leaves of a *Eucalyptus grandis* Hill ex Maiden plantation in Brazil, resulted in soil organic carbon (TC) and soil nitrogen (TN) (0–5 cm soil depth) that were 33% and 43% higher, respectively, seven years after all harvest residues were removed [13]. Similarly, TC and TN (0–10 cm soil depth) increased by 17% and 11%, respectively, in a two-year-old *A. mangium* plantation in Sumatra, Indonesia, when harvest residues were retained rather than removed [21]. This is because decomposing harvest residues can provide a significant source of nutrients to the soil [16,17,23]. For example, harvest residues potentially contributed 176 kg ha^{-1} of N, 15 kg ha^{-1} of P and 276 kg ha^{-1} of K to soil fertility during the first year following harvest in a *Eucalyptus globulus* Labill. plantation in Western Australia [16], and 176, 20, 375, 460, and 92 kg ha^{-1} of N, P, K, Ca, and Mg, respectively, during the first two years following harvest in a *Eucalyptus dunnii* Maiden plantation in Uruguay [17]. Retention of harvest residues, therefore, can conserve nutrients for recycling.

Fast-growing short-rotation plantation species generally require large amount of nutrients to optimize yield [24]. The amount of nutrient uptake by tropical acacia trees is generally highest during the first three years after planting [1,21]; approximately 180, 7, 80, 90, and 23 kg ha^{-1} $year^{-1}$ of N, P, K, Ca, and Mg were respectively accumulated in the above-ground stand biomass of *A. mangium* at two years old [21]. Although the demand for P appears to be low in Vietnam, plantation soils are characteristically very low in soil P [1,25,26] and most forest growers can only afford to apply a small dose of fertilizer at planting [6,26,27]. As deficiency in P could be a concern in successive rotations of acacia [1,21], retention of harvest residues may provide a sufficient source of P as well as other nutrients which, given the limited soil nutrient pool, could support a demand that results in commercial rates of growth.

Harvesting of short-rotation plantations can be associated with the export of large quantities of organic matter and nutrients from the site [1,22,24]. For example, removal of stemwood without bark from a six-year-old *Acacia auriculiformis* A. Cunn. ex Benth. plantation in Southern Vietnam, led to the export of 135.2, 47.3, 115.3, and 15.7 kg ha^{-1} of N, P, K, and Ca, respectively [1]. Retention of bark on-site is potentially important, with the quantities of nutrients extracted by harvesting increasing to 256.5, 55.7, 155.6, and 41.1 kg ha^{-1} of N, P, K, and Ca, respectively, in the same plantation if stemwood with bark had been removed. Similarly, in a 10-year-old *A. mangium* plantation in Sumatra, Indonesia, harvesting stemwood with, rather than without bark, would have increased the export of N, P, K, Ca, and Mg by 55%, 15%, 52%, 97%, and 48%, respectively [21]. Although the bark of acacia trees represents only 9–10% of the total aboveground stand biomass (AGB) when trees are harvested, it contributed 19–24% of N, 7–11% of P, 15–17% of K, 30–36% of Ca, and 7–15% of Mg in the above studies [1,21]. As debarking on site at harvesting can substantially reduce the export of nutrients [1,21], the retention of bark and its even distribution across the site may potentially reduce the costs of fertilizer application in the next rotation.

Given that decomposition and nutrient release from plant materials are largely influenced by environment [28] and substrate quality [29–31], materials of lower nutrient content generally decompose more slowly, immobilize more nutrients during decomposition, and have slower rates of net mineralization than nutrient-rich materials [32]. Acacias are a N-fixing species, and therefore the concentration of N in a given material is generally higher in acacias than in other genera, such as eucalypts [33,34]. This may lead to different activities of microbial decomposers, and thus rates of decomposition and nutrient release. As little is known about the dynamics of nutrient release from the decomposition of acacia harvest residues, particularly bark residue [33–36], it is crucial to measure rates of decomposition and nutrient release from its various components, especially in tropical environments where the area of acacia plantation estates has grown rapidly.

Acacia plantations, including *A. mangium*, *A. auriculiformis*, and their hybrid, currently represent a significant proportion of the commercial forest in Vietnam [37] and are managed on a rotation length of 5–8 years for both pulp and timber production [6]. Most of the resource is currently in the second and third rotation, and their productivity varies from 10 to 25 m^3 ha^{-1} $year^{-1}$ [6,27], depending on site condition as well as management inputs. Sustaining the productivity of these plantations will rely on maintaining, and if possible increasing, the nutrient capital of the current forestry land base [6,37]. The objectives of this study were to evaluate the contribution of harvest residues, particularly the bark component, to nutrient cycling in these intensive short-rotation plantation systems. To this end, decomposition rates of harvest residues of *A. mangium* were quantified. The results from this study are expected to assist with optimizing residue management of short-rotation tropical acacia plantations.

2. Materials and Methods

2.1. Location and Site Description

The study site was located at Phuc, a commune in the Yen Binh district of Yen Bai province in Northern Vietnam (21°51′ N, 105°00′ E; 100 m above sea level). The climate is tropical with four distinct seasons. During the study period, the mean monthly temperatures ranged from 15 to 28 °C (average 22 °C), and the mean monthly rainfall ranged from 6.1 to 375 mm (average 144 mm) (Figure 1).

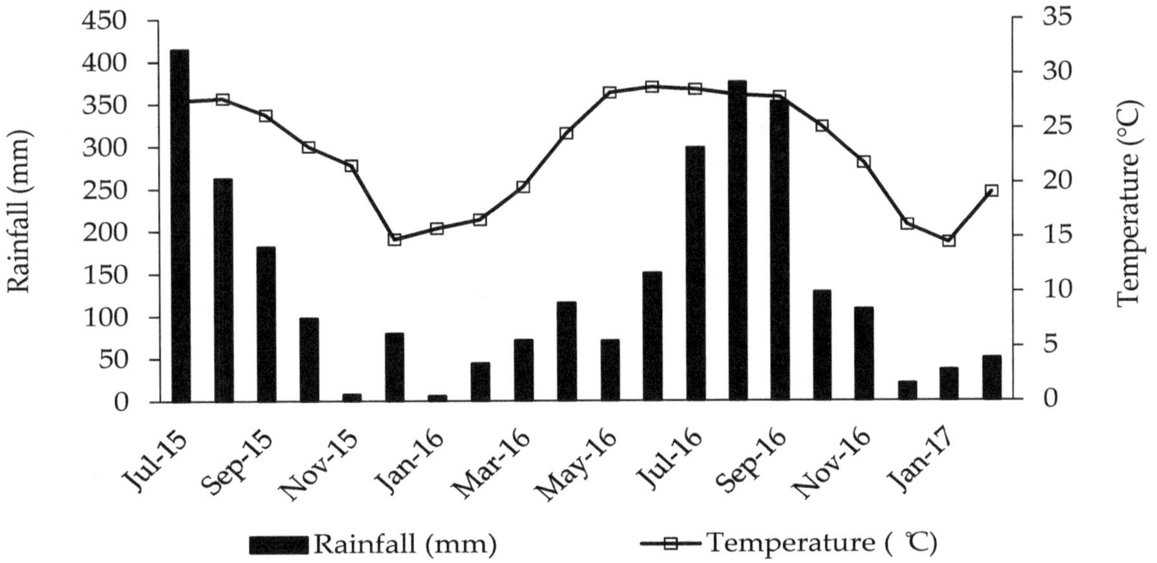

Figure 1. Average monthly temperatures (*continuous line*) and monthly rainfall (*vertical bars*) near the site during the experimental period.

The site is characterized by steep slopes that range from 10° (at the top of the hill) to 30°–40° (in the middle and at the bottom of the hill). The soil is classified as a Ferric (Ferralic) Acrisol [38] with a depth of ~0.5 m (in the middle and bottom of the hill) to 1 m (on top of the hill). The topsoil (0–10 cm depth) is acidic with a pH (1:5 water) of 3.8, a Bray extractable P of 4.04 mg kg^{-1} and soil organic carbon (TC) of 44 g kg^{-1} (Table 1). A more detailed description of soil properties is provided in Bich et al. [39].

Table 1. Soil chemical properties at the experimental site in Northern Vietnam. Standard errors are shown in parentheses (n = 5).

Soil Depth	pH	Total C	Total N	Total P	Bray Extractable P	Exchangeable Cations (cmol$_c$ (+) kg^{-1})		
(cm)	(1:5 Water)	g kg^{-1}	g kg^{-1}	g kg^{-1}	mg kg^{-1}	K	Ca	Mg
0–10	3.79	44.36	2.10	0.09	4.04	0.12	0.70	0.45
	(0.03)	(2.57)	(0.09)	(0.00)	(0.51)	(0.01)	(0.11)	(0.09)
10–30	3.94	27.80	1.54	0.08	2.21	0.08	0.57	0.40
	(0.08)	(2.14)	(0.07)	(0.00)	(0.37)	(0.02)	(0.13)	(0.07)

In the 1980s, the area was converted from secondary forest (degraded natural forest) to a native tree plantation of *Styrax tonkinensis* (Pierre) Craib. Two successive rotations of *A. mangium* were then planted in 2000 and 2008. The second rotation was planted at a spacing of 2 m × 2.5 m and mixed with NPK fertilizer (17 kg ha^{-1} N, 15 kg ha^{-1} P, and 8 kg ha^{-1} K) at planting. During the April 2015 harvest, trees were aged seven years and had a mean annual increment (MAI) of 13.3 ± 2.3 m^3 ha^{-1} year^{-1} based on measurements made on surface area rather than horizontal-projected area (noting that the horizontal projected area is less than the ground surface area because of the slope of the plots). The total aboveground stand biomass (AGB) was 60.8 t ha^{-1}, comprising stemwood (70%), bark (15%), branches (11%), and leaves (4%). The understory vegetation was high and dense, and dominated by woody shrubs and grass.

2.2. *Estimation of Biomass and Nutrients in Harvest Residues, Litter and Understory Vegetation*

In February 2015, 16 representative plots of 750 m^2 were established to determine stand growth and biomass accumulation in the second rotation. Height (H, m) and diameter (DBH, cm) at breast height (1.3 m) of all trees in the plots were measured. A total of 30 trees, covering the range of five diameter classes i.e., 6–9, 10–13, 14–17, 18–21, and 21–24 cm (six trees per diameter class), were randomly sampled for each diameter class to develop predictive models for calculating the biomass and nutrient content of the stand. The method for measuring the components of stand biomass is described in Huong et al. [1]. Briefly, after felling the tree, DBH and H (to a top end diameter of 3 cm) were measured. The tree was divided into five equal-length sections based on the tree height up to 3 cm in diameter over bark. Fresh biomass of stemwood, stem bark (manually stripped), branches, and leaves were weighed at the site and subsamples were then dried to a constant weight at 65 °C.

Forest floor litter and understory vegetation were assessed by sampling all biomass within five randomly located quadrats (16 m^2) in each plot. Litter was separated into woody (branches, stemwood plus bark) and non-woody (leaves, reproductive parts) materials, and the understory into woody and non-woody plants. Each part of the material (woody or non-woody) of the litter and understory vegetation component was weighed and a combined subsample of about 200 g from each part (woody or non-woody) of the components was selected for each plot (total of 32 subsamples per component). The subsamples were oven-dried at 65 °C to a constant weight.

The dry mass ratio of tree, litter, and understory components were used to calculate the total dry mass of each component. Subsamples of biomass components from two trees per diameter class (a total of ten trees) and of litter (16 subsamples of each woody and non-woody part), and understory vegetation (16 subsamples of each woody and non-woody plants) were ground to 0.02 mm particle size for nutrient analyses.

To estimate the biomass of harvest residue components, allometric relationships (equations) between DBH and biomass components of living trees were established (see Supplementary Figure S1) for bark (y = 0.0208 × $DBH^{2.3914}$, r^2 = 0.92), branches (y = 0.0044 × $DBH^{2.8555}$, r^2 = 0.81), and leaves (y = 0.0020 × $DBH^{2.7966}$, r^2 = 0.89). Based on the allometric equations, biomass components were estimated for each individual tree in sample plots and were then summed to give plot totals and then expressed as total biomass per ha.

2.3. Rate of Decomposition and Nutrient Release of Harvest Residues (Branches, Leaves, and Bark)

The experimental site was initially established in a third rotation of the *A. mangium* plantation in June 2015 to examine the effect of residue management treatments and fertilizer application at planting on growth and soil properties. The details of the experimental design are presented in Bich et al. [39]. In brief, the design was 20 plots based on a randomized complete block design with five replications. Each replicate occupied a similar position in the landscape, with plots grouped according to elevation: top-hill, middle-hill, and bottom-hill. This study used only the five plots that combined residue retention and fertilizer as 17 kg ha^{-1} of N, 15 kg ha^{-1} of P, and 8 kg ha^{-1} of K applied in the base of the planting hole at planting. All harvested stemwood without bark down to 3 cm in diameter over bark had been exported from the site. All other harvested tree components, bark, branches, and leaves, as well as forest floor components, were retained as residues and distributed evenly across the site prior to application of treatments. Each plot contained 6 × 6 trees and was surrounded by two rows of buffer trees. Trees had a spacing of 2.5 m × 3 m (1333 trees ha^{-1}).

Decomposition of harvest residues from the *A. mangium* stand excluding litter and understory vegetation was measured by the mesh bag method [40]. Briefly, 10 kg sample of leaves, branches (<3 cm in diameter, ~10 cm lengths), and bark were collected randomly across the site immediately after clear-felling in April 2015 and air-dried at room temperature. The mesh bags—30 cm × 28 cm in size with 2 mm mesh—were used to minimize leakage of small fragments while allowing access by decomposer organisms [41]. Each bag was packed with the air-dried residues, either 40 g leaves, 100 g bark, or 100 g branches. Additional samples were oven dried at 65 °C and weighed to establish moisture content and dry weight before grinding for nutrient analysis.

A total of 21 mesh bags comprising nine of leaves and six each of branches and bark were located randomly within each of the five plots (a total of 105 bags) on 25 July 2015. Bags were positioned on top of the litter layer to simulate the position and microclimate representative of the majority of the harvest residues. For the leaf component, one bag from each plot was collected every 60 days until 540 days after the bags were first placed in the plot (1 sample per plot × 5 plots per collection time × 9 times = 45 samples); for the bark and branch components, one bag of each was collected every 90 days during the same period (1 sample per plot × 5 plots per collection time × 6 times = 30 samples per component). The bags were air dried, and the contents brushed free of soil, insect frass, and other debris, oven dried at 65 °C and weighed to determine dry weight loss. Oven-dried samples were ground (≤0.02 mm particle size) for nutrient analysis.

2.4. Plant Nutrient Analysis

Oven-dried biomass samples of tree components, litter, understory, and decomposing harvest residues from mesh bags were digested in concentrated sulphuric acid with 30% hydrogen peroxide; all nutrients were measured from that digest. Total N was analyzed by Automatic Kjeldahl distillation (UDK 149, PLT Scientific SDN BHD, Puchong Selangor Darul Ehsan, Malaysia), P by spectrophotometry (Jasco 7800 spectrophotometer-JASCO International Co., Ltd, Tokyo, Japan), K by flame photometry (Model 410 Flame Photometer Range—Sherwood Scientific Ltd, Cambridge, UK), and Ca and Mg by atomic absorption spectroscopy [42]. The initial nutrient content of each stand biomass component was calculated from nutrient concentrations multiplied by dry weight while the nutrient content of residues from the mesh bags was calculated as the product of the dry weight remaining and the relevant nutrient concentration.

2.5. Statistical Analysis

Long-term patterns of mass loss (up to 10 years) can be analyzed by a range of decay models [43]. However, given that this decomposition study was only 1.5 years long, a single exponential decay model [44] was used to fit the mass loss of each residue component (bark, leaves, and branches) as a function of time as follows: $M_t = M_0 \times e^{-kt}$, where M_t is residue dry weight at time t, M_0 is the initial residue dry weight at time 0, k is exponential decay coefficient (day^{-1}), and t is time in days. The exponential decay equation was converted to a linear form ($\ln[M_t] = \ln[M_0] - kt$) before the regression procedures were performed. Linear regression of ln [% mass remaining] versus time t was used to determine the decay constant k; the time required for the decomposition of half of the initial residue weight ($t_{0.5}$) was then calculated as $t_{0.5} = 0.693/k$ [44]. Analysis of covariance was conducted to test whether these decomposition rates differed among leaf, branch, and bark; that is, the regression lines of the three separate components were compared to see if they had the same slope. The statistical analyses were conducted with SPSS for Windows version 22.0 (IBM Corp, Armonk, NY, USA, 2013).

3. Results

3.1. Biomass and Nutrient Content of Harvest Residues, Litter and Understory Vegetation

The total initial dry weight of all residues (from harvest and forest floor) following logging was 27.2 t ha^{-1}, comprising 66% harvest residues and 34% forest floor (Table 2). The total initial amounts of N, P, K, Ca, and Mg maintained on the site were 439.6, 14.8, 60.7, 185.0, and 20.1 kg ha^{-1}, respectively. Harvest residues alone accounted for 60% of N, 61% of P, 43% of K, 64% of Ca, and 24% of Mg (Table 2).

Table 2. Initial dry weight and nutrient content of harvest residues and forest floor on the site following harvest of a seven-year-old *A. mangium* plantation in Northern Vietnam. Standard errors are shown in parentheses (n = 16). DM is dry matter.

Components	Dry Weight	Nutrient Content (kg ha^{-1})				
	(t DM ha^{-1})	N	P	K	Ca	Mg
Harvest residue components						
Bark	8.9 (0.3)	133.4 (5.2)	2.6 (0.1)	8.4 (0.3)	76.5 (3.0)	1.2 (0.1)
Branches	6.6 (0.3)	53.9 (1.8)	1.9 (0.1)	4.9 (0.1)	28.9 (1.3)	0.9 (0.1)
Leaves	2.5 (0.1)	76.0 (3.3)	4.5 (0.2)	13.1 (0.6)	13.7 (0.6)	2.8 (0.1)
Subtotal	18.1 (0.7)	263.4 (10)	9.0 (0.3)	26.4 (1.0)	119.2 (4.9)	4.9 (0.2)
Forest floor residue components						
Litter	5.8 (0.6)	128.9 (13.3)	5.1 (0.5)	31.6 (3.3)	45.4 (4.7)	13.7 (1.4)
Understory	3.3 (0.4)	47.3 (5.6)	0.7 (0.1)	2.7 (0.3)	20.4 (2.4)	1.5 (0.2)
Subtotal	9.1 (0.6)	176.2 (12)	5.9 (0.5)	34.4 (3.1)	65.8 (4.3)	15.2 (1.3)
Total residue with bark	27.2 (0.9)	439.6 (15)	14.8 (0.6)	60.7 (3.2)	185.0 (6.3)	20.1 (1.2)
Total residue without bark	18.2 (0.7)	306.1 (12)	12.2 (0.5)	52.3 (3.1)	108.5 (4.5)	18.9 (1.3)

3.2. Decomposition of Harvest Residue Components

A single exponential decay model explained the decomposition of each of the harvest residue components very well (Figure 2; Table 3). The half-lives for leaf, branch, and bark dry weight were 0.47, 1.29, and 3.09 years, respectively (Table 3). The slopes of the regression lines indicate that the decomposition rates of bark and branches were significantly lower than that of leaves ($p < 0.05$; Figure 2; Table 3). The decomposition of leaf dry weight was the fastest throughout the study period with only 10% remaining after 540 days, as compared to 47% and 72% for branches and bark, respectively (Table 3).

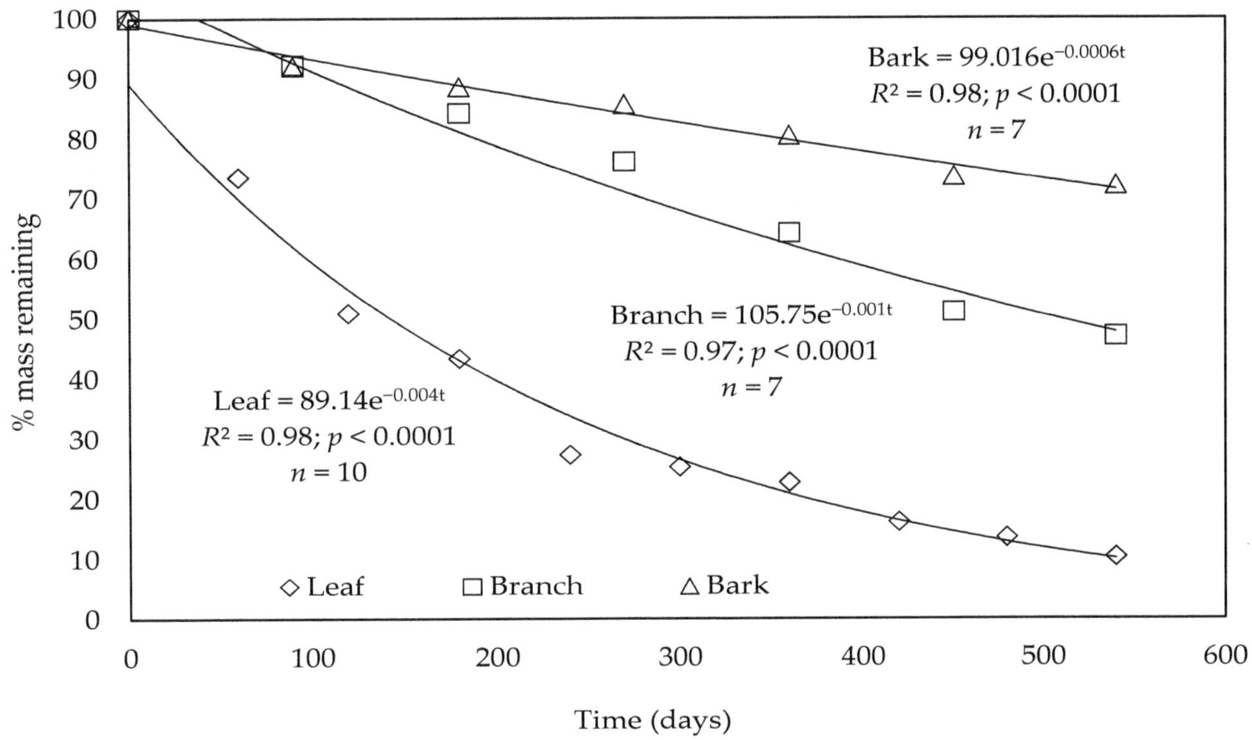

Figure 2. Exponential decay functions fitted to dry weight loss of harvest residue components of *A. mangium* in Northern Vietnam during decomposition for 540 days.

Table 3. Loss of dry weight (%) over the 540-day period, decomposition constant (k, year^{-1}), the proportion of explained variation (R^2), root mean squared error (*RMSE*), and half-life ($t_{0.5}$, year) of harvest residue components of *A. mangium* in Northern Vietnam. $N = 10$ collection times for leaves and 7 for branches and bark.

Component	Dry Weight Loss (%)	k (Year^{-1})	R^2	*RMSE*	*p*-Value	$t_{0.5}$ (Year)
Leaves	90	1.47	0.98	0.10	<0.0001	0.47
Branches	53	0.54	0.97	0.05	<0.0001	1.29
Bark	28	0.22	0.98	0.02	<0.0001	3.09

3.3. Nutrient Release During Decomposition of Harvest Residues

Among the harvest residue components, leaves consistently contained the highest concentration of N, P, K, and Mg, but not Ca during decomposition (Figure 3). Leaves also showed the fastest release of nutrients (Figure 4) as described by the single exponential model (see Supplementary Table S1).

N release from decomposing harvest residue components was similar to their dry weight loss over time (Figure 4), as their N concentration remained relatively stable during the study period (Figure 3a). In contrast, the K and Ca release from all residue components was more rapid; the Mg and P release was less rapid than the dry weight loss (Figure 4).

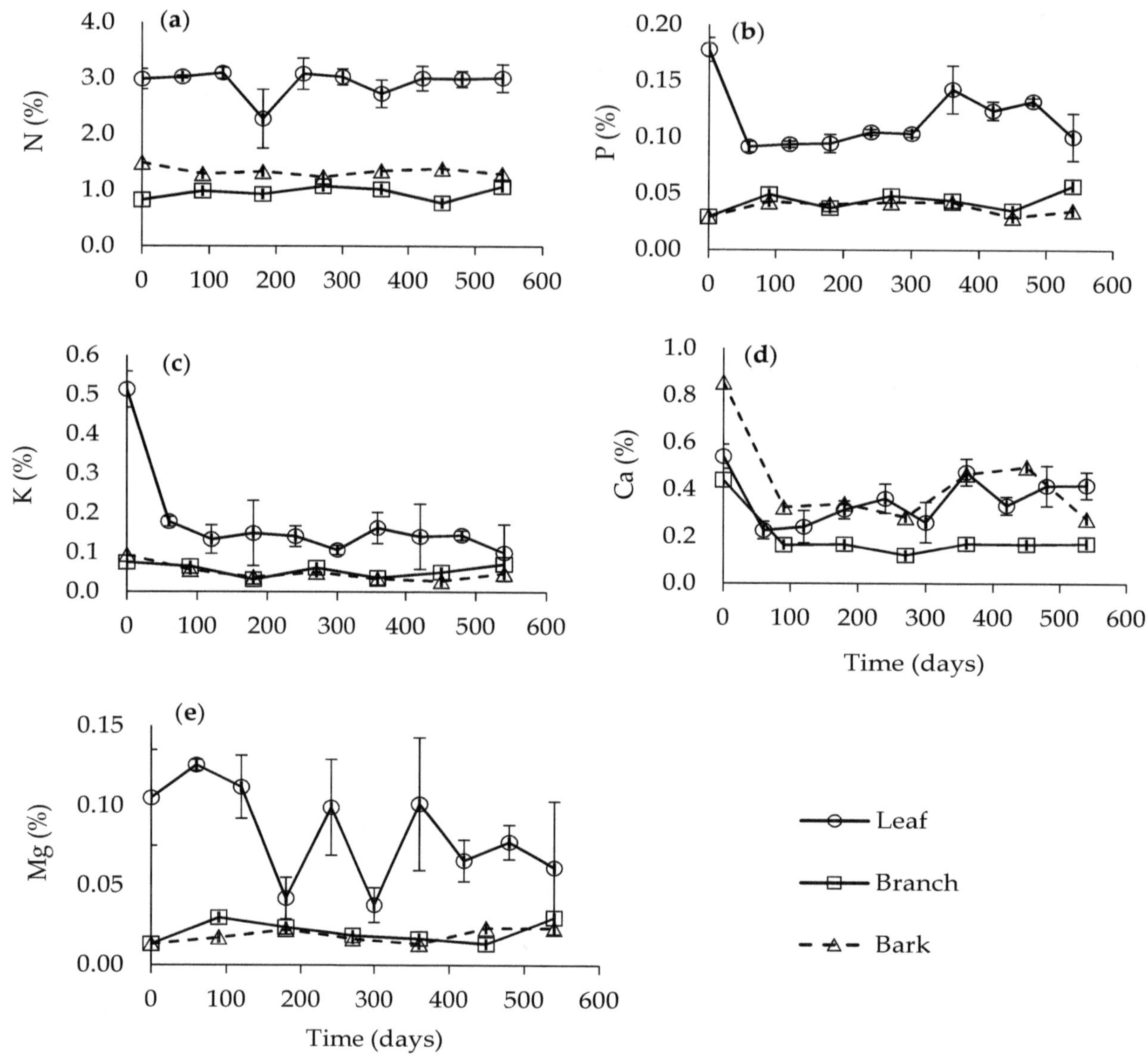

Figure 3. Dynamics of (**a**) N, (**b**) P, (**c**) K, (**d**) Ca, and (**e**) Mg concentration in harvest residues of *A. mangium* during the 540-day period of decomposition in Northern Vietnam. Vertical bars indicate standard errors for each sampling ($n = 5$).

Throughout the study period, the loss of Ca in all the residue components was rapid; 92%, 82%, and 76% was lost in leaves, branches, and bark, respectively (Figure 4). The losses of K were of similar magnitude to that of Ca with 98%, 56%, and 63% being lost in leaves, branches and bark, respectively (Figure 4). In contrast, there was an accumulation of P and Mg, particularly in the bark and branches; P and Mg in branches were 55% and 107% higher, respectively, than their initial content 90 days after establishment (Figure 4b,c).

For all components together, the amount of N, P, K, Ca, and Mg that potentially could be recycled from harvest residues to the soil 18 months following logging was 137.1, 4.7, 20.8, 94.5, and 2.2 kg ha^{-1}, respectively (Table 4). Most of N, P, K, and Mg, respectively 50%, 91%, 62% and 118% was released from decomposing leaves. The value for Mg was >100% because the other components immobilized Mg over the first 18 months. Bark could potentially contribute 61% of the Ca, 35% of N, and 25% of K (Table 4).

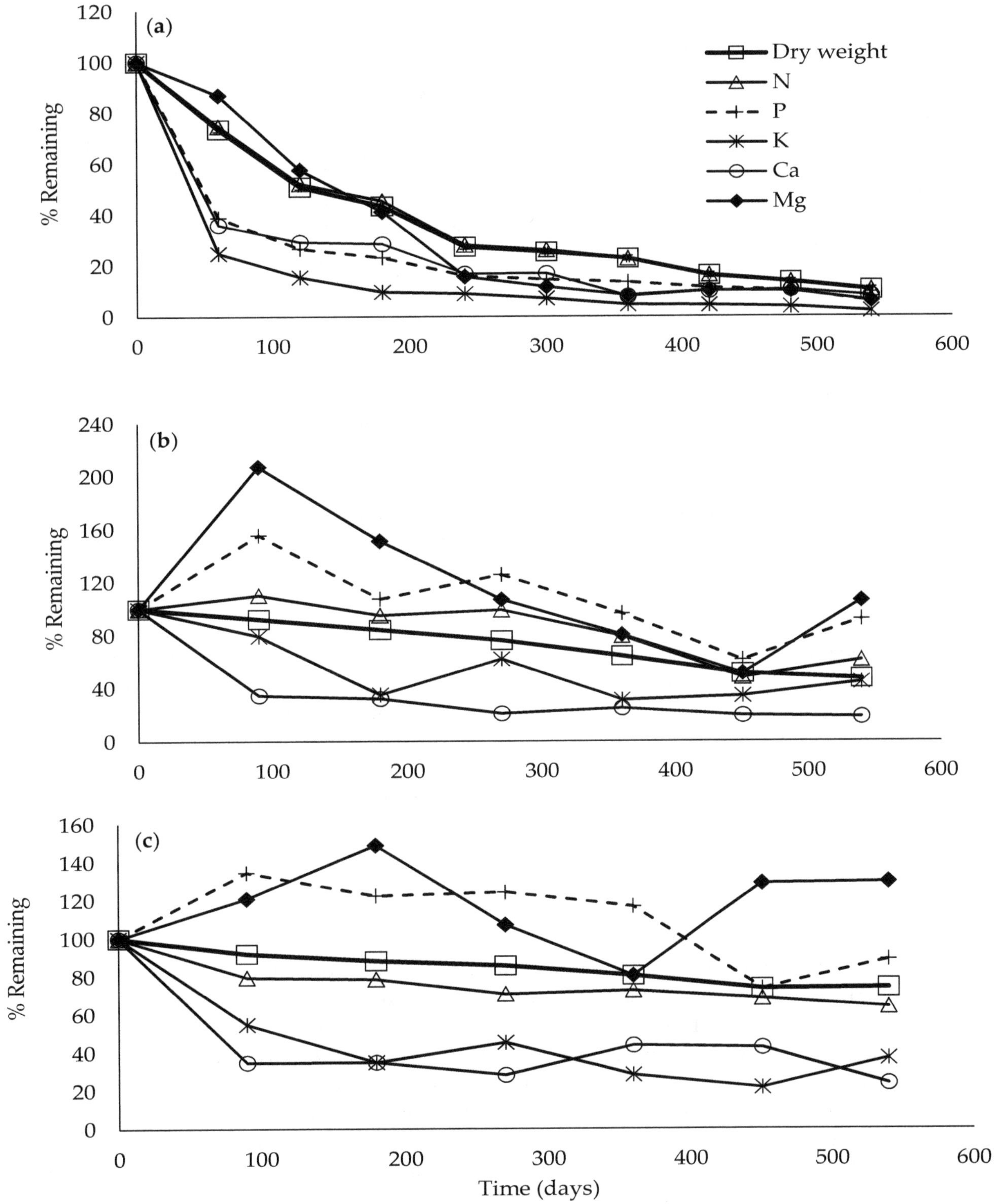

Figure 4. Loss of dry weight and release of nutrients during decomposition for 540 days for (**a**) leaves; (**b**) branches; and (**c**) bark of *A. mangium* in Northern Vietnam.

Table 4. Dry weight loss and nutrient release from different components of decomposing harvest residues 18 months following logging of a seven-year-old *A. mangium* plantation in Northern Vietnam. Standard errors are shown in parentheses ($n = 5$). DM is dry matter.

Components	Dry Weight Loss	Nutrient Release from Decomposing Harvest Residues (kg ha^{-1})				
	(t DM ha^{-1})	N	P	K	Ca	Mg
Leaves	2.3 (0.1)	68.1 (2.0)	4.3 (0.1)	12.8 (0.1)	12.6 (0.4)	2.6 (0.1)
Branches	3.5 (0.3)	20.9 (1.5)	0.2 (0.8)	2.7 (0.1)	23.7 (2.0)	−0.1 (0.0)
Bark	2.3 (0.1)	48.0 (2.8)	0.3 (0.0)	5.3 (0.2)	58.2 (1.5)	−0.4 (0.0)
Total residue with bark	8.1 (0.0)	137.1 (3.1)	4.7 (0.4)	20.8 (0.5)	94.5 (0.6)	2.2 (0.2)
Total residue without bark	5.8 (0.1)	89.0 (2.3)	4.4 (0.4)	15.6 (0.5)	36.3 (0.8)	2.6 (0.2)

4. Discussion

This study has shown that harvest residues retained on site in the form of bark, branches, and leaves of a seven-year-old *A. mangium* plantation accounted for two-thirds of the total initial dry weight and 24–64% of macro-nutrient contents of total residues that included litter and understory vegetation. The export of bark from the site would lead to the significant loss of some nutrients. Decomposition rates and nutrient release from the harvest residues were associated with the nutrient concentration (substrate quality) of each residue component. Despite the limited number of mesh bags used, this study clearly demonstrated large amounts of N, K, and Ca, but not P and Mg being released to the soil over the 1.5 year-study period. These results are now discussed in the context of the contribution of harvest residues to nutrient cycling in short-rotation plantations of acacia in the tropics, especially the role of bark residue.

Although the bark of the *A. mangium* represented just 15% of the total stand AGB, its export from the site would have reduced the total dry weight of all residues by one-third. The contribution of bark residue to this total dry weight was 41%, 31%, and 19% higher than that observed in *A. mangium* in Sumatra [21], *A. auriculiformis* in Southern Vietnam [1], and *E. dunnii* in Uruguay [17], respectively. Potentially the contribution of bark to total AGB in this study and at other sites in Vietnam (also 15%) [2] was proportionately higher than that observed in *A. mangium* in Sumatra (10%) [21,45], *A. auriculiformis*, and its hybrid (*A. mangium* × *A. auriculiformis*) in Vietnam (7–8%) [1,46] and *E. dunnii* in Uruguay (12%) [17]. For eucalypt and pine plantations, bark biomass generally accounts for 7–10% of total stand AGB [47–49]. Thus, retention of bark on site for *A. mangium* plantations tends to be more important in residue retention than for other species.

The bark component potentially conserved 30% of N, 18% of P, 14% of K, 41% of Ca, and 6% of the Mg content of all residues for recycling. For N, this was similar to that in *A. auriculiformis* in Southern Vietnam [1], but 20% and 50% greater than that in *A. mangium* in Indonesia [17,21] and *E. dunnii* in Uruguay [17,21], a difference that is attributable to the higher concentration of N in bark in the Vietnamese (1.5%) than Indonesian (1.1%) and Uruguayan (0.2%) studies. In contrast, K was between 22% and 56% lower than in these other studies [1,17,21] owing to its lower concentration (0.1% vs. 0.2−0.5%). This lower concentration may be associated with soil K being less available in this study than at the other acacia sites [1,21]. Concentration was also the factor that led to the nutrient content of only N (30%) and Ca (41%) approximately matching bark's one-third contribution to the dry weight of all residues. Hence, retention of the bark of *A. mangium* on site at harvesting conserves some nutrients better than others and was a relatively poor source of P, K, and Mg.

The patterns of mass loss during decomposition varied among components, with leaves decaying the fastest, consistent with previous reports for hardwood plantation species [16,17,20,41,50]. Among nutrients in decomposing material, the concentration of N is the most important factor determining the rate of decomposition in a given environment [29,30]. In this study, the decay constant (k) for leaves was 2.7 and 6.7 times higher and the N concentration 2.0 and 3.7 times higher than for branches and bark, respectively. The k of leaves and bark in this study were 50–65%

and 20–50% faster, respectively, than those observed in other studies of *Eucalyptus* plantations in Australia and Uruguay [41]; similarly, the *k* for branches was approximately 60% faster than for *Eucalyptus* species [16,17], and 66–96% faster than that observed for *Pinus radiata* D. Don plantations across New Zealand [51]. These higher *k* values are partly attributable to the higher N concentration in *A. mangium* harvest residues. In addition, higher temperatures and moisture contents in the tropical climate of Vietnam are factors that will cause faster rates of decomposition [52–54]. Thus, harvest residues from tropical acacia plantations release nutrients, especially N, more rapidly than other species.

During the first year of the study, net release of N was found from leaves and bark, while net immobilization of N occurred in branches. Net N release is influenced by the C/N ratio of the decomposing materials [17,29,30,55–57]. If C/N is >40, the net effect is N immobilization during decomposition [54]. In this study, the C/N ratio of branch residue was 65 [2]; therefore, the initial immobilization of its N was expected. N was subsequently mobilized owing to the oxidization of a C and N uptake, resulting in reduction of the C/N ratio during the decomposition process. Conversely, the C/N ratio of the leaf material was initially 17, and for bark it was 35 [2]. By comparison, in *E. dunnii*, the C/N ratios of leaf, branch, and bark were much higher at 36, 134, and 174, respectively [17] due to 13−47% lower N concentration in the eucalypt harvest residues. These lower C/N ratios in acacias than eucalypts were associated with a much faster release of N from *A. mangium* than *E.* hybrid (*E. urophylla* × *E. grandis*) harvest residues [34] and higher rates of N turnover in topsoil under *A. mangium* than *E. grandis* [58]. By the end of the 1.5-year study period, the total amount of N released from decomposing *A. mangium* harvest residues was 137 kg ha^{-1}, 35% of which released from bark. In contrast, bark made no net contribution to the release of N from harvest residues of *E. dunnii* and *E. globulus* for up to two years [16,17] and *E. grandis*, *E. globulus*, *E. dunnii*, and *Pinus taeda* L. for up to six months following harvest [59]. While the quantities of N taken up by *A. mangium* can be as much as 180 kg ha^{-1} yr^{-1} in the first two years of growth [21], *A. mangium* bark residue can potentially make an important contribution to this demand.

Over the 1.5 years of this study, harvest residues released only around 5 kg ha^{-1} of P. Other studies have found that a small amount of P at planting is sufficient to optimize growth of tropical plantation acacias [60,61], but this is in concentrated form near the seedling at planting, so the timing and rates of release of P from residues and its spatial distribution may not meet all of the demand for P early in the growth cycle. However, as the relative importance of the growth response to P declines with increasing stand age [21,61], the amount of P released from residues together with its release from other sources in the soil is likely to meet the demand later in the rotation.

High levels of immobilization of P but not N in the branches and bark during decomposition suggest that P was the more limiting to microbial decomposer activity of organic matter in this system. The addition of P fertilizer has been shown to increase microbial activity in *A. mangium* plantations [62,63]. Tropical plantation soils are characteristically low in soil P because of their high P-fixing capacity [21,25,64] and in this study, P in the topsoil (4.0 mg kg^{-1}) was very low, even for tropical plantation soils [65]. There was also a notable decline in soil available P in this study [66], that has also been found elsewhere in acacia plantations. Hence, the addition of P fertilizer may serve an added purpose—i.e., the promotion of decomposition and nutrient release from acacia harvest residues.

The release of K and Ca was very rapid, especially during the first three months of decomposition, and reflected the high K and Ca concentrations in the harvest residues. Such rapid release of K is commonly found in plantations across a range of environments [16,17,20,59] because K is highly mobile; hence its rapid leaching by rain soon after placement of the mesh bags in a number of related experiments [16,17,67,68]. The rapid release of Ca and decline in Ca concentration in harvest residues is linked to the rapid release rates of Ca oxalate [69,70], which accounts for 20–56% of total Ca in fresh plant tissues [69]. Over the study period, 20.8 and 94.5 kg ha^{-1} of K and Ca, respectively, were released from the residues, with bark contributing 25% of the K and 61% of the Ca, or potentially at least one-quarter and >50% of the demand [21]. As growth responses to Ca and K fertilizers are uncommon in tropical acacia plantations grown with residue retention [1,71], the substantial contribution of residues to the K and Ca nutrient pools, the latter through bark, may explain this lack of response.

In contrast, the release of Mg was very slow, with net immobilization over the first few months, especially in the branch and bark components. The slow release of Mg is because it is dependent on cellular degradation in the residues [17,72]. In Ivory Coast, only 10% of initial Mg was released from decomposing litter of *A. mangium* and *A. auriculiformis* in a one-year study period [36]. In this study, <3 kg ha^{-1} was released, suggesting that *A. mangium* harvest residues may not meet the potential demand for Mg uptake by this species, which can be as much as 9 kg ha^{-1} $year^{-1}$ over a similar period for an equivalent MAI [21].

5. Conclusions

Harvest residues played a crucial role in conserving nutrients for recycling in this *A. mangium* plantation, contributing 66% of the total dry biomass and 24–64% of macronutrients, of all residues, including litter and understory vegetation. Bark removal would have reduced the quantity of all residues by one-third, and a similar proportion of the Ca and N contents. Based on the amounts of nutrients recycled in 1.5 years of the study, we suggest that recycling of N, Ca, and K, but not P and Mg, were potentially able to meet a significant part of the nutrient demand in the next rotation. While slow release should have promoted capture of the nutrients by growing trees in the next rotation, the amount of N and K lost by leaching and P to the high-fixing tropical soil remains unknown. In addition, a high quantity of N, but not P, released may have caused a nutrient imbalance in the soils. As P supply is crucial to the growth of these acacias, the addition of P fertilizer at planting is recommended, both to boost immediate supply and to potentially enhance decomposition rates of the harvest residues.

Supplementary Materials:
Figure S1: The allometric relationships between dry biomass of stand components and tree diameter (DBH, cm). DM is dry matter. $N = 30$. Table S1: Nutrient releases (%) from leaf component over the 540-day period, nutrient release constant (k, $year^{-1}$), regression coefficient (R^2) and root mean squared error ($RMSE$) and haft-life ($t_{0.5}$, year) of nutrient degradation of *A. mangium* harvest residues in Northern Vietnam. N = 10 collection times.

Author Contributions: N.V.B., T.L.D., V.D.H., H.V.T., and D.M. conceived and designed the experiments; N.V.B., H.V.T., and N.V.T. performed the experiments, collected data, and carried out laboratory work; N.V.B, D.M., A.E., and D.R. analyzed the data; N.V.B., A.E., D.M., T.L.D., D.R., K.J.E., and C.M. drafted the manuscript; and all authors contributed to manuscript revision.

Acknowledgments: We thank colleagues from the Silviculture Research Institute (SRI) and Thac Ba Forest Enterprise for assistance with data collection, sampling and maintenance of the field experiments, Chris Beadle for editing the manuscript, Chris Harwood, Jacqui England and two anonymous reviewers for their review of the manuscript.

References

1. Huong, V.D.; Nambiar, E.S.; Quang, L.T.; Mendham, D.S.; Dung, P.T. Improving productivity and sustainability of successive rotations of Acacia auriculiformis plantations in South Vietnam. *South. For.* **2015**, *77*, 51–58. [CrossRef]
2. Hai, V.D.; Trieu, D.T.; Tiep, N.H.; Bich, N.V.; Duong, D.T. *Biomass Accumulation and Carbon Sequestration in Some Major Types of Plantation Forests in Vietnam*; Agriculture Publishing House: Hanoi, Vietnam, 2009.
3. Bao Bac Giang. Burning Tree Bark Residue to Produce Charcoal: Advantages and Disadvantages. Available online: http://baobacgiang.com.vn/bg/nhip-cau-ban-doc/150436/dot-vo-go-lay-than--loi-it--hai-nhieu.html (accessed on 3 July 2018).
4. Bac, V. Producing Compost Fertiliser from Bark of Some Tree Species. Available online: https://www.vietnamplus.vn/san-xuat-phan-bon-huu-co-tu-vo-cay-nguyen-lieu-giay/213711.vnp (accessed on 3 July 2018).
5. Anh, N. Producing Wood Glue from Bark of Acacia Tree. Available online: http://www.khoahocphothong.com.vn/tao-keo-dan-go-tu-vo-keo-la-tram-26294.html (accessed on 3 July 2018).

6. Nambiar, E.K.S.; Harwood, C.E.; Kien, N.D. Acacia plantations in Vietnam: Research and knowledge application to secure a sustainable future. *South. For.* **2015**, *77*, 1–10. [CrossRef]
7. Goncalves, J.L.D.; Alvares, C.A.; Higa, A.R.; Silva, L.D.; Alfenas, A.C.; Stahl, J.; Ferraz, S.F.D.; Lima, W.D.P.; Brancalion, P.H.S.; Hubner, A.; et al. Integrating genetic and silvicultural strategies to minimize abiotic and biotic constraints in Brazilian eucalypt plantations. *For. Ecol. Manag.* **2013**, *301*, 6–27. [CrossRef]
8. Laclau, J.P.; Levillain, J.; Deleporte, P.; Nzila, J.D.; Bouillet, J.P.; Saint Andre, L.; Versini, A.; Mareschal, L.; Nouvellon, Y.; M'Bou, A.T.; et al. Organic residue mass at planting is an excellent predictor of tree growth in *Eucalyptus* plantations established on a sandy tropical soil. *For. Ecol. Manag.* **2010**, *260*, 2148–2159. [CrossRef]
9. Rocha, J.H.T.; Goncalves, J.L.D.; Gava, J.L.; Godinho, T.D.; Melo, E.A.S.C.; Bazani, J.H.; Hubner, A.; Arthur, J.C.; Wichert, M.P. Forest residue maintenance increased the wood productivity of a *Eucalyptus* plantation over two short rotations. *For. Ecol. Manag.* **2016**, *379*, 1–10. [CrossRef]
10. Raison, R.J.; O'Connell, A.M.; Khanna, P.K.; Keith, H. Effect of repeated fiers on nitrogen and phosphorus budgets and cycling processes in forest ecosystems. In *Fire in Mediterranean Ecosystems*; Trabaud, L., Prodon, R., Eds.; Commission of the European Communities: Brussels, Belgium, 1993; pp. 347–363.
11. Mendham, D.S.; O'Connell, A.M.; Grove, T.S.; Rance, S.J. Residue management effects on soil carbon and nutrient contents and growth of second rotation eucalypts. *For. Ecol. Manag.* **2003**, *181*, 357–372. [CrossRef]
12. Deleporte, P.; Laclau, J.P.; Nzila, J.D.; Kazotti, J.G.; Marien, J.N.; Bouillet, J.P.; Szwarc, M.; D'Annunzio, R.; Ranger, J. Effects of slash and litter management practices on soil chemical properties and growth of second rotation eucalypts in the Congo. In *Site Management and Productivity in Tropical Plantation Forests, Proceedings of Workshops, Piracicaba, Brazil, 22–26 November 2004 and Bogor, Indonesia, 6–9 November 2006*; Nambiar, E.K.S., Ed.; Center for International Forestry Research: Bogor, Indonesia, 2008; pp. 5–22.
13. Goncalves, J.L.M.; Wichert, M.C.P.; Gava, J.L.; Masetto, A.V.; Junior, J.C.A.; Serrano, M.I.P.; Mello, S.L.M. Soil fertility and growth of *Eucalyptus grandis* in Brazil under different residue management practices. *South. For.* **2007**, *69*, 95–102. [CrossRef]
14. Raison, R.J.; Khanna, P.K.; Woods, P.V. Mechanisms of element transfer to the atmosphere during vegetation fires. *Can. J. For. Res.* **1985**, *15*, 132–140. [CrossRef]
15. Giardina, C.P.; Sanford, R.L.; Døckersmith, I.C.; Jaramillo, V.J. The effects of slash burning on ecosystem nutrients during the land preparation phase of shifting cultivation. *Plant Soil* **2000**, *220*, 247–260. [CrossRef]
16. Shammas, K.; O'Connell, A.M.; Grove, T.S.; McMurtrie, R.; Damon, P.; Rance, S.J. Contribution of decomposing harvest residues to nutrient cycling in a second rotation *Eucalyptus globulus* plantation in south-western Australia. *Biol. Fertil. Soils* **2003**, *38*, 228–235.
17. Hernández, J.; del Pino, A.; Salvo, L.; Arrarte, G. Nutrient export and harvest residue decomposition patterns of a *Eucalyptus dunnii* Maiden plantation in temperate climate of Uruguay. *For. Ecol. Manag.* **2009**, *258*, 92–99. [CrossRef]
18. De Souza, I.F.; De Barros, N.F.; Da Silva, I.R.; Renier, R.F.; Silva, L.D.; De Novais, R.F. Decomposition of eucalypt harvest residues as affected by management practices, climate and soil properties across southeastern Brazil. *For. Ecol. Manag.* **2016**, *374*, 186–194. [CrossRef]
19. Ferreira, G.W.D.; Soares, E.M.B.; Oliveira, F.C.C.; Silva, I.R.; Dungait, J.A.J.; Souza, I.F.; Vergutz, L. Nutrient release from decomposing *Eucalyptus* harvest residues following simulated management practices in multiple sites in Brazil. *For. Ecol. Manag.* **2016**, *370*, 1–11. [CrossRef]
20. Rocha, J.H.T.; Marques, E.R.G.; Gonçalves, J.L.D.M.; Hübner, A.; Brandani, C.B.; Ferraz, A.D.V.; Moreira, R.M. Decomposition rates of forest residues and soil fertility after clear-cutting of *Eucalyptus grandis* stands in response to site management and fertilizer application. *Soil Used Manag.* **2016**, *32*, 289–302. [CrossRef]
21. Hardiyanto, E.; Nambiar, S.E. Productivity of successive rotations of *Acacia mangium* plantations in Sumatra, Indonesia: Impacts of harvest and inter-rotation site management. *New For.* **2014**, *45*, 557–575. [CrossRef]
22. Achat, D.L.; Deleuze, C.; Landmann, G.; Pousse, N.; Ranger, J.; Augusto, L. Quantifying consequences of removing harvesting residues on forest soils and tree growth—A meta-analysis. *For. Ecol. Manag.* **2015**, *348*, 124–141. [CrossRef]
23. Versini, A.; Zeller, B.; Derrien, D.; Mazoumbou, J.-C.; Mareschal, L.; Saint-André, L.; Ranger, J.; Laclau, J.-P. The role of harvest residues to sustain tree growth and soil nitrogen stocks in a tropical *Eucalyptus* plantation. *Plant Soil* **2014**, *376*, 245–260. [CrossRef]

24. Folster, H.; Khanna, P.K. Dynamics of Nutrient Supply in Plantation Soils. In *Management of Soil, Nutrients and Water in Tropical Plantation Forests*; Nambiar, E.K.S., Brown, A.G., Eds.; ACIAR: Canberra, Australia, 1997; pp. 339–373.
25. Sam, D.D.; Binh, N.N. *Assessment of Productivity of Forest Lands in Vietnam*; Statistical Publishing House: Hanoi, Vietnam, 2001.
26. Dong, T.L.; Doyle, R.; Beadle, C.L.; Corkrey, R.; Quat, N.X. Impact of short-rotation acacia hybrid plantations on soil properties of degraded lands in central Vietnam. *Soil Res.* **2014**, *52*, 271–281. [CrossRef]
27. Hung, T.T.; Almeida, A.C.; Eyles, A.; Mohammed, C. Predicting productivity of *Acacia* hybrid plantations for a range of climates and soils in Vietnam. *For. Ecol. Manag.* **2016**, *367*, 97–111. [CrossRef]
28. Meentemeyer, V. Macroclimate the lignin control of litter decomposition rates. *Ecology* **1978**, *59*, 465–472. [CrossRef]
29. Ge, X.; Zeng, L.; Xiao, W.; Huang, Z.; Geng, X.; Tan, B. Effect of litter substrate quality and soil nutrients on forest litter decomposition: A review. *Acta. Ecol. Sin.* **2013**, *33*, 102–108. [CrossRef]
30. Krishna, M.P.; Mohan, M. Litter decomposition in forest ecosystems: a review. *Energy Ecol. Environ.* **2017**, *2*, 236–249. [CrossRef]
31. Murphy, K.L.; Klopatek, J.M.; Klopatek, C.C. The effects of litter quality and climate on decomposition along an elevational gradient. *Ecol. Appl.* **1998**, *8*, 1061–1071. [CrossRef]
32. Vitousek, P. Nutrient cycling and nutrient use efficiency. *Am. Nat.* **1982**, *119*, 553–572. [CrossRef]
33. Bachega, L.R.; Bouillet, J.-P.; de Cássia Piccolo, M.; Saint-André, L.; Bouvet, J.-M.; Nouvellon, Y.; de Moraes Gonçalves, J.L.; Robin, A.; Laclau, J.-P. Decomposition of *Eucalyptus grandis* and *Acacia mangium* leaves and fine roots in tropical conditions did not meet the home field advantage hypothesis. *For. Ecol. Manag.* **2016**, *359*, 33–43. [CrossRef]
34. Tchichelle, S.V.; Epron, D.; Mialoundama, F.; Koutika, L.S.; Harmand, J.M.; Bouillet, J.P.; Mareschal, L. Differences in nitrogen cycling and soil mineralisation between a eucalypt plantation and a mixed eucalypt and *Acacia mangium* plantation on a sandy tropical soil. *South. For.* **2017**, *79*, 1–8. [CrossRef]
35. Hardiyanto, E.; Anshori, S.; Sulistyono, D. Early results of site management in *Acacia mangium* plantations at PT. Musi Hutan Persada, South Sumatra, Indonesia. In *Site Management and Productivity in Tropical Plantation Forests, Proceedings of Workshop, Congo, July 2001 and China, February 2003*; Nambiar, E., Ranger, J., Tiarks, A., Toma, T., Eds.; Center for International Forestry Research: Bogor, Indonesia, 2004; pp. 93–108.
36. Ngoran, A.; Zakra, N.; Ballo, K.; Kouame, C.; Zapata, F.; Hofman, G.; Van Cleemput, O. Litter decomposition of *Acacia auriculiformis* Cunn. Ex Benth. and *Acacia mangium* Willd. under coconut trees on quaternary sandy soils in Ivory Coast. *Biol. Fertil. Soils* **2006**, *43*, 102–106. [CrossRef]
37. Harwood, C.E.; Nambiar, E.K.S. Productivity of acacia and eucalypt plantations in Southeast Asia. 2. trends and variations. *Int. For. Rev.* **2014**, *16*, 249–260. [CrossRef]
38. Food and Agriculture Organization. *World Reference Base for Soil Resources*; Food and Agriculture Organization of the United Nations: Rome, Italy, 1998.
39. Bich, N.V.; Mendham, D.; Evans, K.J.; Dong, T.L.; Hai, V.D.; Thanh, H.V.; Mohammed, C. Effect of residue management and fertiliser application on the productivity of young eucalyptus hybrid and *Acacia mangium* planted on steep slopes in Northern Vietnam. *South. For.* **2018**, in press.
40. Bärlocher, F. Leaf Mass Loss Estimated by Litter Bag Technique. In *Methods to Study Litter Decomposition: A Practical Guide*; Graça, M.A.S., Bärlocher, F., Gessner, M.O., Eds.; Springer: Dordrecht, The Netherlands, 2005; pp. 37–42.
41. O'Connell, A.M. Decomposition of slash residues in thinned regrowth eucalypt forest in Western Australia. *J. Appl. Ecol.* **1997**, *34*, 111. [CrossRef]
42. Berry, W.L.; Johnson, C.M. Determination of calcium and magnesium in plant material and culture solutions, using atomic-absorption spectroscopy. *Appl. Spectrosc.* **1966**, *20*, 209–211. [CrossRef]
43. Harmon, M.E.; Silver, W.L.; Fasth, B.; Chen, H.; Burke, I.C.; Parton, W.J.; Hart, S.C.; Currie, W.S.; Lidet. Long-term patterns of mass loss during the decomposition of leaf and fine root litter: An intersite comparison. *Glob. Chang. Biol.* **2009**, *15*, 1320–1338. [CrossRef]
44. Olson, J.S. Energy storage and balance of producers and decomposers in ecological systems. *Ecology* **1963**, *44*, 322–330. [CrossRef]

45. Siregar, S.T.H.; Nurwahyudi; Mulawarman. Effects of inter-rotation management on site productivity of *Acacia mangium* in Riau Province, Sumatra, Indonesia. In *Site Management and Productivity in Tropical Plantation Forests, Proceedings of Workshops, Piracicaba, Brazil, 22–26, November 2004 and Bogor, Indonesia, 6–9 November 2006*; Nambiar, E.K.S., Ed.; Center for International Forestry Research (CIFOR): Bogor, Indonesia, 2008; pp. 93–106.
46. Hai, V.D.; Trieu, D.T.; Thang, H.V.; Bich, N.V.; Cam, N.V.; Dinh, P.X.; Phuong, V.; Dien, P.V.; Tam, D.D.; Tiep, N.H. *The Research on Development of Schima wallichi and Schima superba Tree Species in Vietnam. Final Project Report No. VAFS 43/2010*; Vietnamese Academy of Forest Sciences: Hanoi, Vietnam, 2010; p. 120.
47. Santana, R.C.; Barros, N.F.; Comerford, N.B. Above-ground biomass, nutrient content, and nutrient use efficiency of eucalypt plantations growing in different sites in Brazil. *N. Z. J. For. Sci.* **2000**, *30*, 225–236.
48. Du Toit, B.; Dovey, S.B.; Fuller, G.M.; Job, R.A. Effects of harvesting and site management on nutrient pools and stand growth in a South African euclypt plantation. In *Site Management and Productivity in Tropical Plantation Forests, Proceedings of Workshops, Congo, July 2001 and China, February 2003*; Nambiar, E.K.S., Ranger, J., Tiarks, A., Toma, T., Eds.; Center for International Forestry Research (CIFOR): Bogor, Indonesia, 2004; pp. 31–43.
49. Li, X.; Yi, M.J.; Son, Y.; Park, P.S.; Lee, K.H.; Son, Y.M.; Kim, R.H.; Jeong, M.J. Biomass and carbon storage in an age-sequence of Korean Pine (*Pinus koraiensis*) plantation forests in Central Korea. *J. Plant Biol.* **2011**, *54*, 33–42. [CrossRef]
50. Palviainen, M.; Finér, L.; Kurka, A.-M.; Mannerkoski, H.; Piirainen, S.; Starr, M. Decomposition and nutrient release from logging residues after clear-cutting of mixed boreal forest. *Plant Soil* **2004**, *263*, 53–67. [CrossRef]
51. Garrett, L.G.; Kimberley, M.O.; Oliver, G.R.; Pearce, S.H.; Paul, T.S.H. Decomposition of woody debris in managed Pinus radiata plantations in New Zealand. *For. Ecol. Manag.* **2010**, *260*, 1389–1398. [CrossRef]
52. Zhang, X.Y.; Wang, W. Control of climate and litter quality on leaf litter decomposition in different climatic zones. *J. Plant Res.* **2015**, *128*, 791–802. [CrossRef] [PubMed]
53. Song, X.Z.; Jiang, H.; Peng, C.H.; Zhou, G.M.; Chang, S.X.; Yu, S.Q.; Ma, Y.D.; Peng, S.L.; Wei, X.H. Leaf litter decomposition along the temperate-tropical transect (East China): The influence of stand succession, litter quality and climate. *Pol. J. Ecol.* **2012**, *60*, 265–276.
54. Parton, W.; Silver, W.L.; Burke, I.C.; Grassens, L.; Harmon, M.E.; Currie, W.S.; King, J.Y.; Adair, E.C.; Brandt, L.A.; Hart, S.C.; et al. Global-scale similarities in nitrogen release patterns during long-term decomposition. *Science* **2007**, *315*, 361–364. [CrossRef] [PubMed]
55. Moore, T.R.; Trofymow, J.A.; Prescott, C.E.; Titus, B.D. Nature and nurture in the dynamics of C, N and P during litter decomposition in Canadian forests. *Plant Soil* **2011**, *339*, 163–175. [CrossRef]
56. Stevenson, F.J.; Cole, M.A. *Cycles of Soils: Carbon, Nitrogen, Phosphorus, Sulfur, Micronutrients*, 2nd ed.; John Wiley and Sons Ltd.: New York, NY, USA, 1999; p. 427. ISBN 0471320714.
57. Heal, O.; Anderson, J.; Swift, M. Plant litter quality and decomposition: An historical overview. In *Driven By Nature. Plant Litter Quality and Decomposition*; Cadisch, G., Giller, K., Eds.; CAB International: Wallingford, CT, USA, 1997; pp. 3–30.
58. Voigtlaender, M.; Laclau, J.P.; Goncalves, J.L.D.; Piccolo, M.D.; Moreira, M.Z.; Nouvellon, Y.; Ranger, J.; Bouillet, J.P. Introducing *Acacia mangium* trees in *Eucalyptus grandis* plantations: Consequences for soil organic matter stocks and nitrogen mineralization. *Plant Soil* **2012**, *352*, 99–111. [CrossRef]
59. Sánchez, G.; Pino, A.d.; Hernández, J. Decomposition of *Eucalyptus* sp. and *Pinus taeda* harvest residues under controlled temperature and moisture conditions. *Open J. For.* **2018**, *8*, 18.
60. Harwood, C.E.; Nambiar, E.K.S.; Dinh, P.X.; Toan, L.X.; Quang, L.T. Managing wood production from small grower *Acacia* hybrid plantations on eroded soils in central Vietnam. *Aust. For.* **2017**, *80*, 286–293. [CrossRef]
61. Mendham, D.S.; Hardiyanto, E.B.; Wicaksono, A.; Nurudin, M. Nutrient management of contrasting *Acacia mangium* genotypes and weed management strategies in South Sumatra, Indonesia. *Aust. For.* **2017**, *80*, 127–134. [CrossRef]
62. Mori, T.; Ohta, S.; Ishizuka, S.; Konda, R.; Wicaksono, A.; Heriyanto, J.; Hardjono, A. Effects of phosphorus and nitrogen addition on heterotrophic respiration in an *Acacia mangium* plantation soil in South Sumatra, Indonesia. *Tropics* **2013**, *22*, 83–87. [CrossRef]

63. Mori, T.; Wachrinrat, C.; Staporn, D.; Meunpong, P.; Suebsai, W.; Matsubara, K.; Boonsri, K.; Lumban, W.; Kuawong, M.; Phukdee, T.; et al. Contrastive effects of inorganic phosphorus addition on soil microbial respiration and microbial biomass in tropical monoculture tree plantation soils in Thailand. *ANRES* **2016**, *50*, 327–330. [CrossRef]
64. Nambiar, E.K.S.; Brown, A.G. *Management of Soil, Nutrients and Water in Tropical Plantation Forests*; ACIAR Monograph No.43; Australian Centre for International Agricultural Research (ACIAR): Canberra, Australia, 1997; p. 571.
65. Tiarks, A.; Ranger, J. Soil properties in tropical plantation forests: Evaluation and effects of site management: A summary. In *Site Management and Productivity in Tropical Plantation Forests, Proceedings of workshops, Piracicaba, Brazil, 22–26 November 2004 and Bogor, Indonesia, 6–9 November 2006*; Nambiar, E.K.S., Ed.; Center for International Forestry Research: Bogor, Indonesia, 2008; pp. 191–204.
66. Bich, N.V.; Eyles, A.; Mendham, D.; Evans, K.J.; Dong, T.L.; Hai, V.D.; Thang, H.V.; Thinh, N.V.; Mohammed, C. Effect of harvest residue management on soil properties of eucalyptus hybrid and Acacia mangium plantations planted on steep slopes of northern Vietnam. **2018**. Manuscript in preparation.
67. Ranjbar, F.; Jalali, M. Calcium, magnesium, sodium, and potassium release during decomposition of some organic residues. *Commun. Soil Sci. Plant. Anal.* **2012**, *43*, 645–659. [CrossRef]
68. Osono, T.; Takeda, H. Potassium, calcium, and magnesium dynamics during litter decomposition in a cool temperate forest. *J. For. Res.* **2004**, *9*, 23–31. [CrossRef]
69. Dauer, J.M.; Perakis, S.S. Calcium oxalate contribution to calcium cycling in forests of contrasting nutrient status. *For. Ecol. Manag.* **2014**, *334*, 64–73. [CrossRef]
70. O'Connell, A.M.; Malajczuk, N.; Gailitis, V. Occurrence of calcium oxalate in karri (Eucalyptus diversicolor F. Muell.) forest ecosystems of south western Australia. *Oecologia* **1983**, *56*, 239–244. [CrossRef] [PubMed]
71. Hardiyanto, E.B.; Wicaksono, A. Inter-rotation site management, stand growth and soil properties in *Acacia mangium* plantations in South Sumatra, Indonesia. In *Site Management and Productivity in Tropical Plantation Forests, Proceedings of Seventh Workshops, Piracicaba, Brazil, 22–26 November 2004 and Bogor, Indonesia, 6–9 November 2006*; Nambiar, E.K.S., Ed.; CIFOR: Bogor, Indonesia, 2008; pp. 107–122.
72. O'Connell, A.M.; Grove, T.S. Biomass production nutrient uptake and nutrient cycling in the Jarrah (*Eucalyptus marginata*) and Karri (*Eucalyptus diversicolor*) forests of South-Western Australia. In *Nutrition of Eucalypts*; Attiwill, P.M., Mark, A.A., Eds.; CSIRO Publishing: Collingwood, Australia, 1996; pp. 155–189.

2

Surface CO_2 Exchange Dynamics across a Climatic Gradient in McKenzie Valley: Effect of Landforms, Climate and Permafrost

Natalia Startsev [1], Jagtar S. Bhatti [1,*] and Rachhpal S. Jassal [2]

1 Natalia Startsev. Canadian Forest Service, Northern Forestry Centre, 5320 122 Street, Edmonton, AB T6H 3S5, Canada; natalia.startsev@canada.ca
2 Biometeorology and Soil Physics Group, University of British Columbia, Vancouver, BC V6T 1Z4, Canada; rachhpal.jassal@ubc.ca
* Correspondence: jagtar.bhatti@canada.ca

Academic Editors: Scott X. Chang and Xiangyang Sun

Abstract: Northern regions are experiencing considerable climate change affecting the state of permafrost, peat accumulation rates, and the large pool of carbon (C) stored in soil, thereby emphasizing the importance of monitoring surface C fluxes in different landform sites along a climate gradient. We studied surface net C exchange (NCE) and ecosystem respiration (ER) across different landforms (upland, peat plateau, collapse scar) in mid-boreal to high subarctic ecoregions in the Mackenzie Valley of northwestern Canada for three years. NCE and ER were measured using automatic CO_2 chambers (ADC, Bioscientific LTD., Herts, England), and soil respiration (SR) was measured with solid state infrared CO_2 sensors (Carbocaps, Vaisala, Vantaa, Finland) using the concentration gradient technique. Both NCE and ER were primarily controlled by soil temperature in the upper horizons. In upland forest locations, ER varied from 583 to 214 g $C{\cdot}m^{-2}{\cdot}year^{-1}$ from mid-boreal to high subarctic zones, respectively. For the bog and peat plateau areas, ER was less than half that at the upland locations. Of SR, nearly 75% was generated in the upper 5 cm layer composed of live bryophytes and actively decomposing fibric material. Our results suggest that for the upland and bog locations, ER significantly exceeded NCE. Bryophyte NCE was greatest in continuously waterlogged collapsed areas and was negligible in other locations. Overall, upland forest sites were sources of CO_2 (from 64 $g{\cdot}C{\cdot}m^{-2}{\cdot}year^{-1}$ in the high subarctic to 588 g $C{\cdot}m^{-2}{\cdot}year^{-1}$ in mid-boreal zone); collapsed areas were sinks of C, especially in high subarctic (from 27 $g{\cdot}$ $C{\cdot}$ m^{-2} $year^{-1}$ in mid-boreal to 86 $g{\cdot}C{\cdot}m^{-2}{\cdot}year^{-1}$ in high subarctic) and peat plateaus were minor sources (from 153 $g{\cdot}$ $C{\cdot}$ $m^{-2}{\cdot}$ $year^{-1}$ in mid-boreal to 6 $g{\cdot}C{\cdot}m^{-2}{\cdot}year^{-1}$ in high subarctic). The results are important in understanding how different landforms are responding to climate change and would be useful in modeling the effect of future climate change on the soil C balance in the northern regions.

Keywords: net carbon exchange; ecosystem respiration; upland forest; bogs; collapse scar; permafrost

---

1. Introduction

Northern regions have experienced considerable climate change during the last few decades, affecting the substantial pool of soil carbon stored in the permafrost [1–3]. Regions exposed to these changes are in many respects fragile and may undergo significant ecological alterations including permafrost thawing, potential release of fossil methane, and decomposition of previously frozen peat. More than one third of the planetary soil organic C is stored in the permafrost area [4,5], therefore assessing and forecasting of the carbon (C) storage dynamics in this pool has a major impact on the understanding of global C balance. Particular interest in this case lies in the monitoring of the changes in the surface CO_2 flux, and climate-driven changes in the permafrost status and peat accumulation rates. These changes are sensitive to local landforms and soil microclimate variations.

Effects that warming, and lengthening of the growing season, have on the C pool in the northern areas are multifaceted. On one hand, increases in temperature, depth of active soil layer, and length of growing season increase the rate of C loss from the soil; on the other hand, the same factors, aided by potential CO_2 fertilization effects, can increase the assimilation rate of the plants and hence biomass production [6]. A large amount of C stored in the soil in northern regions is evidence that assimilation rates continuously exceed of ecosystem respiration. The primary reason for this positive net C exchange (NCE) is lower decomposition rates of soil organic C due to low soil temperatures, soil waterlogging, and permafrost. The increase in mean annual air temperature affected soil temperatures, increasing the depth of the active layer, and escalating organic matter decomposition and thus ecosystem respiration, thereby shifting the NCE balance from carbon accumulation to degradation of stored carbon [7]. The implications of this shift include a release of vast amounts of carbon to the atmosphere in response to a comparatively minor increase in mean annual soil temperature [1,8].

Soil microbial activity response to changes in soil temperature varies depending on the temperature range. In the frozen soil layers at the temperature well below zero and with biological activity nearly non-existent [5,9], an increase of a few degrees Celsius would not change the carbon balance significantly as long as the soil stays frozen. However, in soil layers above freezing point, similar increase in soil temperature would cause an exponential increase in decomposition and ecosystem respiration [7,10]. The most appreciable effect of the climate change on the carbon balance can be expected in the areas of the zonal boundaries where relatively small changes in the soil temperature signify the difference between the frozen and thawed state influencing soil water regime, permafrost degradation, and ground subsidence, thereby resulting in a nearly instantaneous rise in CO_2 production.

Thawing of the permafrost follows a complex pattern of more intricate localized conditions than ecoclimatic gradient [11]. Areas affected by the permafrost frequently have a mosaic of landforms with locations of distinctly different soil moisture and temperature regimes in close proximity. Some of the most common landforms in the permafrost-affected areas include forested uplands, lichen-dominated peat plateaus, and collapse features of different origins. They differ in vegetation, allocation of C storage, soil types, hydrological regime, and consequently in the C balance. The effect of the landform and depth of the saturated soil layer on CO_2 fluxes is not fully understood and existing studies offer conflicting information. For example, Waddington and Roulet [12] suggested that CO_2 fluxes from the collapsed areas exceeded those from peat plateaus, while Mitsch and Gosselink [13] showed an opposite relationship.

Uplands generally have better drainage, mineral horizons are present within the active layer, and most of the CO_2 assimilation is performed by the tree crowns, while shaded ground covering the bryophyte layer plays only a minor role in total ecosystem assimilation [14]. Organic layers are frequently coarse mor to moder, and although comparatively thick, contain a limited and relatively constant amount of C [15]. This suggests that these areas are historically C neutral, neither sinks nor sources of CO_2, but that the assimilation and respiration functions of the ecosystem are largely stratified: tree crowns responsible for the bulk of total assimilation, and decomposition taking place at the ground surface, making the soil surface a strong source of CO_2 while the ecosystem as a whole remains carbon neutral [1]. Climate warming might increase or decrease total pools of stored carbon as higher temperatures would promote faster decomposition of soil C, greater tree growth rate would provide more litter fall and more carbon stored in standing biomass [3]. It is reasonable to expect that in either case, these ecosystems would reach another state of equilibrium with new, somewhat higher, decomposition rates of organic matter on the ground compensating for the greater assimilation rate of the crowns and undergrowth.

Peat plateaus landforms are composed of perennially frozen peat and appear to have developed under non-permafrost conditions which subsequently became elevated and permanently frozen [16].

These areas contain a large amount of carbon stored in the peat, indicating that historically they underwent a period of C accumulation [10]. However, measurements indicate that peat plateaus at present are either minor C sources or C neutral, or alternate between minor C sources and minor C sinks from year to year [17]. Most studies suggest that C balance of peat plateaus can be positive or negative varying from year to year [12,18]. This indicates that either warming has already shifted the NCE or that present peat plateaus have reached a stable stage of ecosystem development. Further warming is expected to increase respiration [5], and the depth of active layer thereby exposing stored C to degradation, or destabilization and collapse of the permafrost to create collapsed areas in place of peat plateaus [10,19].

Collapsed areas are also known as thermokarst form where permafrost thawed and ground surface subsided. These are soils composed of water-saturated peat with the water table close to the surface or above it, initially forming fast growing sphagnum communities [16]. Thermokarst features of the landform are predominantly sources of carbon emissions in the form of methane [12]. There are indications that high assimilation rates and low soil respiration due to waterlogging create conditions for net C sequestration [13]. Several studies suggested that collapsed areas are the greatest sinks of CO_2 in northern regions due to active sphagnum growth under conditions of constant supply of water, sunlight, and warmer conditions, while at the same time producing the greatest amounts of methane [20–23]. Carbon accumulation in these areas is conditioned by the presence of the high water table, which depends on the presence of permafrost in the surrounding areas that prevents water from draining away from the collapsed areas [24]. Further permafrost thaw would cause drainage of collapsed areas thereby affecting conservation of soil organic matter and supporting sphagnum growth, thus converting them from sinks to major sources of CO_2.

In Canada, the Mackenzie valley region in the northwest has undergone the most warming (1.7 °C) over the last century [25]. As a result, changes in the permafrost distribution have affected forest and peatland ecosystems in the Mackenzie valley region. It is uncertain how the above mentioned changes will affect the distribution, composition and C source/sink capacity of forests and peatlands, i.e., balance between organic matter production, heterotrophic mineralization, and greenhouse gases (GHG) emissions [2]. Limited data and understanding of how changing environmental conditions and permafrost thawing influences the C cycle of forests and peatlands over short timescales poses an even greater challenge to predict the changes in the C sink/source relationships and GHG dynamics under a changing climate. The goal is to advance and further develop C monitoring and assessment along diverse landscape (forest–peatland) and ecoregion (boreal–subarctic–arctic) gradients across the Mackenzie valley. This study was conducted with the following objectives: (1) To measure soil CO_2 fluxes, including NCE, along a climatic gradient and across different landforms; (2) To assess effects of permafrost, temperature and solar radiation on NCE and soil CO_2 production; and (3) To develop a statistical relationship to estimate ecosystem respiration (ER) along latitudinal gradient. Considering the difference in the soil surface carbon balance along the climatic gradient and between most common landforms, spanning from strong C sources to sinks, the task of evaluating and modeling of CO_2 emissions and assimilation in changing climate requires an approach that would take all of these variables into account.

2. Methods

2.1. Site Description

The study was conducted at four locations along a north-south climatic gradient in the Mackenzie Valley, Western Canada (Figure 1), from mid boreal with isolated patches of permafrost near Anzac, northern Alberta (AZ), to boreal forest with sporadic discontinuous permafrost zone in the area near Fort Simpson (FS), low subarctic with extensive discontinuous permafrost in the proximity of Norman Wells (NW), and high subarctic with extensive continuous permafrost in the Inuvik (IN) area. The mean values of the main climatic parameters for the sites (Table 1) indicate mean annual temperature and growing degree days decreasing but mean annual snow pack and precipitation increasing with latitude. Each study site was chosen to include a gradient from upland to peatland conditions and to a collapsed scar (Figure 2). Individual study plots were located within the upland forest (UL), peat plateau (PP; in Anzac site, a lichen-populated peat bog was used as ecologically comparable substitute), and a collapse feature (CS; in Anzac site, collapse feature was represented by an internal lawn; in Fort Simpson and Norman Wells, collapse scars; and in Inuvik a polygonal trench). Internal lawns are characteristically less than 50 cm lower than the surrounding bog and contain dead stands of *Picea mariana* and a *Sphagnum* ground cover [16].

Figure 1. Map of the site locations.

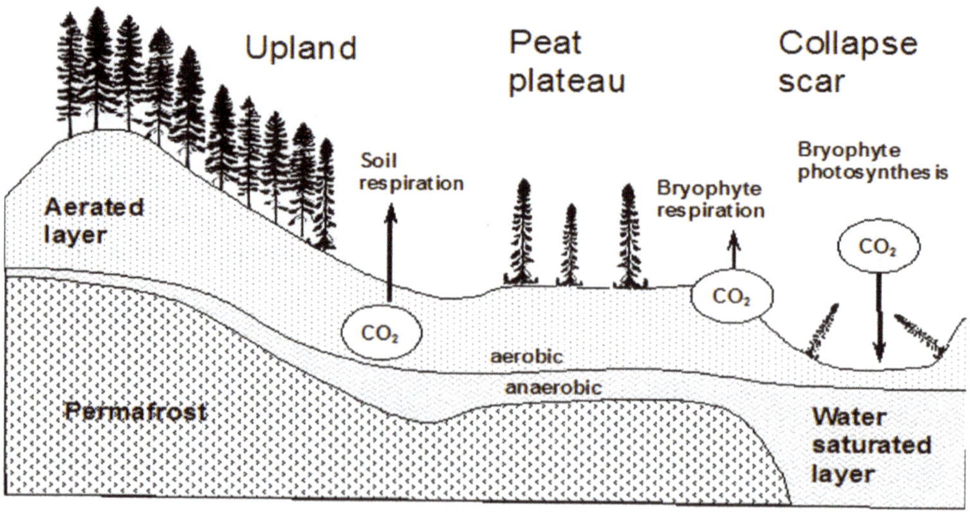

Figure 2. Schematic of the site landform features and key fluxes of CO_2. The arrows only represent the major fluxes not their magnitude at any point in time.

Table 1. Locations, ecozone permafrost status, and climatic description of the study sites.

	Latitude (°)	Longitude (°)	Ecozone	Permafrost status	Mean Annual Air Temperature (°C)	Growing Degree Days above 0 °C, (Days)	Mean Annual Snow Pack, (cm)	Mean Annual Precipitation, (mm)
Anzac (AZ)	56.40	−111.03	Mid-boreal	Isolated patches (0%–10%)	1.49	2251	32	464
Fort Simpson (FS)	61.63	−121.40	High-boreal	Sporadic discontinuous (10%–50%)	−2.17	2064	54	361
Norman Wells (NW)	65.21	−127.01	Low Subarctic	Extensive discontinuous (50%–90%)	−4.89	1817	48	318
Inuvik (IN)	68.32	−133.43	High Subarctic	Continuous (90%–100%)	−6.9	1295	60	259

Study Plots Establishment

At each site and each landform location, a study plot was established to facilitate measurements and minimize site disturbance. Board walks were laid in a cross pattern for easy reach to permanent sampling points; 10 cm in diameter plastic tubes (10 cm in length) were driven into the soil to serve as permanent collars for CO_2 measurements. In addition, 10 cm in diameter and 40 cm long plastic tubes were installed for the measurements of the water table.

2.2. *Vegetation Description*

In Anzac and Fort Simpson, uplands were characterized by a dense *Picea mariana* stand with a feathermoss understorey consisting primarily of *Hylocomium splendens* and *Pleurozium schreberi*. Peat plateaus had a sparse *Picea mariana* canopy with *Ledum groenlandicum* and *Vaccinium vitis-idaea* dominating the shrub layer, and a ground layer of lichen species, including *Cladina mitis*, *C. rangiferina*, and *C. stellaris*. The Norman Wells upland had a sparse *Picea mariana* canopy with *Ledum groenlandicum*, *Arctostaphylos rubra*, and *Vaccinium uliginosum* dominating the shrub layer, and a ground layer of primarily moss species, including *Tomenthypnum nitens*, *Hylocomium splendens*, and *Aulacomnium palustre*. The Inuvik upland plot was dominated by an open canopy of *Picea mariana* with a shrub layer of *Vaccinium vitis-idaea* and *Ledum decumbens*. *Petasites* sp. and *Rubus chamaemorus* were predominant in the herbaceous layer, while *Cladina mitis*, *Sphagnum angustifolium*, *Peltigera* spp. and *Cladonia* spp. comprised the ground layer.

Both Norman wells and Inuvik peat plateaus were treeless, with vegetation in Norman Wells dominated by a ground layer of *Cladina mitis* with patches of *Sphagnum fuscum*. The vascular vegetation consisted of a mix of *Ledum decumbens* with *Vaccinium vitis-idaea*, *Andromeda polifolia*, *Myrica gale*, and *Vaccinium uliginosum* in the shrub layer, and an herbaceous layer of scattered *Carex* sp. and *Rubus chamaemorus*. The Inuvik peat plateau was vegetated by *Ledum decumbens*, *Andromeda polifolia*, *Betula glandulosa*, and *Vaccinium vitis-idaea* dominating the shrub layer, a herbaceous layer of *Rubus chamaemorus* and *Carex* spp., and a ground layer comprised primarily of the lichen species *Flavocetraria nivalis* and *Cladina mitis*. Collapse scars were treeless at all sites, with vegetation dominated by a moss layer of *Sphagnum* sp. and *S. fuscum* with shrub cover including *Chamaedaphne calyculata* and *Andromeda polifolia*, and a herbaceous layer of *Smilacina trifolia*, *Eriophorum* sp. and *Carex* sp.

2.3. *Soil Description*

Soils at the study sites differed most notably with landforms and the presence of permafrost in the soil profile. Upland soils had comparatively better drainage and varied from moderately to imperfectly drained. The water table was within the upper 40 cm of the soil profile at the peat plateaus. Collapse features of the landforms did not have permafrost with the exception of the Inuvik site where the collapsed area was represented by a polygonal trough with ice-wedge at the depth of 50 cm. Upland soils were represented by Gleyed Dystric Brunisol in Anzac, Orthic Dystric Brunisol in Fort Simpson, Orthic Static Cryosol in Norman Wells, and Gleysolic Turbic Cryosol in Inuvik. Peat plateaus at all sites had a cover of Fibric organic cryosol except for Anzac where it was Typic Mesic Fibrosol developed in the absence of permafrost. The soil in collapse scars was classified as Hydric Fibrisol at all sites except Inuvik, where Mesic Organic Cryosol was found in the polygonal trench.

2.4. *CO_2 Flux Measurements*

2.4.1. Intra-Site Spatial Variability of CO_2 Fluxes

To determine spatial variability within a site, we measured NCE and ER using the PP system (EGM-4 portable CO_2 Gas Analyzer, PP System Inc., Amesbury, MA, USA) at three different locations, namely upland, peat plateau, and collapse scars. At each study location, six permanent gas flux sampling collars were installed within easy reach of the boardwalk in such a way that they were not shaded by the boardwalk. In order to improve the representation of spatial variability in hummocky areas, both hummocks and hollows were included as sample points. The measurements began in the fall of 2007 and continued throughout the growing seasons of 2008 and 2009. Surface CO_2 fluxes were measured using the PP system. Surface NCE and ER were measured using a clear and an opaque cover, respectively. The rate of assimilation of the soil cover vegetation was obtained from the difference between NCE and ER.

Spatially explicit measurements were taken during daytime, usually in the early afternoon, during the warmest part of the day when temperature-driven microbiological activity would have reached its maximum. In order to calculate daily flux values, a relationship was established between hourly fluxes and soil temperature (Table 2) and used to extrapolate flux values for all twenty-four hours of the day. To estimate net carbon exchange (NCE) at the soil surface, exponential regression (1):

$$\mathrm{NCE} = ae^{(bT)} \quad (1)$$

was used where NCE is the net carbon exchange ($g{\cdot}CO_2{\cdot}m^{-2}{\cdot}h^{-1}$) and T is the soil temperature at the 5-cm depth (°C), a is the reference NCE at zero °C, and coefficient b represents the temperature sensitivity. For the purpose of modeling the temperature effect on ER, an exponential regression (2):

$$\mathrm{ER} = ae^{(bT)} \quad (2)$$

was used where ER is the ecosystem respiration ($g{\cdot}CO_2{\cdot}m^{-2}{\cdot}h^{-1}$) and T is soil temperature at the 5-cm depth (°C), a is reference respiration at zero °C, and coefficient b represents the temperature sensitivity. Soil moisture has insignificant effect on the soil CO2 fluxes. Values of Q_{10}, which is the relative increase in respiration for a 10 °C increase in temperature, were obtained as Equation (3):

$$Q_{10} = \mathrm{b}(1 + 10\,\mathrm{T}^{-1}) \quad (3)$$

Similarly, the effect of solar irradiation on the net assimilation rate (NAR) was evaluated using a log function (Equation (4)):

$$\mathrm{NAR} = a'\ln(Q) - b' \quad (4)$$

where NAR is surface CO_2 net assimilation rate ($g{\cdot}CO_2{\cdot}m^{-2}{\cdot}h^{-1}$) and Q is PAR in $\mu mol{\cdot}photons{\cdot}m^{-2}{\cdot}s^{-1}$, and a' and b' are the coefficients.

Table 2. Parameter values to estimate the effect of soil temperature T (°C) at the 5-cm depth on net carbon exchange (NCE) ($g{\cdot}CO_2{\cdot}m^{-2}{\cdot}h^{-1}$) using equation NCE = $ae^{(bT)}$ and coefficient of determination for landforms across different ecoregions (sites).

Site	Location	a ($g{\cdot}CO_2{\cdot}m^{-2}{\cdot}h^{-1}$)	b ($°C^{-1}$)	R^2
Mid-boreal	Upland	0.128	0.17	0.91
	Peat Plateau	0.018	0.28	0.75
	Collapse scar	0.072	0.08	0.79
High-boreal	Upland	0.151	0.09	0.84
	Peat Plateau	0.048	0.11	0.91
	Collapse scar	0.012	0.15	0.82
Low Subarctic	Upland	0.035	0.15	0.82
	Peat Plateau	0.034	0.11	0.76
	Collapse scar	0.047	0.11	0.68
High Subarctic	Upland	0.023	0.16	0.97
	Peat Plateau	0.010	0.27	0.87
	Collapse scar	0.011	0.356	0.82

2.4.2. Temporal Variation in CO_2 Fluxes

Surface NCE was also measured for extended periods of time (up to 45 days) with clear chambers using ACE automatic CO_2 exchange systems (ADC BioScientific Ltd., Hertfordshire, UK). ACE units were equipped with a built-in photosynthetically active radiation (PAR) silicone photocell sensor and a soil temperature probe, and were powered by gel-cell batteries and solar panels. The ACE units were deployed simultaneously in upland, and peat plateau locations with one unit in each plot for comparison of local CO_2 fluxes, surface temperatures, and illumination intensities between landform positions.

We also measured soil CO_2 concentrations using Vaisala solid-state infrared CO_2 sensors (Vaisala Oyj, Helsinki, Finland) at three depths (5, 20, and 40 cm) with Vaisala model GMP222 (measurement range 0–10,000 $\mu mol \cdot CO_2 \cdot mol^{-1}$) at the 5 cm depth, and model GMP221 (measurement range 0–20,000 $\mu mol \cdot CO_2 \cdot mol^{-1}$) at the 20 and 40 cm depths. The sensors were protected with Teflon socks and installed at different depths as explained in Jassal et al. [26], and connected to a data logger (Campbell Scientific CR1000) recording soil CO_2 concentrations every two hours. Soil respiration was obtained with the gradient technique using Equation (5) (see below). To obtain diffusivity values, soil physical properties, such as bulk density and porosity, and variations in soil temperature and moisture content were also measured. The gradient approach allowed for the calculation of the contribution of different soil layers to the total surface flux.

Soil temperature and volumetric water content (θ_v) were measured continuously starting spring 2008 to fall 2010 using HOBO U12 units equipped with temperature sensors (Model TMC6-HD, Onset Computer Corp., Bourne, MA, USA) and soil moisture sensors (Watermark Sensor type WMSM, Delta-T Device Ltd., Cambridge, UK), respectively. The temperature sensors were installed at the 5, 25, and 50 cm depths while the moisture sensors were installed at the 5 cm depth. Soil temperature and moisture readings were also taken every two hours.

2.4.3. Calculating Soil CO_2 Fluxes

CO_2 flux was calculated from the Fick's first law of diffusion (Equation (5)):

$$F = -D_s \frac{dC}{dz} \tag{5}$$

where F is CO_2 flux in $\mu mol \cdot m^{-1} \cdot s^{-1}$, D_s is diffusion coefficient in the soil ($m^2 \cdot s^{-1}$), C is soil CO_2 concentration at the given depth and dC/dz is vertical gradient of CO_2 concentrations in the soil between two measurement depths. The value of D_s was calculated as $D_s = \zeta D_a$ where ζ is soil gas tortuosity factor and D_a is CO_2 diffusivity in the air. The latter is calculated from the standard diffusivity value (D_{a0}) at 20 °C (293.15 K) and 101.3 kPa using the following equation (Equation (6)):

$$D_a = D_{a0} \left(\frac{T}{293.15}\right)^{1.75} \left(\frac{P}{101.3}\right) \tag{6}$$

where T is temperature (K) and P is pressure (kPa). The value of D_{a0} is empirically given as 14.7 $m^2 \cdot s^{-1}$. We used the Millington-Quirk model [27] for calculation of tortuosity (ζ) as it was proven to provide the best correlation to the measured tortuosity [28] (see Equation (7)).

$$\zeta = \frac{\alpha^{\frac{10}{3}}}{\phi^2} \tag{7}$$

where α is air-filled porosity and ϕ is total porosity. The calculated soil CO_2 fluxes, using the gradient technique, for the uppermost soil layer (0–5 cm) were validated by plotting them against soil respiration values measured using the PP system on bare soil.

2.5. Statistical Analysis

Annual NCE and ER values were calculated for each site and landform position for the growing season (1 June to 30 September) of 2008, 2009, and 2010 by summing all daily values. As outlined in Section 2.4.1, collars were installed along three arms of the boardwalk, with two pseudo-replications on each arm. Arms were treated as replications. Re-measurements of the collars each month were initially treated as repeated measures. However, that did not permit for the account of variations in the temperature and light.

The latest calculation treats re-measurements as replicated measurements using ANCOVA testing for month × temperature covariance. Barlett's test was used to test homoscedasticity and Pearson's chi-square test for normality for comparison of the groups by site and by plot. In case, the assimilation in the two plots variability ratio did not meet the criteria for homoscedasticity, the weighed variables method in analysis was employed.

Comparison of temporal variations of NCE and ER between different sites and landform positions was conducted using analysis of Variance (ANOVA) with repeated measures coupled with Tukey's Studentized Range Test in SAS (SAS Institute Inc., Cary, NC, USA, 2004) [29]. Interaction effects were elucidated using the *pdiff* option in SAS. To determine the significance of trends in NCE, ER, and assimilation rate and landforms variables, Pearson product-moment correlations were carried out in SAS. Significance was tested using p-value = 0.05.

3. Results

3.1. Weather Monitoring

During three years of the study, weather at all four sites was close to their climatic normals (Table 1). The temperature and precipitation at different sites are presented in Figure 3. Year 2007 was somewhat warmer at all sites except for Norman Wells where the annual temperature was the same as the long-term average.

The annual mean temperature in 2007 in Anzac and Inuvik was warmer than their long-term means by 0.2 and 1.2 °C, respectively. 2008 was also somewhat warmer at all sites except Norman Wells. Year 2009 was slightly cooler at all sites except Inuvik where the mean annual temperature was about 1 °C higher than normal. Though the length of the warm (above freezing) season was significantly longer in the southern sites with mild winters, there were no significant differences in the mid-summer temperatures between the sites. Soil temperature generally followed air temperature with a maximum in July, though it did not warm up to the same degree in the north as in the south during shorter summers.

Precipitation along the valley typically is between 300 to 400 mm, tapering off along the Mackenzie Delta to 220 mm at Inuvik. Individual daily rainfalls are typically light with few exceeding 5 mm; however, heavy rains can happen during summer months with rainfall occasionally exceeding 30 mm for sites in Fort Simpson and Newman Wells. After October, most of the precipitation falls as snow.

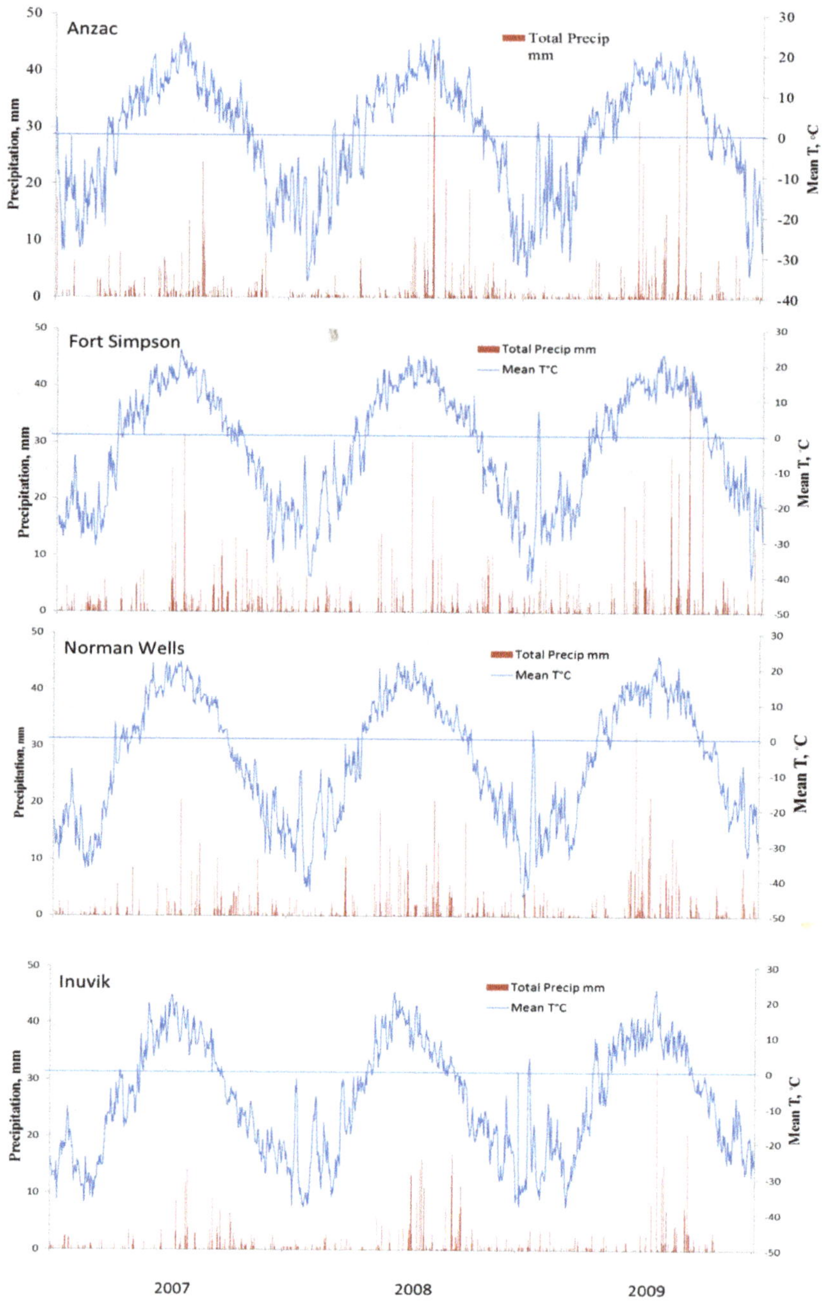

Figure 3. Seasonal patterns of soil temperature (T_s) and precipitation (Ppt) during the three year period at Anzac (AB), Fort Simpson (NWT), Norman Wells (NWT) and Inuvik (NWT).

3.2. Diurnal Variations in NCE

Diurnal measurements of surface NCE indicated the surface was a source of CO_2 with respiration dominating the gas exchange balance at most of the study sites. Hourly measurement of NCE corresponded closely to diurnal temperature variations. Figure 4 shows the distinct diurnal pattern in measured NCE in collapse scars, peat plateau, and upland areas at the Fort Simpson site with upland showing maximum (9.0 μmole$\cdot$m$^{-2}\cdot$s^{-1}) NCE followed by peat plateau (3.9 μmole$\cdot$m$^{-2}\cdot$s^{-1}) and collapse scars (0.3 μmole$\cdot$m$^{-2}\cdot$s^{-1}), also suggesting that photosynthetic activity of the ground cover bryophytes played a notable role only in collapsed areas. Both upland and tree-covered peat plateau areas had higher surface CO_2 fluxes (emissions) during the daytime, but collapse scar exhibited a double-peak pattern with higher respiration during nighttime, which began to decrease with sunrise and reached negative values (becoming carbon sink) during morning hours. During mid-day when the temperature was at the peak, collapse scar NCE turned positive again followed by another period of negative NCE showing the prevalence of assimilation during late afternoon until sunset.

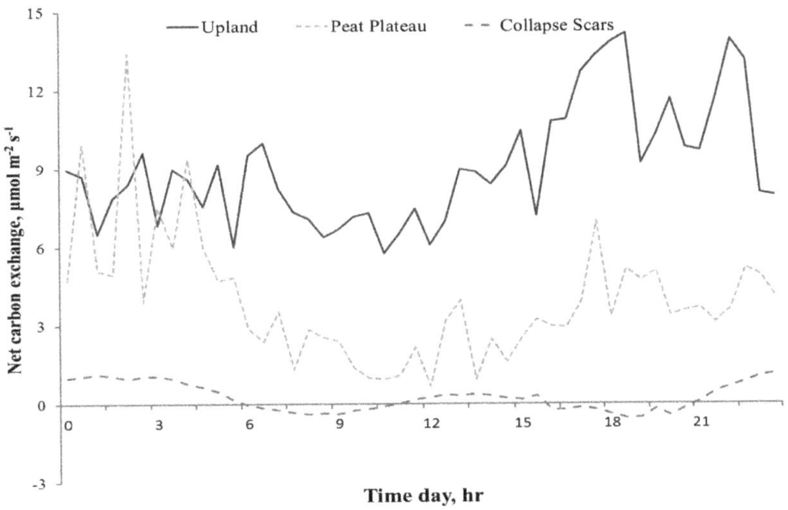

Figure 4. Role of landform on diurnal variation in NCE in upland, peat plateau, and collapse scar on 25 June 2009, at Fort Simpson (NWT) Canada.

3.3. Statistical Model of Surface CO_2 Fluxes against Temperature

Net carbon exchange (NCE) was primarily controlled by soil temperature at the 5 cm depth as the exponential model (Equation (1), NCE = $ae^{(bT)}$) explained between 68% to 97% variation in observed fluxes for upland, peat plateau, and collapse scar sites along the latitudinal gradient (Table 2). Adding soil moisture factor into the model did not improve it any further. The model showed variation in coefficients a and b, and Q_{10} values, based on landform along the north-south gradient. At peat plateau and collapsed areas, the effect of temperature was less obvious (Tables 3 and 4) where soil moisture would have a dominant control. There was also a good agreement between ER measurements and soil temperature at the 5 cm depth at different sites and landforms. The values of the parameters in Equation (2) (ER = $ae^{(bT)}$) are presented in Table 3. The value of the a coefficient was higher for all the upland sites as compared to peat plateau. The ER-temperature and NCE temperature responses did not differ dramatically among landform units suggesting that ER was a major component of the NCE. The $Q_{10\ values}$ estimated for different landform positions along the latitudinal gradient are also shown in Table 4. The response of ER to the temperature varied between ecoregions and landform positions with Q_{10} values being the greatest in the peat plateau of south most sites (5.6–11.5) followed by uplands (2.9–5.4) and collapsing scar (1.2–4.0), and decreasing with increases in latitude (Table 4). Furthermore, we also found that ER was negatively affected by water table and permafrost depths.

Table 3. Parameter values to estimate the effect of soil temperature T (°C) at the 5-cm depth on surface respiration ($g{\cdot}CO_2{\cdot}m^{-2}{\cdot}h^{-1}$) using equation ER = $ae^{(bT)}$ and coefficient of determination for landforms across different ecoregions (sites).

Site	Location	a ($g{\cdot}CO_2{\cdot}m^{-2}{\cdot}h^{-1}$)	b ($°C^{-1}$)	R^2
Mid-boreal	Upland	0.061	0.251	0.68
	Peat Plateau	0.010	0.339	0.83
High-boreal	Upland	0.146	0.080	0.63
	Peat Plateau	0.001	0.216	0.91
Low Subarctic	Upland	0.008	0.208	0.65
	Peat Plateau	0.001	0.519	0.58
High Subarctic	Upland	0.010	0.183	0.56
	Peat Plateau	0.002	0.299	0.46

Table 4. Q_{10} values for surface respiration in upland (UL), peat plateau (PP) and collapse scars (CS) across different ecoregions (sites).

Site	Uplands	Peat Plateau	Collapse Scar
Mid-boreal	5.38	11.49	3.97
High-boreal	3.41	11.36	3.28
Low Subarctic	3.16	5.45	2.18
High Subarctic	2.86	5.53	1.22

3.4. Time Series of NCE and ER

The temporal variations in surface NCE and ER are shown in Figures 5 and 6, respectively, depicting most pronounced differences among sites. Surface NCE and ER followed a bell-shaped curve during the year closely corresponding to the seasonal variation in soil temperature. The peak NCE and ER values were observed during the summer months of July and August (Figure 5). The greatest amount of CO_2 was released in the southernmost site; however, from mid-boreal to low-arctic, the rate of CO_2 release from uplands dropped sharply due to a lower seasonal temperature in the north. A comparison of Figures 5 and 6 show that in upland locations, ER was the major component of NCE, exceeding assimilation rate by ground cover vegetation three-fold. Therefore, the response of NCE to ecoclimatic factors was, to a large extent, dictated by the response of ER. NCE did not vary much between peat plateaus with extensive and continuous permafrost in low or high subarctic regions. Results show that total CO_2 emissions were greatest in the upland location, least in the collapse scar with peat plateau occupying the intermediate position (Figure 5).

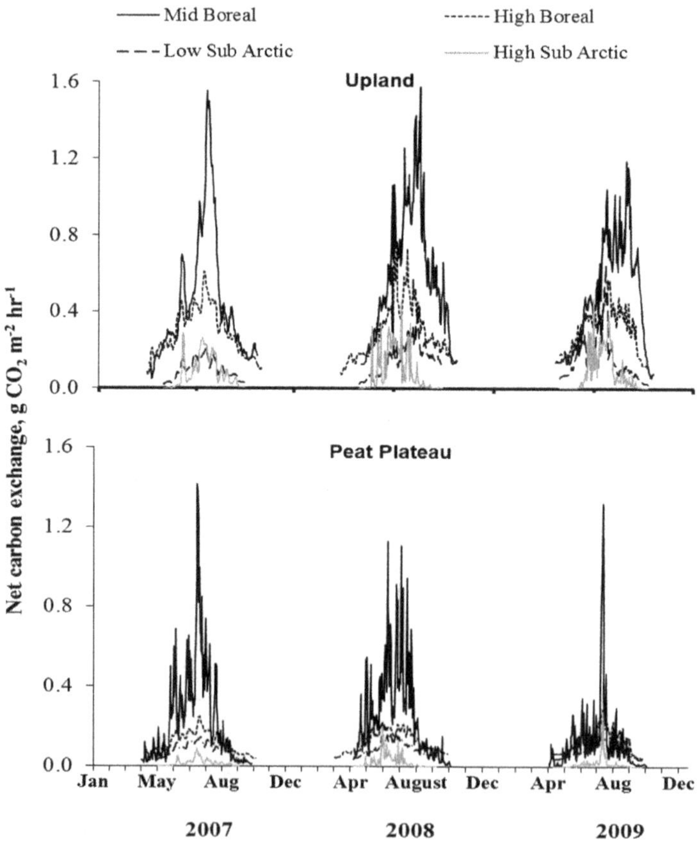

Figure 5. Temporal variation in net carbon exchange (NCE) for different landforms at sites within different ecoregions.

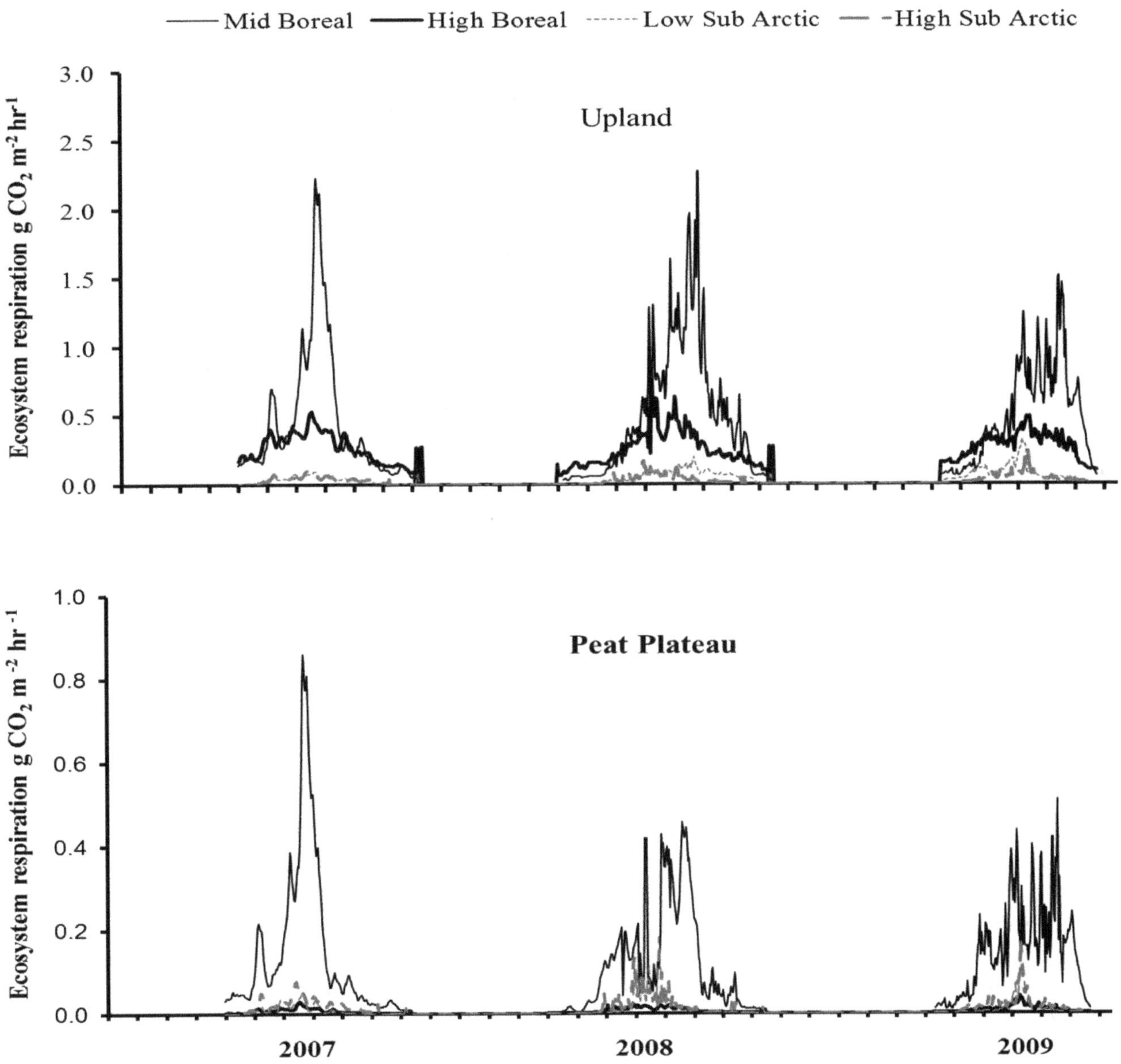

Figure 6. Temporal variation in ecosystem respiration (ER) at upland and peat plateau landforms at sites within different ecoregions.

Calculated flux contributions from different soil layers using the CO_2 gradient technique showed that in both peat plateau (PP) and upland (UL) areas, the greatest contribution (about 75%) of soil respiration (SR) originated in the upper 5 cm layer (Figure 7). Combined contribution of the lower horizons did not exceed 25% of the total flux. Both in upland and peat plateau, soil temperature was a strong driver of CO_2 release in SR. In peat plateau, the lower layer (21 to 40 cm) contributed about only 5% of SR, which could be due to the presence of the water table. Peat plateau areas exhibited greater variability in the flux with a pronounced increase in the CO_2 emissions after rain events, which can be attributed to increased decomposition [26] or displacement of soil-stored CO_2 [30].

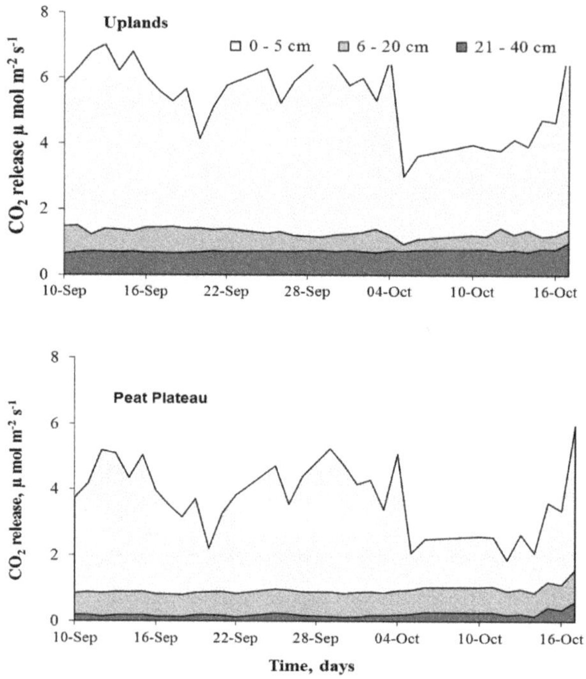

Figure 7. Comparative contribution of different soil layers to soil CO_2 efflux in uplands (top panel) and peat plateau (bottom panel) during 2008 at Fort Simpson (NWT) Canada.

3.5. *Effect of Light on Surface Assimilation Rate*

The surface assimilation rate was calculated as the difference between measured hourly NCE and ER, and was correlated to the simultaneously measured PAR values. Surface assimilation was significantly correlated with PAR at different landforms along the climatic gradient (Table 5). Correlation obeyed logarithmic relationship and was most significant in the areas of unimpeded solar radiation. In collapsed scar areas, seasonal assimilation was consistently higher (Table 6) than in either uplands or peat plateaus by as much as three times due to the abundance of sphagnum. In peat plateaus, assimilation rate was highly variable due to the frequent drying out of the lichens, which dominated the ground vegetation. With less shading in the peat plateau as compared to uplands (forest areas), significantly higher assimilation was observed (Table 6). During dry days, the surface assimilation was not measurable, and when surface was moist, assimilation by lichens was comparatively low (less than 0.05 $g{\cdot}CO_2{\cdot}m^{-2}{\cdot}h^{-1}$) but increased logarithmically with illumination.

Table 5. Parameter values to estimate the effect of light on surface assimilation ($g{\cdot}CO_2{\cdot}m^{-2}{\cdot}h^{-1}$) using Equation: $F'(CO_2) = a' \times \ln(Q)^{-b'}$ and coefficient of determination for landforms across different ecoregions (sites).

Site	Location	A'	b'	R^2
	Upland	0.41	0.15	0.57
Mid boreal	Peat Plateau	0.45	0.17	0.76
	Collapse Scar	1.27	0.26	0.81
	Upland	0.21	0.07	0.71
High boreal	Peat Plateau	0.20	1.14	0.55
	Collapse Scar	0.17	0.53	0.67
	Upland	0.18	0.93	0.69
Low Subarctic	Peat Plateau	0.11	0.61	0.93
	Collapse Scar	0.15	0.75	0.74
	Upland	0.16	0.64	0.66
High Subarctic	Peat Plateau	0.02	0.03	0.71
	Collapse Scar	0.38	0.12	0.87

Table 6. Cumulative seasonal (1 June to 30 September 2009) net assimilation rate at different landform positions (in g·C·m^{-2}·season^{-1}).

Site	Upland	Peat Plateau	Collapse Scar
Low boreal	82.6 ± 4.01 Aa	139.2 ± 12.4 Ba	211.3 ± 9.68 Ca
High boreal	62.0 ± 6.58 Ab	92.0 ± 8.92 Bb	198.0 ± 5.24 Ca
Low Subarctic	56.7 ± 3.74 Ac	63.3 ± 7.69 Ac	127.4 ± 3.92 Cb
High Subarctic	38.12 ± 4.62 Ad	54.6 ± 8.16 Bc	86.5 ± 6.15 Cc

Capital letter represent the $p < 0.05$ between land form and lower case letter shows the $p < 0.05$ between sites.

3.6. Annual ER and NCE

A comparison of seasonal NCE among different sites (Figure 8) showed that net carbon exchange at the surface was a significantly larger source at the upland sites (38 to 576 g·C·m^{-2}·year^{-1}) as compared to peat plateau (8 to 159 g·C·m^{-2}·year^{-1}) while collapse scars (−85 to −23 g·C·m^{-2}·year^{-1}) were C sinks at all locations along the transect. In general, annual NCE decreased with mean annual temperature, which decreased with increase in latitude, with high sub-arctic site showing a net assimilation integrated over upland, peat plateau, and collapse scar areas. The upland and peat plateau sites were, on average, net CO_2 sources to the atmosphere, while the collapse scar sites were, on average, net CO_2 sinks. The NCE measurements pertain only to the ground surface and short vegetation and do not include CO_2 uptake by trees and large shrubs. ER varied along the latitudinal gradient, for upland sites from 6.0 to 538 g·C·m^{-2}·year^{-1}, for peat plateau from 2.0 g·C·m^{-2}·year^{-1} to 187g·C·m^{-2}·year^{-1}, followed by collapse scars with 60 to 214 g·C·m^{-2}·year^{-1} (Figure 9). Low soil temperature along with high water table at both upland and peat plateau sites in low and high subarctic regions resulted in lower ER. Since the water table was at the surface at all the collapse scar sites, ER was negligible. In mid and high-boreal zone, ER at the upland sites was two- to three-fold higher as compared to peat plateau sites with and without collapse scar.

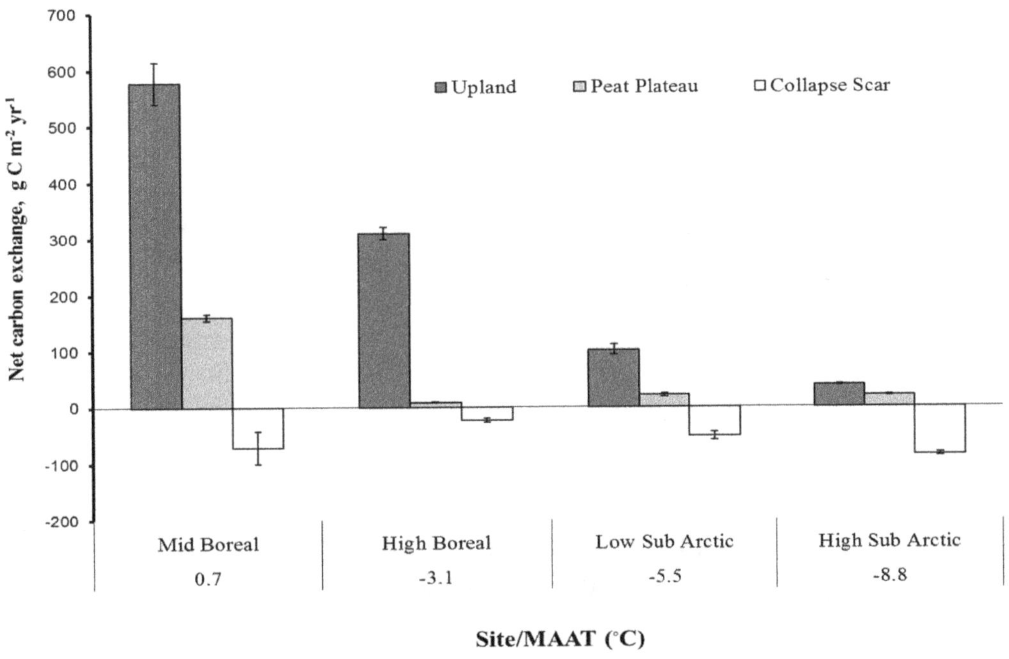

Figure 8. Seasonal net carbon exchange (NCE) for different sites and landform positions. MATT (°C) is mean annual air temperature.

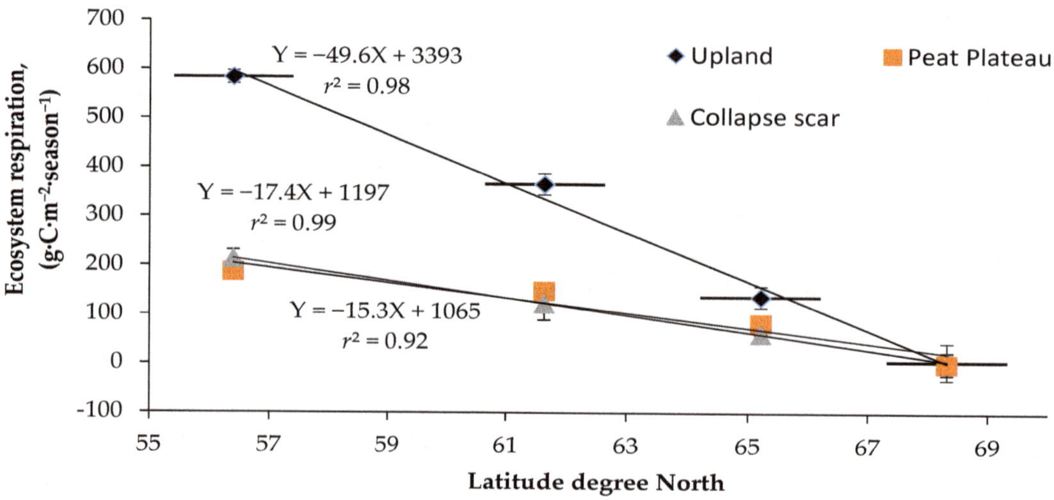

Figure 9. Cumulative seasonal surface fluxes for different landforms at sites along the latitudinal gradient.

4. Discussion

CO_2 exchange from the surface of permafrost-affected areas is determined by a combination of several factors, including soil microclimate [31–34], landform features [2,35–37], water table depth [38], extent of permafrost [2,39,40], and vegetation type [41,42]. There are two important contributors to the CO_2 flux: soil respiration and assimilation, and respiration of the ground cover vegetation. They vary differently in response to physiology and ecoclimate. NCE, assimilation rate and ER by surface vegetation were significantly affected by the local landform positions (i.e., uplands, peat plateau, and collapse scars), and different sites along latitude north–south gradient.

4.1. Spatial Variation in Net Carbon Exchange (NCE)

Net ecosystem exchange in a subarctic forest was lower than that of boreal forest due to extremely cold soils and high water table in the north. Bisbee [41] reported NCE to be 562 $g{\cdot}C{\cdot}m^{-2}{\cdot}year^{-1}$ for southern sites while Wang et al. [43] reported it to be 226–413 $g{\cdot}C{\cdot}m^{-2}{\cdot}year^{-1}$ for northern sites in the BOREAS study, which are similar to the NCE observed in the present study for mid- and high-boreal forest zones. In peat plateau, a presence of permafrost significantly decreased NCE from 160 $g{\cdot}C{\cdot}m^{-2}{\cdot}year^{-1}$ to 20–50 $g{\cdot}C{\cdot}m^{-2}{\cdot}year^{-1}$ among areas with sporadic and continuous permafrost. Similar NCE values were observed for a nutrient rich fen and sub-arctic palsa in Scandinavian peatlands [43,44]. However, it is not always possible to compare our seasonal NCE estimates with other studies, because of differences in climate, duration of the study period and variations in snow-cover period. Many studies have observed that spring conditions greatly influence seasonal and annual CO_2 exchange, where earlier snow melts and/or warmer spring temperatures almost universally enhance accumulated CO_2 uptake in high-boreal and subarctic regions [40,45,46]. Collapsed areas usually behaved as CO_2 sinks due to high productivity of bryophytes and restricted respiration in flooded soils. The length of the thawed period was significantly shorter in the northern sites, limiting both decomposition and growth periods, but this effect was apparently compensated by the increase in bryophyte assimilation due to longer daylight conditions in subarctic regions. It is likely that the combination of these two factors was responsible for similar values of NCE in collapsed areas at different sites as total assimilation varying between 20 and 70 $g{\cdot}C{\cdot}m^{-2}{\cdot}year^{-1}$. Our NCE estimates for collapse scar sites were in the range of values reported for low Arctic tundra sites [47–49] and also observed large inter-annual variability in NCE

4.2. Assimilation Rates across Different Landforms

Another important factor that determines ecosystem carbon exchange is photosynthetic activity (i.e., CO_2 assimilation rate) of other ground cover vegetation [50]. Assimilation of CO_2 by the bryophytes is affected by light intensity, moisture, temperature, and species composition. All these factors varied with ecozone and landform. Upland forests of the southern sites had greater shading than northern uplands and therefore had a limited assimilation rate of bryophyte layer, though the total length of the growing season was longer and the canopy provided some protection from drying. While assimilation rate of the ground cover vegetation was inhibited by shading, relatively good drainage of the upper soil layers, stable soil moisture levels, and abundant input of organic debris from tree crowns provided favorable conditions for high microbial respiration. Bryophyte layer of peat plateaus had a large proportion of lichens that are less effective in CO_2 assimilation [51] but are prone to drying, especially in non-treed northern sites. *Sphagnum* present at peat plateau sites may have been moisture-limited because *Sphagnum* moss photosynthesis is highly sensitive to moisture [52,53]. Tuittila et al. [54] measured the optimum water table level for photosynthesis by *S. angustifolium* and observed decreased photosynthesis rate at lower water table levels in bogs. Most effective bryophyte assimilation took place in the collapsed areas due to unimpeded illumination, earliest warming in the spring, unlimited supply of soil moisture, and relatively effective photosynthesis of mosses [55].

4.3. Spatial Variation in Ecosystem Respiration (ER)

Variation in soil temperature at the 5-cm depth explained most of the variability in ER. Q_{10} values for ER observed in the study (Tables 3 and 4) were within the range of values reported for boreal forest soils [56–58]. Many other studies on boreal forest stands [58,59] have shown that Ts (soil temperature) exerts a major influence on ER. Although, ER in uplands was primarily driven by soil temperature, high water table present in the collapse scar sites inhibited respiration [60]. Therefore the highest ER was observed in the southern-most location in the relatively dry upland soils. In the northern sites, it was limited by short frost-free periods, and frequently by water saturation in peat plateau and collapse scar sites [61]. Spatial variability in ER showed a strong relationship with latitudinal gradient which is a good proxy for the combined influence of radiation, length of growing season, temperature, precipitation, and vegetation cover (Figure 8). The different relationships for upland and peat plateau showed that the vegetation and climate were the most critical factors in regulating the spatial variability in ER with latitude [62,63]. As temperature, length of growing season and vegetation productivity decrease from south to north, ER decreases with latitude.

4.4. Quantitative Relationships between NCE and ER and Climate Variables

Soil temperature and moisture are important parameters in regulating surface CO_2 fluxes in boreal ecosystems [60,64–66]. Various empirical models have been used to quantify the relationship between ER and temperature, e.g., simple linear by Bisbee, [41] and Wang et al. [43] exponential by Lavigne et al., [64] and Rayment and Jarvis, [65]. We found that NCE and ER were highly correlated to soil temperature with an exponential relationship for all the sites. Increasing temperature can increase surface CO_2 fluxes through (a) activating dormant microbes and increasing microbial species richness, thus mineralization of C; (b) increasing root respiration by increasing photosynthates translocation from the aboveground part of the plant to roots; (c) accelerating gas exchange between ecosystems and the atmosphere [60]. However, the temperature sensitivity of NCE and ER is not constant, but tends to be modified by seasonal changes in soil moisture, root biomass, litter inputs, microbial populations, and other seasonally fluctuating conditions and processes [67]. Although temperature represented a significant control on the variability of the surface CO_2 fluxes in this study, other studies have found that water table depth or timing of the snowmelt play a more important role in predicting the CO_2 uptake capacity in many peatlands. For example, in a minerogenic peatland in Sweden, Peichl et al. [58] found that the growing season NCE was strongly correlated with pre-growing season

T_{air}, (air temperature) and suggested that less root damage would occur during warmer winters and access to water by the vegetation would be easier. Similarly, in a mild maritime climate ombrotrophic bog, winter temperature would be a main control on the inter-annual variability in NEE [61].

Another important driver of ER is the nature and source of the decomposing material. Landform is closely associated with vegetation type. While upland locations at all sites had black spruce stands, collapsed areas had only ground cover vegetation, and peat plateaus had trees in boreal ecozones but no trees in subarctic sites. Accordingly, these locations varied in the amount of plant debris input and its quality. Startsev and Lieffers [68] showed that decomposing litter generally trapped in the bryophyte layer contributed to the total CO_2 production, and provided bryophytes with a source of easily available nutrients and carbohydrates. In our study, the total amount of plant debris input was greater in the southern treed areas where trees were bigger and more productive, which was also an important factor in southern treed peat plateaus. In collapsed scar areas and northern subarctic treeless peat plateaus, the only source of decomposing organic matter was the annual production of ground vegetation and small shrubs, which provided considerably less decomposable material.

The type of ground vegetation could be another major variable to determine ER as upland sites were dominated by feathermoss in mid and high-boreal ecozones while peatland was dominated by the lichens. We found that Q_{10} values were highest at the peat plateau sites dominated by lichens. Bergeron et al. [62] also observed that Q_{10} was highest under lichen followed by sphagnum and lowest for feathermoss dominated understory vegetation. ER of bryophyte layer was also affected by temperature. The observed decrease in Q_{10} values from 5.4 and 5.5 to 2.2 and 1.2, upon thawing of peat plateau in low subarctic and high subarctic region respectively, were close to the range reported by previous studies [69,70].

Permafrost is another factor inhibiting respiration rates. ER was most active in the upper soil horizons (Figure 6). It appears that at upland and peat plateau sites, the depth of active layer was sufficient to support microbiological activity throughout the growing season, and therefore presence of permafrost had only a limited effect on ER. On the other hand, at collapse scar sites, accompanying factors such as length of time over which soil was frozen, the mean soil temperature throughout the year, and the presence of perched water table in the soil profile had a major impact on ER [38]. In collapse scar sites, lower ER could be attributed to slow decomposition rates of deep organic matter, low O_2 availability in saturated soils, low temperature, and substrate recalcitrance [36,37,71]. Although winter surface CO_2 fluxes were not measured in this study, Wang et al. [43] reported that in high-boreal forest, winter SR accounted for 5%–19% of the annual ER, which agrees well with the value reported in similar boreal forests [72] while Kim et al. [39] observed that wintertime soil CO_2 efflux contributed 24% of the annual CO_2 efflux in Alaska.

4.5. Implicatiuons of Changing Climate

Existing climatic records [73] indicate that climate change over the last five decades has increased mean annual air temperature at the study sites by 0.5–2 °C. Climatic change in is likely to affect NCE in several direct and indirect ways. Longer growing seasons, elevated CO2, and increased nutrients released from decomposing organic carbon may all stimulate plant growth [74] to increase C assimilation. Temperature-driven increases in soil respiration would likely be pronounced in the drained mid-boreal uplands where Q_{10} values are the largest. Higher respiration rates of drained upland soils combined with large input of organic debris from the tree crowns and limited assimilation of shaded bryophytes determined that uplands were consistent sources of CO_2, especially in the southern sites [43]. Low productivity of bryophytes in peat plateaus, in association with limited soil respiration due to greater water-saturated layer and permafrost, resulted in peat plateaus being near-neutral or minor sources of CO_2 with little difference across climatic gradient. Degradation of peat plateau to collapse scars will result in a dramatic change in hydrology and surface vegetation at the local level, and may result in a net increase in NCE, which would offset increase in ER depending on the extent of permafrost thawing. Accelerated collapsing of the peat plateau would subject soil to waterlogging, thereby decreasing aerobic respiration. Furthermore, with receding of permafrost, the connectivity of collapsing features would drain the landscape [75,76] resulting in increased active layer depth, thereby making the previously frozen organic matter available for decomposition. However, an increase in the rate of surface assimilation by bryophytes is to be expected in the expanding collapsed areas. With rising water table, the *Sphagnum* spp. found on hummocks tends to increase productivity [77]. However, lower water table and increase in mean annual air temperature would result in significant increase in surface CO_2 emissions, while only minor increase in surface assimilation rates can be expected [61]. As to whether collapse scar sites in boreal and subarctic regions are an annual source or sink for CO_2, we cannot know without estimating wintertime fluxes. Even small wintertime effluxes would offset growing season CO_2 sinks. There is clearly a pressing need for future research in this area.

5. Conclusions

This study elucidated the changes in the surface CO_2 exchange along the longitudinal gradient as influenced by the local landform and soil microclimate variations. Both ER and NCE decreased with latitude with the highest values observed at the upland and the lowest at peat plateau locations while collapse scar sites were C sinks. The strong relationship between ER and latitude can be used to estimate regional ER and the relationship can be refined with more data, especially wintertime measurements. Both NCE and ER were primarily controlled by soil temperature in the upper horizons along the climatic gradient. Environmental conditions and vegetation type determine the C source-sink relationship. With continuously waterlogged conditions at the collapsed scar sites, bryophyte NCE was highest, and was negligible at upland and peat plateau locations. On an average, upland forest sites and peat plateaus were respectively major and minor sources of CO_2, collapsed scar sites were sinks of CO_2.

Acknowledgments: This project was made possible with the generous support of the Government of Canada Program for International Polar Year, and Natural Resources Canada, Canadian Forest Service. Many thanks are also due to Steve Gooderham, Mike Gravel, Tom Lakusta and Paul Rivard, from the Forest Management Division, Government of the Northwest Territories. Without the valuable contributions of field and laboratory staff, this project would not have been possible and we would like to acknowledge the hard work of: Patrick Hurdle, Thierry Verem Sanders, Ruth Errington, and Catherine McNalty.

Author Contributions: J.S.B. conceived the idea and designed the experiments; N.S. performed the experiments; N.S. and J.S.B. analyzed the data; J.S.B., N.S. and R.S.J. wrote the paper.

References

1. Abbott, B.W.; Jones, J.B.J.; Schuur, E.; Chapin, F.; Bowden, W.B.; Bret-Harte, M.S.; Epstein, H.E.; Flannigan, M.D.; Harms, T.K.; Hollingsworth, T.N.; et al. Biomass offsets little or none of permafrost carbon release from soils, streams, and wildfire: An expert assessment. *Environ. Res. Lett.* **2016**, *11*, 034014. [CrossRef]
2. Schuur, E.G.; McGuire, A.D.; Schädel, C.; Grosse, G.; Harden, J.W.; Hayes, D.J.; Hugelius, G.; Koven, C.D.; Kuhry, P.; Lawrence, D.M.; et al. Climate change and the permafrost carbon feedback. *Nature* **2015**, *520*, 171–179. [CrossRef] [PubMed]
3. Schuur, E.A.; Bockheim, G.J.; Canadell, J.G.; Euskirchen, E.; Field, C.B.; Goryachkin, S.V.; Hagemann, S.; Kuhry, P.; Lafleur, P.M.; Lee, H.; et al. Vulnerability of Permafrost Carbon to Climate Change: Implications for the Global Carbon Cycle. *Bioscience* **2008**, *58*, 701–714. [CrossRef]
4. Hugelius, G.; Strauss, J.; Zubrzycki, S.; Harden, J.W.; Schuur, E.A.G.; Ping, C.-L.; Schirrmeister, L.; Grosse, G.; Michaelson, G.J.; Koven, C.D.; et al. Estimated stocks of circumpolar permafrost carbon with quantified uncertainty ranges and identified data gaps. *Biogeosciences* **2014**, *11*, 6573–6593. [CrossRef]

5. Tarnocai, C.; Canadell, J.G.; Schuur, E.A.G.; Kuhry, P.; Mazhitova, G.; Zimov, S.A. Soil organic carbon pools in the northern circumpolar permafrost region. *Glob. Biogeochem. Cycles* **2009**, *23*. [CrossRef]
6. Raich, J.W.; Schlesinger, W.H. The global carbon dioxide flux in soil respiration and its relationship to vegetation and climate. *Tellus B* **1992**, *44*, 81–99. [CrossRef]
7. Zhuang, Q.; McGuire, A.D.; O'Neill, K.P.; Harden, J.W.; Romanovsky, V.E.; Yarie, J. Modeling soil thermal and carbon dynamics of a fire chronosequence in interior Alaska. *J. Geophys. Res.* **2003**, *107*. [CrossRef]
8. Kurz, W.A.; Shaw, C.H.; Boisvenue, C.; Stinson, G.; Metsaranta, J.; Leckie, D.; Dyk, A.; Smyth, C.; Neilson, E.T. Carbon in Canada's boreal forest—A synthesis. *Environ. Rev.* **2014**, *21*, 260–292. [CrossRef]
9. Price, D.T.; Alfaro, R.I.; Brown, K.J.; Flannigan, M.D.; Fleming, R.A.; Hogg, E.H.; Girardin, M.P.; Lakusta, T.; Johnston, M.; McKenney, D.W.; et al. Anticipating the consequences of climate change for Canada's boreal forest ecosystems. *Environ. Rev.* **2013**, *21*, 322–365. [CrossRef]
10. Harden, J.W.; Koven, C.D.; Ping, C.L.; Hugelius, G.; McGuire, A.D.; Camill, P.; Jorgenson, T.; Kuhry, P.; Michaelson, G.J.; O'Donnell, J.A.; et al. Field information links permafrost carbon to physical vulnerabilities of thawing. *Geophys. Res. Lett.* **2012**, *39*, L15704. [CrossRef]
11. Thie, J. Distribution and Thawing of Permafrost in the Southern Part of the Discontinuous Permafrost Zone in Manitoba. *Arctic* **1974**, *27*, 189–200. [CrossRef]
12. Waddington, J.M.; Roulet, N.T. Carbon balance of a boreal patterned peatland. *Glob. Chang. Biol.* **2000**, *6*, 87–98. [CrossRef]
13. Mitsch, W.J.; Gosselink, J.G. *Wetlands*, 4th ed.; John Wiley and Sons: Hoboken, NJ, USA, 2007; p. 582.
14. Drewitt, G.B.; Black, T.A.; Nesic, Z.; Humphreys, E.R.; Jork, E.M.; Swanson, R.; Ethier, G.J.; Griffis, T.; Morgenstern, K. Measuring forest-floor CO_2 fluxes in a Douglas-fir forest. *Agric. For. Meteorol.* **2002**, *110*, 299–317. [CrossRef]
15. Green, R.N.; Trowbridge, R.L.; Klinka, K. Towards a taxonomic classification of humus forms. *For. Sci.* **1993**, *39*, 1–49.
16. Halsey, L.A.; Vitt, D.H.; Zoltai, S.C. Disequilibrium response of permafrost in boreal continental western Canada to climate change. *Clim. Chang* **1995**, *30*, 57–73. [CrossRef]
17. Moore, T.R.; Roulet, N.T.; Waddington, J.M. Uncertainty in Predicting the Effect of climatic Change on the Carbon Cycling of Canadian Peatlands. *Clim. Chang.* **1998**, *40*, 229–245. [CrossRef]
18. Kayranli, B.; Scholz, M.; Mustafa, A.; Hedmark, A. Carbon storage and fluxes within freshwater wetlands: A critical review. *Wetlands* **2010**, *30*, 111–124. [CrossRef]
19. Grosse, G.; Harden, J.; Turetsky, M.R.; McGuire, A.D.; Camill, P.; Tarnocai, C.; Frolking, S.; Schuur, E.A.G.; Jorgenson, T.; Marchenko, S.; et al. Vulnerability of high-latitude soil organic carbon in North America to disturbance. *J. Geophys. Res.* **2011**, *116*, G00K06. [CrossRef]
20. Liblik, L.; Moore, T.R.; Bubier, J.L.; Robinson, S.D. Methane emissions from wetlands in the discontinuous permafrost zone: Fort Simpson, NWT, Canada. *Glob. Biogeochem. Cycles* **1997**, *11*, 485–494. [CrossRef]
21. Turetsky, M.R.; Wieder, R.K.; Vitt, D.H. Boreal peatland C fluxes under varying permafrost regimes. *Soil Biol. Biochem.* **2002**, *34*, 907–912. [CrossRef]
22. Prater, J.L.; Chanton, J.P.; Whiting, G.J. Variation in methane production pathways associated with permafrost decomposition in collapse scar bogs of Alberta, Canada. *Glob. Biogeochem. Cycle* **2007**, *21*, GB4004. [CrossRef]
23. Flanagan, L.B.; Syed, K.H. Simulation of both photosynthesis and respiration in response to warmer and drier conditions in a boreal peatland ecosystem. *Glob. Chang. Biol.* **2011**, *17*, 2271–2872. [CrossRef]
24. Elberling, B.; Michelsen, A.; Schädel, C.; Schuur, E.A.G.; Christiansen, H.H.; Berg, L.; Tamstorf, M.P.; Sigsgaard, C. Long-term CO_2 production following permafrost thaw. *Nat. Clim. Chang.* **2013**, *3*, 890–894. [CrossRef]
25. Skinner, W.; Maxwell, B. Climatic Patterns, Trends and Scenarios in the Arctic. In Proceedings of the Sixth Biennial AES/DIAND Meeting on Northern Climate & Mid-Study Workshop of the Mackenzie Basin Impact Study, Yellowknife, NT, Canada, 10–14 April 1994; pp. 125–137.
26. Jassal, R.; Black, A.; Novak, M.; Morgenstern, K.; Nesic, Z.; Gaumont-Guay, D. Relationship between soil CO_2 concentrations and forest-floor CO_2 effluxes. *Agric. For. Meteorol.* **2005**, *130*, 176–192.
27. Millington, R.J.; Quirk, J.M. *Transport in Porous Media*; Elsevier: Amsterdam, The Netherlands, 1960; pp. 97–106.
28. Sallam, A.; Jury, W.A.; Letey, J. Measurement of gas diffusion coefficient under relatively low air-filled porosity. *Soil Sci. Am. J.* **1984**, *48*, 3–6. [CrossRef]

29. SAS Institute Inc. *The SAS System for Windows*, 9th ed.; SAS Institute Inc.: Cary, NC, USA, 2004.
30. Lee, X.; Wu, H.J.; Sigler, J.; Oishi, C.; Siccama, T. Rapid and transient response of soil respiration to rain. *Glob. Chang. Biol.* **2004**, *10*, 1017–1026. [CrossRef]
31. Bond-Lamberty, B.; Thomson, A. Temperature-associated increases in the global soil respiration record. *Nature* **2010**, *464*, 597–582. [CrossRef] [PubMed]
32. Bronson, D.R.; Gower, S.T.; Tanner, M.; Linder, M.; van Herk, I. Response of soil surface CO_2 flux in a boreal forest to ecosystem warming. *Glob. Chang. Biol.* **2008**, *14*, 856–867. [CrossRef]
33. Davidson, E.A.; Belk, E.; Boone, R.D. Soil water content and temperature as independent or confounded factors controlling soil respiration in a temperate mixed hardwood forest. *Glob. Chang. Biol.* **1998**, *4*, 217–227. [CrossRef]
34. Davidson, E.A.; Jassens, I.A. Temperature sensitivity of soil carbon decomposition and feedback to climate change. *Nature* **2006**, *440*, 165–173. [CrossRef] [PubMed]
35. Schuur, E.A.G.; Abbott, B.W.; Bowden, W.B.; Brovkin, V.; Camill, P.; Canadell, J.G.; Chanton, J.P.; Chapin, F.S., III; Christensen, T.R.; Ciais, P.; et al. Expert assessment of vulnerability of permafrost carbon to climate change. *Clim. Chang.* **2013**, *119*, 359–374. [CrossRef]
36. Clymo, R.S. The limits to peat bog growth. *Philos. Trans. Soc. R. Ser. B* **1984**, *3113*, 605–654. [CrossRef]
37. Wickland, K.P.; Striegl, R.G.; Neff, J.C.; Sachs, T. Effects of permafrost melting on CO_2 and CH_4 exchange of a poorly drained black spruce lowland. *J. Geophys. Res.* **2006**, *111*, G02011. [CrossRef]
38. Strachan, I.B.; Pelletier, L.; Bonneville, M.-C. Interannual variability in water table depth controls net ecosystem carbon dioxide exchange in a boreal peatland. *Biogeochemistry* **2016**, *127*, 99–111. [CrossRef]
39. Kim, Y.; Kim, S.D.; Enomoto, H.; Kushida, K.; Kondoh, M.; Uchida, M. Latitudinal distribution of soil CO_2 efflux and temperature along the Dalton Highway, Alaska. *Polar Sci.* **2013**, *7*, 162–173. [CrossRef]
40. Aurela, M.; Laurila, T.; Tuovinen, J.-P. The timing of snow melt controls the annual CO_2 balance in a subarctic fen. *Geophys. Res. Lett.* **2004**, *31*, LI6119. [CrossRef]
41. Bisbee, K.E.; Gower, S.T.; Norman, J.M.; Nordheim, E.V. Environmental controls on ground cover species composition and productivity in a boreal black spruce forest. *Oecologia* **2001**, *129*, 261–270. [CrossRef]
42. Yuan, W.P.; Luo, Y.Q.; Richardson, A.D.; Oren, R.; Luyssaert, S.; Janssens, I.A.; Ceulemans, R.; Zhou, X.H.; Grunwald, T.; Aubinet, M.; et al. Latitudinal patterns of magnitude and interannual variability in net ecosystem exchange regulated by biological and environmental variables. *Glob. Chang. Biol.* **2009**, *15*, 2905–2920. [CrossRef]
43. Wang, C.; Bond-Lamberty, B.; Gower, S.T. Carbon distribution of a well- and poorly-drained black spruce fire chronosequence. *Glob. Chang. Biol.* **2003**, *9*, 1–14. [CrossRef]
44. Aurela, M.; Lohila, A.; Tuovinen, J.-P.; Hatakka, J.; Riutta, T.; Laurila, T. Carbon dioxide exchange on a northern boreal fen. *Boreal Environ. Res.* **2009**, *14*, 699–710.
45. Griffis, T.J.; Rouse, W.R.; Waddington, J.M. Interannual variability of net ecosystem CO_2 exchange at a subarctic fen. *Glob. Biogeochem. Cycles* **2000**, *14*, 1109–1121. [CrossRef]
46. Lafleur, P.M.; Humphrey, E.R. Spring warming and carbon dioxide exchange over low Arctic tundra in central Canada. *Glob. Chang. Biol.* **2007**, *14*, 740–756. [CrossRef]
47. Kwon, H.-J.; Oechel, W.C.; Zulueta, R.C.; Hastings, S.J. Effects of climate variability on carbon sequestration among adjacent wet sedge tundra and moist tussock tundra ecosystems. *J. Geophys. Res.* **2006**, *111*, G03014. [CrossRef]
48. Groendahl, L.; Friborg, T.; Soegaard, H. Temperature and snowmelt controls on interannual variability in carbon dioxide exchange in the high Arctic. *Theor. Appl. Climatol.* **2007**, *88*, 111–125. [CrossRef]
49. Humphreys, E.R.; Lafleur, P.M. Does earlier snowmelt lead to greater CO_2 sequestration in two low Arctic tundra ecosystems? *Geophys. Res. Lett.* **2011**, *38*, L09703. [CrossRef]
50. Tuba, Z.; Csintalan, Z.; Proctor, M.C.F. Photosynthetic Responses of a Moss, *Tortula ruralis*, ssp. *ruralis*, and the Lichens *Cladonia convoluta* and *C. furcata* to Water Deficit and Short Periods of Desiccation, and Their Ecophysiological Significance: A Baseline Study at Present-Day CO_2 Concentration. *New Phytol.* **1996**, *133*, 353–361. [CrossRef]
51. Lund, M.; Lafleur, P.M.; Roulet, N.T.; Lindroth, A.; Christensen, T.R.; Aurela, M.; Chojnicki, B.H.; Flanagan, L.B.; Humphreys, E.R.; Laurila, T.; et al. Variability in exchange of CO_2 across 12 northern peatland and tundra sites. *Glob. Chang. Biol.* **2010**, *16*, 2436–2448. [CrossRef]

52. McNeil, P.; Waddington, J.M. Moisture controls on *Sphagnum* growth and CO_2 exchange on a cut-over bog. *J. Appl. Ecol.* **2003**, *40*, 354–367. [CrossRef]
53. Tuittila, E.-S.; Vasander, H.; Luine, J. Sensitivity of C sequestration in reintroduced *Sphagnum* to water-level variation in a cutaway peatland. *Restor. Ecol.* **2004**, *12*, 483–493. [CrossRef]
54. Moore, T.R.; Bubier, J.L.; Bledzki, L. Litter decomposition in temperate peatland ecosystems: The effect of substrate and site. *Ecosystems* **2007**, *10*, 949–963. [CrossRef]
55. Euskirchen, E.S.; Edgar, C.W.; Turetsky, M.R.; Waldrop, M.P.; Harden, J.W. Differential response of carbon fluxes to climate in three peatland ecosystems that vary in the presence and stability of permafrost. *J. Geophys. Res. Biogeosci.* **2014**, *119*, 1576–1595. [CrossRef]
56. Gaumont-Guay, D.; Black, T.A.; Barr, A.G.; Jassal, R.S.; Nesic, Z. Biophysical controls on rhizospheric and heterotrophic components of soil respiration in a boreal black spruce stand. *Tree Physiol.* **2008**, *28*, 161–171. [CrossRef] [PubMed]
57. Gaumont-Guay, D.; Black, T.A.; Griffis, T.J.; Barr, A.G.; Jassal, R.S.; Nesic, Z. Interpreting the dependence of soil respiration on soil temperature and water content in a boreal aspen stand. *Agric. For. Meteorol.* **2006a**, *140*, 220–235. [CrossRef]
58. Bergeron, O.; Margolis, H.A.; Coursolle, C. Forest floor carbon exchange of a boreal black spruce forest in eastern North America. *Biogeosciences,* **2009**, *6*, 1849–1864. [CrossRef]
59. Tang, J.W.; Dennis, D.B.; Qi, Y.; Xu, L.K. Assessing soil CO_2 efflux using continuous measurement of CO_2 profiles in soils with small solid-state sensors. *Agric. For. Meteo* **2003**, *118*, 207–220. [CrossRef]
60. Gaumont-Guay, D.; Black, T.A.; Griffis, T.J.; Barr, A.G.; Jassal, R.A.; Nesic, Z. Influence of temperature and drought on seasonal and interannual variations of soil, bole and ecosystem respiration in a boreal aspen stand. *Agr. Forest Meteorol.* **2006**, *140*, 203–219. [CrossRef]
61. Helfter, C.; Campbell, C.; Dinsmore, K.J.; Drewer, J.; Coyle, M.; Anderson, M.; Skiba, U.; Nemitz, E.; Billett, M.F.; Sutton, M.A. Drivers of long-term variability in CO_2 net ecosystem exchange in a temperate peatland. *Biogeosciences* **2015**, *12*, 1799–1811. [CrossRef]
62. Raich, J.W.; Potter, C.S.; Bhagawati, D. Interannual variability in global soil respiration, 1980–94. *Glob. Chang. Biol.* **2002**, *8*, 800–812. [CrossRef]
63. Rodeghiero, M.; Cescatti, A. Main determinants of forest soil respiration along an elevation/temperature gradient in the Italian Alps. *Glob. Chang. Biol.* **2005**, *11*, 1024–1041. [CrossRef]
64. Lavigne, M.B.; Ryan, M.G.; Anderson, D.E.; Baldocchi, D.D.; Crill, P.M.; Fitzjarrald, D.R.; Goulden, M.L.; Gower, S.T.; Massheder, J.M.; McCaughey, J.H.; et al. Comparing nocturnal eddy covariance measurements to estimates of ecosystem respiration made by scaling chamber measurements at six coniferous boreal sites. *J. Geophys. Res.* **1997**, *102*, 28977–28985. [CrossRef]
65. Rayment, M.B.; Jarvis, P.G. Temporal and spatial variation of soil CO_2 efflux in a Canadian boreal forest. *Soil Biol. Biochem.* **2000**, *32*, 35–45. [CrossRef]
66. Xu, M.; Qi, Y. Soil-surface CO_2 efflux and its spatial and temporal variations in a young ponderosa pine plantation in northern California. *Glob. Chang. Biol.* **2001**, *7*, 667–677. [CrossRef]
67. Peichl, M.; Öquist, M.; Löfvenius, M.O.; Ilstedt, U.; Sagerfors, J.; Grelle, A.; Lindroth, A.; Nilsson, M.B. A 12-year record reveals pre-growing season temperature and water table level threshold effects on the net carbon dioxide exchange in a boreal fen. *Environ. Res. Lett.* **2014**, *9*, 055006. [CrossRef]
68. Startsev, N.A.; Lieffers, V.J. Emission of nitrogen gas, nitrous oxide, and carbon dioxide on rehydration of dry feathermosses. *Soil Sci. Soc. Am. J.* **2007**, *71*, 214–218. [CrossRef]
69. Bubier, J.L.; Crill, P.M.; Moore, T.R.; Savage, K.; Varner, R.K. Seasonal patterns and controls on net ecosystem CO_2 exchange in a boreal peatland complex. *Glob. Biogeochem. Cycles* **1998**, *12*, 703–714. [CrossRef]
70. Lu, X.; Fan, J.; Yan, Y.; Wang, X. Responses of Soil CO_2 Fluxes to Short-Term Experimental Warming in Alpine Steppe Ecosystem, Northern Tibet. *PLoS ONE* **2013**, *8*. [CrossRef] [PubMed]
71. Yavitt, J.B.; Lang, G.E.; Wieder, R.K. Control of carbon mineralization to CH_4 and CO_2 in anaerobic, *Sphagnum-* derived peat from Big Run Bog. West Virginia. *Biogeochemisty* **1987**, *4*, 141–157. [CrossRef]
72. Winston, G.C.; Sundquist, E.T.; Stephens, B.B.; Trumbore, S.E. Winter CO_2 fluxes in a boreal forest. *J. Geophys. Res.* **1997**, *102*, 795–804. [CrossRef]
73. Environment and Climate Change Canada. Adjusted and Homogenized Canadian Climate Data (AHCCD). Available online: http://www.ec.gc.ca/dccha-ahccd/ (accessed on 14 November 2016).

74. Sistla, S.A.; Moore, J.C.; Simpson, R.T.; Gough, L.; Shaver, G.R.; Schimel, J.P. Long-term warming restructures Arctic tundra without changing net soil carbon storage. *Nature* **2013**, *497*, 615–618. [CrossRef] [PubMed]
75. Quinton, W.L.; Hayashi, M.; Chasmer, L.E. Permafrost-thaw-induced land-cover change in the Canadian subarctic: Implications for water resources. *Hydrol. Process.* **2013**, *25*, 152–158. [CrossRef]
76. Baltzer, J.L.; Veness, T.; Chasmer, L.E.; Sniderhan, A.E.; Quinton, W.L. Forests on thawing permafrost: Fragmentation, edge effects, and net forest loss. *Glob. Chang. Biol.* **2014**, *20*, 824–834. [CrossRef] [PubMed]
77. Bridgham, S.D.; Pastor, J.; Dewey, B.; Weltzin, J.F.; Updegraff, K. Rapid carbon response of peatlands to climate change. *Ecology* **2008**, *89*, 3041–3048. [CrossRef]

3

Storage and Climatic Controlling Factors of Litter Standing Crop Carbon in the Shrublands of the Tibetan Plateau

Xiuqing Nie [1,2,3], Dong Wang [1,3], Lucun Yang [2,4] and Guoying Zhou [2,4,*]

1 Key Laboratory of Tree Breeding and Cultivation of the State Forestry Administration, Research Institute of Forestry, Chinese Academy of Forestry, Beijing 100091, China; niexiuqing123@163.com (X.N.); dwang@caf.ac.cn (D.W.)
2 Key Laboratory of Tibetan Medicine Research, Northwest Institute of Plateau Biology, Chinese Academy of Science, Xining 810008, China; Yanglucun@nwipb.cas.cn
3 Research Institute of nature protected Area, Chinese Academy of Forestry, Beijing 100091, China
4 Qinghai Key Laboratory of Qing-Tibet Biological Resources, Xining 810008, China
* Correspondence: zhougy@nwipb.cas.cn

Abstract: Litter is an important component of terrestrial ecosystems and plays a significant role in carbon cycles. Quantifying regional-scale patterns of litter standing crop distribution will improve our understanding of the mechanisms of the terrestrial carbon cycle, and thus enable accurate predictions of the responses of the terrestrial carbon cycle to future climate change. In this study, we aimed to estimate the storage and climatic controlling factors of litter standing crop carbon in the Tibetan Plateau shrublands. We investigated litter standing crop carbon storage and its controlling factors, using a litter survey at 65 shrublands sites across the Tibetan Plateau from 2011–2013. Ordinary least squares regression analyses were conducted to estimate the relationships between litter standing crop carbon, longitude, and latitude. Multiple linear regressions were used to evaluate relationships among litter standing crop carbon, mean annual temperature (MAT), mean annual precipitation (MAP), and aboveground biomass. The litter standing crop carbon storage was 10.93 Tg C, 7.40 Tg C, and 3.53 Tg C in desert shrublands and alpine shrublands, respectively. Litter standing crop carbon decreased with longitude, and was stable with increasing latitude. Most (80%) of the litter standing crop was stored in branches, with only 20% stored in foliage in the shrublands on the Tibetan Plateau. The conversion coefficient was 0.44 for litter standing crop to litter standing crop carbon, and 0.39 and 0.45 for foliage and branch litter standing crop to foliage and branch litter standing crop carbon, respectively. Aboveground biomass can accelerate more inputs of litter and has a positive effect on litter standing crop carbon. MAT had a positive effect on litter standing crop carbon due to stimulating more input of aboveground biomass. However, MAP had a negative relationship with litter standing crop carbon by enhancing litter decomposition.

Keywords: litter standing crop carbon; conversion coefficient; climatic factors; Tibetan Plateau; shrublands

---

1. Introduction

Litter is an important component of terrestrial ecosystems. Its production and accumulation contribute to carbon sequestration and soil fertility [1]. The litter standing crop depends on the rates of litter production and litter decomposition. If the litter production rate is greater than the litter decomposition rate, litter becomes a sink of atmosphere CO_2; if the litter decomposition rate is greater than the litter production rate, it becomes a carbon source [2]. Although soil acts as the largest carbon pool [3], the litter standing crop pool also accounts for a non-ignorable amount of emissions; the global litter standing crop carbon pool has been estimated to be 43 Pg C for all forest ecosystem carbon stocks [4], and the litter standing crop carbon pool in both forest and grassland ecosystems in China has been estimated as being 0.52 Pg C [5], which means litter plays a significant role in carbon cycles [1,6,7]. Most researchers have focused on litter production and litter decomposition [8,9], while few studies have considered litter standing crop carbon in shrublands [2], especially on the Tibetan Plateau.

As a result of global climatic changes, the Tibetan Plateau has been experiencing warming, and the mean annual temperature (MAT) has increased by 0.05 °C every year [10]. Meanwhile the mean annual precipitation (MAP) has increased by 10.2 mm every ten years [10]. Furthermore, climatic conditions, such as changes in precipitation and temperature, are predicted to not only continue, but also perhaps to accelerate in the future [11,12]. It has been demonstrated that climatic factors such as MAP and MAT greatly affect litter standing crop carbon [5]. Climatic factors also indirectly influence litter decomposition through their effects on biotic factors [2]. It has been demonstrated that MAT and MAP have positive effects on litter production in forests in China [13]. At the regional scale, a similar trend was observed in Tiantong national forest park [14]. However, other studies have indicated that litter production in northeastern China's forests is influenced by temperature, but not significantly affected by precipitation [15]. Therefore, the effects of climatic factors on litter production remain uncertain. Compared with climatic effects on litter production, its effects on litter standing crop have been less researched in the shrublands on the Tibetan Plateau.

The conversion coefficient of litter standing crop to litter standing crop carbon is an important parameter in the carbon budget [16]. Generally, researchers have used 0.50 to convert biomass to biomass carbon [17]. However, the parameter of 0.50 overestimated carbon by 22% when estimating the litter carbon budget in shrublands across south China [16]. Furthermore, different components play important roles in shaping the magnitude of the conversion coefficient [18]. Although the conversion coefficient of litter standing crop to litter standing crop carbon has been estimated in shrublands of southern China [16], components such as foliage and branch standing crop to foliage and branch standing crop carbon have not been considered. Furthermore, although biomass and soil organic carbon have been explored in the Tibetan Plateau shrublands [19–21], it was difficult to assess the total carbon storage in shrublands ecosystem without litter standing crop carbon. It has been demonstrated that the biophysical processes that control soil carbon on the Tibetan Plateau may differ from those in other regions [20,22]. We hypothesized that the conversion coefficient in the Tibetan Plateau shrublands was different from that in the southern China shrublands. We also aimed to explore the litter standing crop carbon responses to changes of MAT, MAP and aboveground biomass in the shrublands on the Tibetan Plateau.

2. Materials and Methods

2.1. Study Area

This study investigated shrublands on the northeastern Tibetan Plateau (Figure 1). Shrublands are one of the most widely distributed ecosystems on the plateau [23]. Shrublands compose woody vegetation, with coverage of more than 30% and a mean height below 5 m [24] and are primarily classified as desert shrublands or alpine shrublands [23]. Desert shrublands are mostly distributed in drier areas, such as the Qaidam Basin, and are characterized by species that can endure severe drought, such as *Kalidium foliatum* (Pall.) Moq, *Salsola abrotanoides* Bunge, and *Sympegma regelii* Bunge. Compared with alpine shrublands, only a few types of super-xerophytic herbs grow in desert shrublands. Brown desert soil and grey brown desert soil are mainly soil types in desert shrublands ecosystems, while soil in desert shrublands is very thin and more infertile than that in alpine shrublands [23]. In contrast, alpine shrublands are located in mountains with a cold and semiarid environment. Chestnut soil, grey cinnamon soil, and alpine shrubby meadow soil are mainly soil types in alpine shrublands ecosystems [23]. Representative species include *Rhododendron thymifolium* Maxim, *R. capitatum* Maxim, and *Sibiraea laevigata* (Linn.) Maxim. Mesophyte herbs, including *Kobresia* spp., *Carex* spp., and

Oxytropis spp., are located in alpine desertblands. MAP and MAT in the study regions range from 17.6 to 764.4 mm and −5.6 to 8.9 °C, respectively [25].

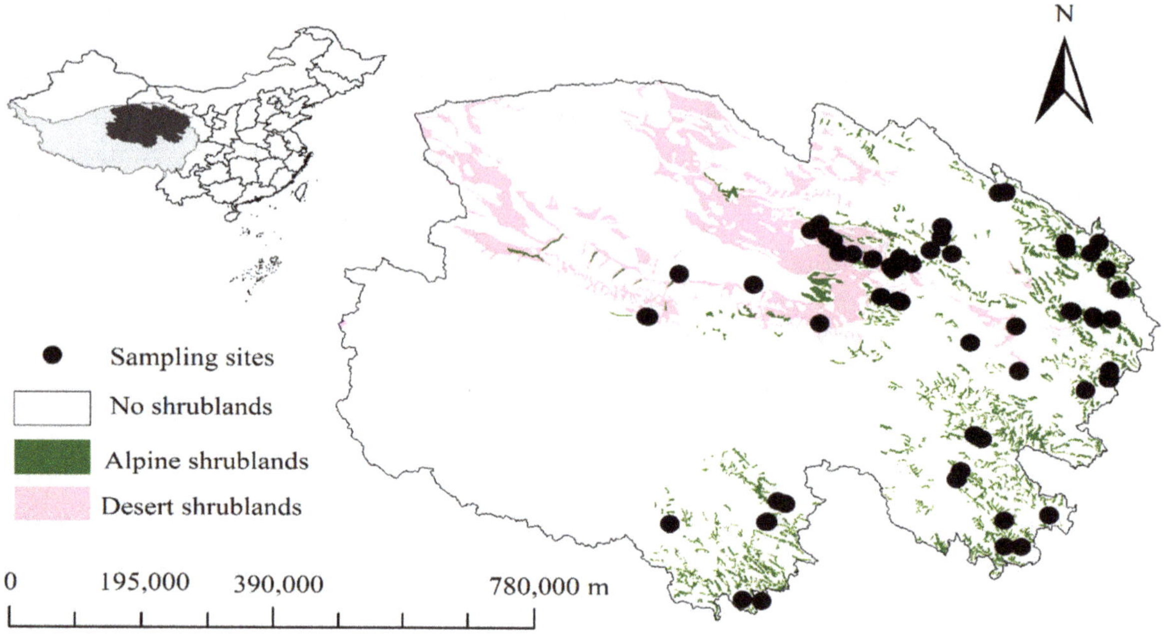

Figure 1. Distribution of 65 sampling sites in the shrublands of the Tibetan Plateau, which was based on the vegetation of China (1:100 0000) [26].

2.2. Field Investigation and Laboratory Measurements

In line with China's vegetation atlas (1:1000000) [26] and field investigations, we systematically selected 195 plots in 65 sites across the northeastern Tibetan Plateau shrublands in July and August from 2011 to 2013 (Figure 1). We selected three plots (1 m × 1 m) at each site and the distance of two plots was between 5 m and 50 m. At each site, all of the litter, including older litter and decomposed litter in the three plots, was collected to determine the litter standing crop [27]. Also, all shrubland plants in three plots were harvested to measure aboveground biomass. The samples of litter and aboveground biomass were oven dried at 65 °C to a constant weight, and weighed to the nearest 0.01 g. The litters were primarily foliage or branches. Carbon content of litter was determined using an elemental analyzer (2400 II CHNS; Perkin-Elmer, Boston, MA, USA).

2.3. Climate Data

To investigate the potential effects of climate factors on the litter standing crop carbon, MAT and MAP were extracted for each site from the WorldClim database (http://www.Worldclim.org/) with a spatial resolution of 1 × 1 km^2 [28,29].

2.4. Data Analysis

Ordinary least squares regression analyses were used to estimate the relationships between litter standing crop carbon, longitude, and latitude. Multiple linear regressions were conducted to evaluate relationships among litter standing crop carbon, MAT, MAP, and aboveground biomass. All analyses were conducted using SPSS 22.0 (SPSS Inc., Chicago, USA). Graphs were created in SigmaPlot 12.5 (Systat Software, Inc., Point Richmond, CA, USA).

3. Results

3.1. Size and Storage of Litter Standing Crop in Tibetan Plateau Shrublands

The litter standing crop ranged from 0.04 kg m^{-2} to 0.48 kg m^{-2}. The branch standing crop ranged from 0.02 kg m^{-2} to 0.48 kg m^{-2}, and foliage was no more than 0.25 kg m^{-2}. The mean litter standing crop was 0.23 kg m^{-2} in the shrublands overall (Table 1), and it was 0.25 kg m^{-2} and 0.19 kg m^{-2} in desert shrublands and alpine shrublands, respectively (Table 1). The mean values for foliage and branch standing crops in the shrublands were 0.04 kg m^{-2}, and 0.19 kg m^{-2}, respectively (Table 1). Most litter standing crop was stored in the branch standing crop, accounting for 80% of all litter standing crop overall (Figure 2a), and 76% and 87% in alpine shrublands and desert shrublands, respectively (Figure 2b,c).

Table 1. Size, storage and carbon storage of foliage, branch, and litter standing crops in the shrublands of the Tibetan Plateau.

Shrubland Type	Area (10^4 km^2)	Litter Standing Crop Density (kg m^{-2})			Litter Standing Crop Storage (Tg)			Litter Standing Crop Carbon Storage (Tg C)		
		Foliage	Branch	Litter	Foliage	Branch	Litter	Foliage	Branch	Litter
Desert shrublands	6.69	0.03	0.22	0.25	2.11	14.63	16.74	0.82	6.58	7.40
Alpine shrublands	4.29	0.05	0.14	0.19	1.94	6.15	8.09	0.76	2.77	3.53
Overall	10.98	0.04	0.19	0.23	4.05	20.78	24.83	1.58	9.35	10.93

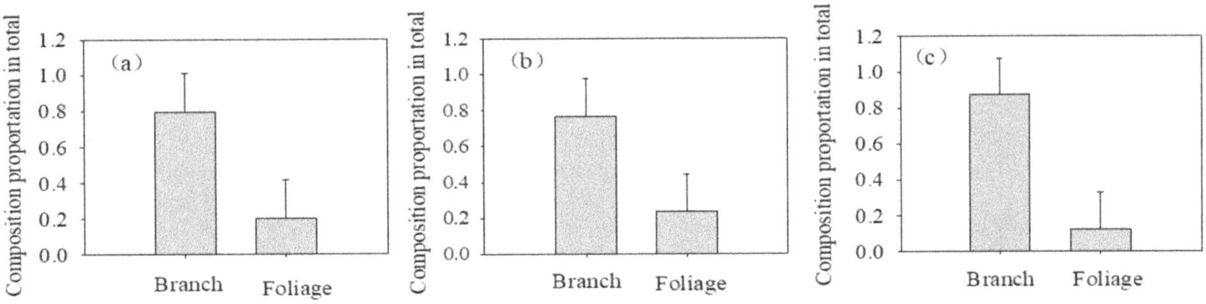

Figure 2. The proportions of branch and foliage standing crop in the litter standing crop on the Tibetan Plateau, (**a**) all shrublands, (**b**) alpine shrublands, (**c**) desert shrublands.

Overall, the litter standing crop storage in the shrublands was 24.83 Tg; 16.74 Tg and 8.09 Tg in desert shrublands and alpine shrublands, respectively. Storages in the foliage and branch standing crops were 4.05 Tg and 20.78 Tg, respectively (Table 1).

3.2. Conversion Coefficient, Storage, and Distribution of Litter Standing Crop Carbon

The conversion coefficient of litter standing crop to litter standing crop carbon was 0.44 in the shrublands of the Tibetan Plateau, which supported our hypothesis that the conversion coefficient was different from that in the shrublands of southern China. The conversion coefficients for foliage and branch standing crop to foliage and branch standing crop carbon were 0.39 and 0.45, respectively. Litter standing crop carbon was 10.93 Tg C in the northeastern Tibetan Plateau shrublands, and they were 7.40 Tg C and 3.53 Tg C in desert shrublands and alpine shrublands, respectively (Table 1). Furthermore, the conversion coefficients of branch standing crop to branch standing crop carbon were 0.42 and 0.46 in desert shrublands and alpine shrublands, respectively. Meanwhile, the conversion coefficients of foliage standing crop to foliage standing crop carbon were 0.27 and 0.40 in desert shrublands and alpine shrublands, respectively.

Spatially, the litter standing crop carbon was stable with increased latitude (Figure 3a), but it decreased with longitude in total shrublands (Figure 3b). Similarly, the significant trend also existed in the alpine shrublands (Figure 3c). In the desert shrublands, the range of longitude was between 94.51° and 98.73°. The litter standing crop carbon has a decreasing trend with an increasing longitude in desert shrublands (Figure 3d). Specifically, the highest litter standing crop carbon was in a low longitude (94.51°–95.91°), while the lowest litter standing crop carbon was in high MAT (97.32°–98.73°). In medium MAT (95.91°–97.32°), the litter standing crop carbon was moderate (Figure 3d).

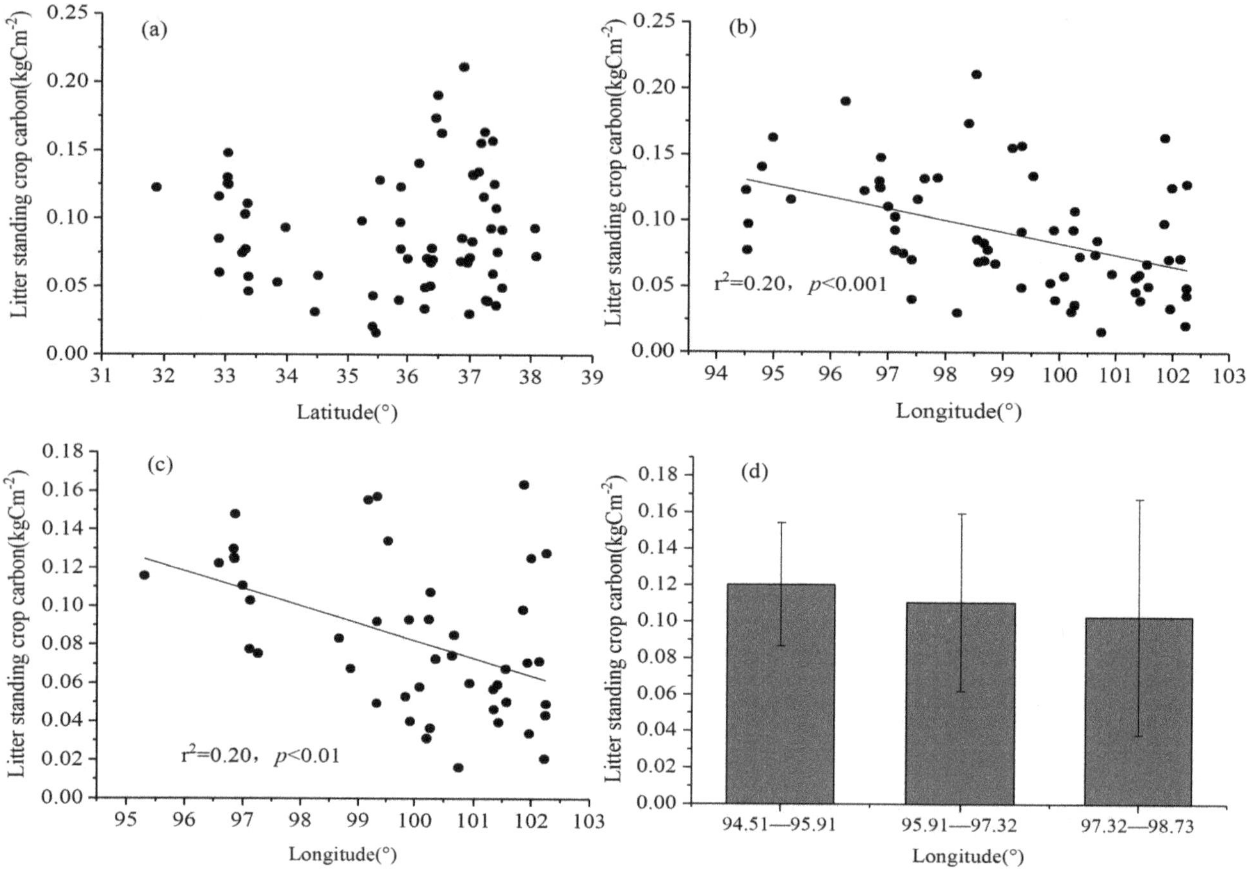

Figure 3. Relationships between litter standing crop carbon and latitude and longitude, (**a**) and (**b**) total shrublands, (**c**) alpine shrublands, (**d**) desert shrublands.

3.3. *Effects of MAP and MAT on Litter Standing Crop Carbon*

Litter standing crop carbon significantly decreased with MAP (Figure 4a) ($p < 0.01$). In the alpine shrublands, the range of MAP was between 160.83 mm and 572.50 mm. The litter standing crop carbon showed a negative trend with MAP in alpine shrublands. Specifically, the highest litter standing crop carbon was in low MAP (160.83 mm–298.06 mm), while the lowest litter standing crop carbon in high MAP (435.28–572.50 mm), and in medium MAP (298.06 mm–435.28 mm), the litter standing crop carbon was moderate (Figure 4b). Similarly, and the range of MAP was between 35.83 mm and 265.00 mm in desert shrublands. The highest litter standing crop carbon was in low MAP (35.83 mm–112.22 mm), while the lowest litter standing crop carbon in high MAP (188.61 mm–265.00 mm), and in medium MAP (112.22 mm–188.61 mm), the litter standing crop carbon was moderate (Figure 4c).

Litter standing crop carbon increased with MAT ($p < 0.01$) (Figure 4d). In the alpine shrublands, the range of MAT was between −2.83 °C and 3.20 °C. The litter standing crop carbon showed a positive trend with MAT in alpine shrublands. Specifically, the highest litter standing crop carbon was in high MAT (1.18 °C–3.20 °C), while the lowest litter standing crop carbon in low MAT(−2.85 °C–−0.83 °C), and in medium MAT (−0.83 °C–1.18 °C), the litter standing crop carbon was moderate (Figure 4e). Similarly to desert shrublands, and the range of MAT was between −1.14 °C and 5.47 °C. The litter standing crop carbon showed a positive trend with MAT in desert shrublands. Specifically, the highest litter standing crop carbon was in high MAT (3.65 °C–5.47 °C), while the lowest litter standing crop carbon was in low MAT (−1.14 °C–0.42 °C), and in medium MAP (0.42 °C–3.65 °C), the litter standing crop carbon was moderate (Figure 4f). Generally, multiple linear regressions showed that in total shrublands, litter standing crop carbon = −0.001 MAP + 0.01 MAT + 0.25 ($r = 0.40$, $p < 0.01$).

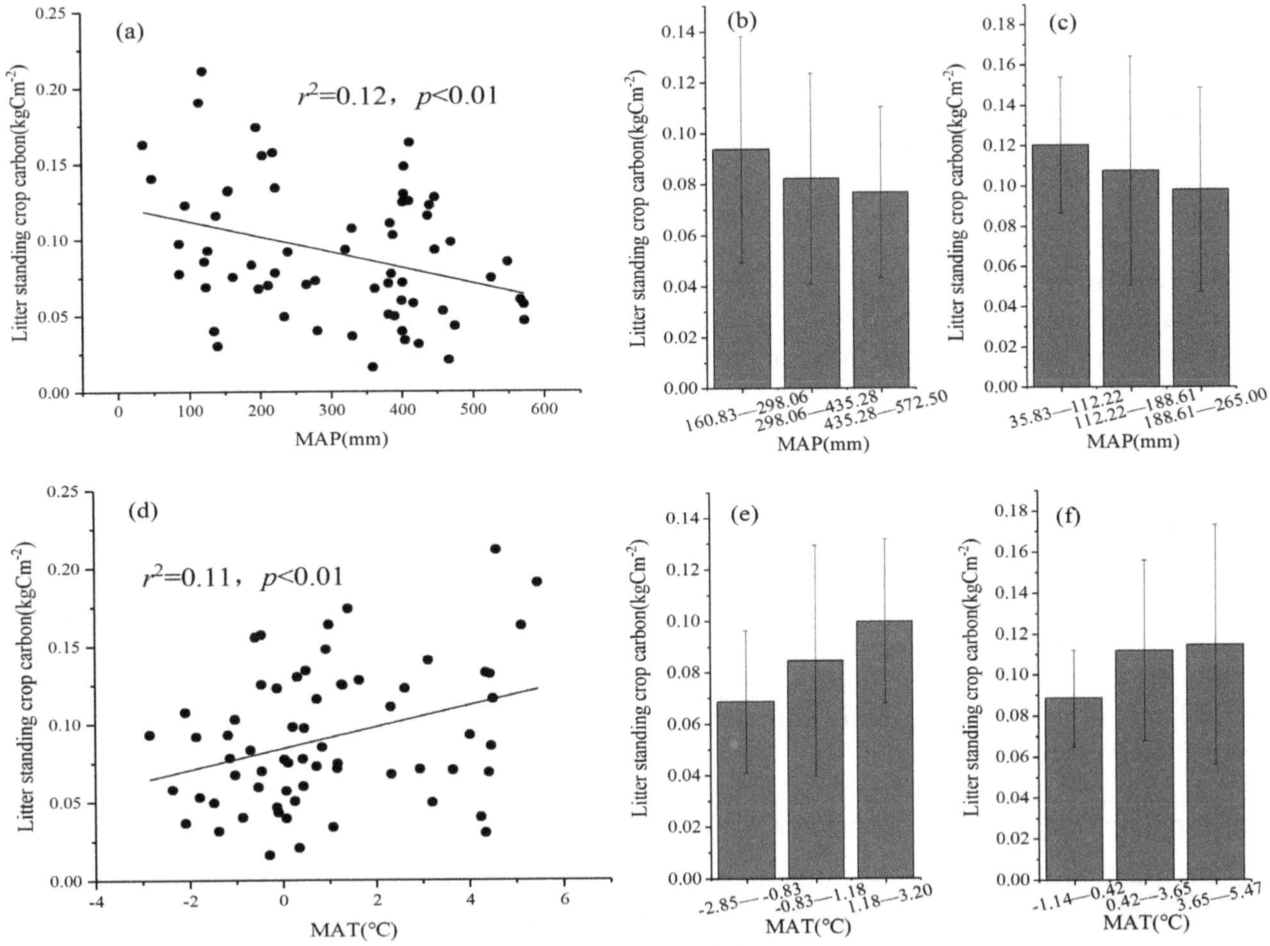

Figure 4. Relationships among mean annual precipitation (MAP) and mean annual temperature (MAT) and litter standing crop carbon, (**a**,**d**) total shrublands, (**b**,**e**) alpine shrublands, (**c**,**f**) desert shrublands.

3.4. *Effects of Aboveground Biomass and Climatic Factors on Litter Standing Crop Carbon*

Increasing aboveground biomass can stimulate litter standing crop carbon (Figure 5a) ($p < 0.01$), specifically, the positive trend also existed in both alpine shrublands and desert shrublands (Figure 5b). Generally, multiple linear regressions showed that litter standing crop carbon = −0.001 MAP + 0.01 MAT + 0.0001Aboveground biomass + 0.11 ($r = 0.42$, $p < 0.01$). Furthermore, the relationships among litter standing crop carbon, aboveground biomass, MAT, and MAP were shown in Figure S1.

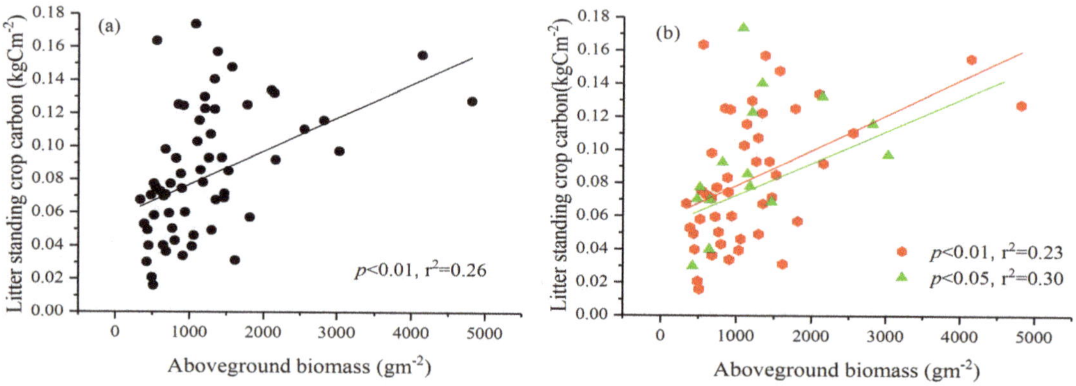

Figure 5. Relationships between aboveground biomass and litter standing crop carbon, (**a**) total shrublands, (**b**) red circle means alpine shrublands, green triangle means desert shrublands.

4. Discussion

4.1. Size of Litter Standing Crop

The density of the litter standing crop in the Tibetan Plateau shrublands was 0.23 kg m^{-2}, nearly half of the litter standing crop in the forests (0.45 kg m^{-2}) of the Tibetan Plateau [5], and was more than three times as much as the litter standing crop of the grasslands (0.05 kg m^{-2}) of the Tibetan Plateau [5]. The mean aboveground biomass in the shrublands was 1.10 kg m^{-2} [19], less than that of 4.22 kg m^{-2} in the forests [30], and greater than that of 0.08 kg m^{-2} in the grasslands of the plateau [31]. As aboveground biomass made the greatest contribution to litter standing crop in ecosystem [1], the litter standing crop in shrublands was in the range between forest and grasslands.

The litter standing crop in the shrublands of south China was 0.32 kg m^{-2} [16], larger than our result of 0.23 kg m^{-2} for the Tibetan Plateau. This difference may result from the greater amount of aboveground biomass in the shrublands across south China, compared to the Tibetan Plateau shrublands. Aboveground biomass can shape the input resource of litter standing crop [1] and significantly accumulate litter standing crop. It has been demonstrated that aboveground biomass in shrublands in the south region of China was 1.44 kg m^{-2} [32], greater than the 1.10 kg m^{-2} of the Tibetan Plateau shrublands [19].

4.2. Controlling Factors of Litter Standing Crop Carbon

Climatic factors can significantly shape litter standing crop [33,34]. Generally, precipitation not only stimulates vegetation growth in arid regions [35], and furthermore increases input of litter standing crop, but also affects the leaching of chemical substances from litter and consequently limits the litter decomposition rate [2]. So precipitation has been shown to have a positive effect on litter production [13] and stimulates accumulation of litter standing crop. However, precipitation has restrained effects on accumulation of litter standing crop carbon in the Tibetan Plateau shrublands. This result was different from those for other regions and biomes. For forests, precipitation did not have a significant effect on litter standing crop in forests of the eastern Inner Mongolia Plateau [18] or forests in China [5]. Litter standing crop carbon was stable with increasing precipitation in southern regions of China shrublands [16]. It should be noted that precipitation can also enhance litter decomposition by maximizing decomposer activity [11,36], which can reduce litter standing crop storage. Litter production and litter decomposition determine litter standing crop [37]. The restrain effects on the accumulation of litter standing crop carbon from precipitation may exceed the positive effects; consequently, increasing MAP decreases litter standing crop carbon in the shrublands, both alpine shrublands and desert shrublands, on the Tibetan Plateau.

Temperature is considered to be the most significant factor determining litter decomposition [38,39]. Rising temperature increases the litter decomposition rate [2]. In our study, MAT contributed to the accumulation of litter standing crop carbon in the shrublands on the Tibetan Plateau, which was different from other reported findings, specifically, that temperature had negative effects on accumulation of litter standing crop carbon in shrubland ecosystems in southern China [15] and in forests in China [5], whereas temperature was not a significant controlling factor for litter standing crop carbon in grasslands in China [5]. One of the most significant climate characteristics on the Tibetan Plateau is cold, and in the Tibetan Plateau shrublands, increasing temperature has been shown to play a significant role in shaping the amount of aboveground biomass [40]. Temperature contributes to the accumulation of aboveground biomass [40], which shapes amount of litter input, and stimulates more inputs into the litter standing crop; consequently, the litter standing crop in the shrublands of the Tibetan Plateau shows an increasing trend with MAT.

The climatic controlling factors on the litter standing crop carbon in the Tibetan Plateau shrublands, such as MAT and MAP, were different from those in other regions, which may indicate that the biophysical processes that control carbon on the Tibetan Plateau may differ from those in other regions [20,22]. In the global warming scenario, MAT is increasing by 0.05 °C every year [10,41], which, combined with the positive relationship between litter standing crop carbon and temperature, contributes to the carbon sink trend in the Tibetan Plateau shrublands.

4.3. Distribution and Conversion Coefficient of Litter Standing Crop to Litter Standing Crop Carbon

Spatially, the litter standing crop carbon decreased with longitude in shrublands, which may result from climatic changes with different regions. With an increasing longitude, the MAP has an increasing trend (Figure S2a), and consequently, increasing precipitation limited accumulation of litter standing crop carbon (Figure 4a–c). At the same time, MAT has a decreasing trend with longitude (Figure S2b), and decreasing temperature also limited the accumulation of litter standing crop carbon (Figure 4d–e). The climatic changes with longitude can explain the distribution of litter standing crop carbon in shrublands on the Tibetan Plateau.

It has been estimated that the conversion coefficient for litter standing crop to litter standing crop carbon is 0.41 in the shrublands of southern China [16], which is lower than our result of 0.44. Different vegetation types result in various rates of litter production and litter decomposition due to climatic conditions [33,34], and eventually, shape the conversion coefficient for litter standing crop to litter standing crop carbon. Cold weather is the most significant climatic characteristics of the Tibetan Plateau [42], due to its high elevation. On the one hand, lower temperature suppresses decomposer activity [11,43], which contributes to the weak decomposition of litter on the Tibetan Plateau. On the other hand, plants tend to produce organic matter with a higher carbon content [44–46]. Therefore, the conversion coefficient for litter standing crop to litter standing crop carbon in shrublands was smaller on the Tibetan Plateau compared with in shrublands of southern China. Researchers use a coefficient of 0.50 to convert biomass to biomass carbon in woody ecosystems such as forests [47,48] and shrublands [17], while the coefficient is smaller (0.45) in herbaceous ecosystems such as grasslands [49,50]. Our results reveal that the conversion coefficient for branch standing crop to branch standing crop carbon was 0.45, larger than the conversion coefficient of 0.36 for foliage standing crop to foliage standing crop carbon, which indicates that the conversion coefficient in woody ecosystems is larger than in herbaceous ecosystems. Studies of forest and grassland ecosystem support this trend, with a conversion coefficient of 0.46 in forests and 0.40 in grasslands in China [5]. Therefore, it is better to use specific conversion coefficients for different ecosystems to estimate litter standing crop carbon. Furthermore, even with in the same ecosystem, such as shrublands, it is better to use a specific parameter to assess carbon pools, specifically, 0.41 for the shrublands of southern China [16] and 0.44 for the Tibetan Plateau shrublands. To conduct a more accurate estimation of the carbon pool, it is necessary to use conversion coefficients for the different components of litter standing crop (in this case, 0.45 and 0.36, respectively, for branch and foliage standing crop in the shrublands of the Tibetan Plateau). Meanwhile, in the Tibetan Plateau, conversion coefficients of branch standing crop to branch standing crop carbon and foliage standing crop to foliage standing crop carbon were 0.46 and 0.42,

0.40, and 0.27 greater in alpine shrublands than in desert shrublands, respectively, which furthermore means that in the neighborhood regions (such as the Tibetan Plateau), the given conversion coefficient needs to be considered for more accurate estimation.

5. Conclusions

We used data from 65 sites to explore the storage, distribution, and environmental controls of litter standing crop carbon in the shrublands of the Tibetan Plateau. The litter standing crop carbon was 10.93 Tg C in the shrublands of the plateau. Most of the litter standing crop was stored in the branch standing crop, which accounted for 80% of all litter standing crop. The conversion coefficient was 0.44 for litter standing crop to litter standing crop carbon. Spatially, the litter standing crop showed a decreasing trend with increasing longitude, and was stable with latitude. Litter standing crop carbon decreased with increasing MAP, but MAT and aboveground biomass have had positive effects on litter standing crop carbon in the shrublands on the Tibetan Plateau.

Author Contributions: Conceptualization, G.Z. and X.N.; Methodology, G.Z., X.N.; Investigation, L.Y.; Writing—Original Draft Preparation X.N. and D.W.; Project Administration, G.Z.

Acknowledgments: We thank Wenzhu Song, Zebing Zhong, Hechun Liu, and Yi Ning for facilitating our field surveys on the Tibetan Plateau and laboratory assistance. This study is funded by the National Program on Basic Work Project of China (2015FY11030001-5), Qinghai Province Natural Science Fund Project (2019-ZJ-910), Qinghai Province International Exchange and Cooperation Project (2019-HZ-807) and Strategic Priority Research Program of CAS (XDA0505030304).

References

1. Descheemaeker, K.; Muys, B.; Nyssen, J.; Poesen, J.; Raes, D.; Haile, M.; Deckers, J. Litter production and organic matter accumulation in exclosures of the Tigray highlands, Ethiopia. *For. Ecol. Manag.* **2006**, *233*, 21–35. [CrossRef]
2. Peng, S.; Liu, Q. The dynamicis of forest litter and its responses to global warming. *Acta Ecol. Sin.* **2002**, *9*, 1534–1544.
3. Bojko, O.; Kabała, C. Loss-on-ignition as an estimate of total organic carbon in the mountain soils. *Pol. J. Soil Sci.* **2014**, *47*, 71–79.
4. Pan, Y.; Birdsey, R.A.; Fang, J.; Houghton, R.; Kauppi, P.E.; Kurz, W.A.; Phillips, O.L.; Shvidenko, A.; Lewis, S.L.; Canadell, J.G. A large and persistent carbon sink in the world's forests. *Science* **2011**, *333*, 988–993. [CrossRef] [PubMed]
5. Wen, D.; He, N. Spatial patterns of litter density and their controlling factors in forests and grasslands of China. *Acta Ecol. Sin.* **2016**, *36*, 2876–2884.
6. Marty, C.; Houle, D.; Gagnon, C. Variation in stocks and distribution of organic C in soils across 21 eastern Canadian temperate and boreal forests. *Forest Ecol. Manag.* **2015**, *345*, 29–38. [CrossRef]
7. Domke, G.M.; Perry, C.H.; Walters, B.F.; Woodall, C.W.; Russell, M.B.; Smith, J.E. Estimating litter carbon stocks on forest land in the United States. *Sci. Total Environ.* **2016**, *557*, 469–478. [CrossRef]
8. Liu, L.; Shen, G.; Chen, F.; Luo, L.; Xie, Z.; Yu, J. Dynamic characteristics of litterfall and nutrient return of four typical forests along the altitude gradients in Mt. Shennongjia, China. *Acta Ecol. Sin.* **2012**, *32*, 2142–2149. [CrossRef]
9. McLaren, J.R.; Buckeridge, K.M.; Weg, M.J.; Shaver, G.R.; Schimel, J.P.; Gough, L. Shrub encroachment in Arctic tundra: Betula nana effects on above-and belowground litter decomposition. *Ecology* **2017**, *98*, 1361–1376. [CrossRef]
10. Yang, Y. Ecological processes in alpine ecosystems under changing environment. *Chin. J. Plant Ecol.* **2018**, *42*, 1–5.
11. Taylor, P.G.; Cleveland, C.C.; Wieder, W.R.; Sullivan, B.W.; Doughty, C.E.; Dobrowski, S.Z.; Townsend, A.R. Temperature and rainfall interact to control carbon cycling in tropical forests. *Ecol. Lett.* **2017**, *20*, 779–788. [CrossRef] [PubMed]

12. Wohl, E.; Barros, A.; Brunsell, N.; Chappell, N.A.; Coe, M.; Giambelluca, T.; Goldsmith, S.; Harmon, R.; Hendrickx, J.M.; Juvik, J. The hydrology of the humid tropics. *Nat. Clim. Chang.* **2012**, *2*, 655. [CrossRef]
13. Jia, B.; Zhou, G.; Liu, Y.; Jiang, Y. Spatial pattern and environment controls of annual litterfall production in natural forest ecosystems in China. *Sci. Sin. Vitae* **2016**, *11*, 1304–1311.
14. Zhao, Y.; Wang, X.; Zhang, Y. Relationship between litterfall production and meteorological factors in evergreen broad-leaved forest in Tiantong national forest park. *J. Beijing Univ. Agric.* **2017**, *32*, 73–77.
15. Zhang, X.; Wang, X.; Zhu, B.; Zong, Z.; Peng, C.; Fang, J. Litter fall production in relation to environmental factions in northeast China's forests. *Chin. J. Plant Ecol.* **2008**, *32*, 1031–1040.
16. Ge, J.; Xiong, G.; Li, J.; Xu, W.; Zhao, C.; Lu, Z.; Li, Y.; Xie, Z. 2017 Litter standing crop of shrubland ecosytem in southern China. *Chin. J. Plant Ecol.* **2017**, *41*, 5–13.
17. Hu, H.; Wang, Z.; Liu, G.; Fu, B. Vegetation carbon storage of Major shrublands in China. *Chin. J. Plant Ecol.* **2006**, *30*, 539–544.
18. Chen, S.; Liu, H.; Guo, D. Litter stocks and chemical quality of natural birch forests along temperature and precipitatioan gradients in eastern Inner Mongolia China. *Chin. J. Plant Ecol.* **2010**, *34*, 1007–1015.
19. Nie, X.; Yang, Y.; Yang, L.; Zhou, G. Above- and Belowground Biomass Allocation in Shrub Biomes across the Northeast Tibetan Plateau. *PLoS ONE* **2016**, *11*, e0154251. [CrossRef]
20. Nie, X.; Peng, Y.; Li, F.; Yang, L.; Xiong, F.; Li, C.; Zhou, G. Distribution and controlling factors of soil organic carbon storage in the northeast Tibetan shrublands. *J. Soils Sediments* **2019**, *19*, 322–331. [CrossRef]
21. Nie, X.; Yang, L.; Li, F.; Xiong, F.; Li, C.; Zhou, G. Storage, patterns and controls of soil organic carbon in the alpine shrubland in the Three Rivers Source Region on the Qinghai-Tibetan Plateau. *Catena* **2019**, *178*, 154–162. [CrossRef]
22. Yang, Y.; Fang, J.; Tang, Y.; Ji, C.; Zheng, C.; He, J.; Zhu, B. Storage, patterns and controls of soil organic carbon in the Tibetan grasslands. *Glob. Chang. Biol.* **2008**, *14*, 1592–1599. [CrossRef]
23. Zhou, X.; Wang, Z.B.; Du, Q. *Vegetation of Qinghai Province*; Qinghai People's Publishing House: Xining, China, 1987; pp. 53–72.
24. Wu, Z. *Vegetation of China*; Science Press: Beijing, China, 1980; p. 430.
25. Zhang, Z. *Geography of Qinghai Plateau*; Science Press: Beijing, China, 2009; pp. 25–26.
26. Chinese Academy of Science. *Vegetation Altas of China*; Science Press: Beijing, China, 2001.
27. Technical Manual Writing Group of Ecosystem Carbon Sequestration Project. *Observation and Investigation for Carbon Sequestration in Terrestrial Ecosystems*; Science Press: Beijing, China, 2015; pp. 153–155.
28. Yang, X.; Tang, Z.; Ji, C.; Liu, H.; Ma, W.; Mohhamot, A.; Shi, Z.; Sun, W.; Wang, T.; Wang, X.; et al. Scaling of nitrogen and phosphorus across plant organs in shrubland biomes across Northern China. *Sci. Rep.* **2014**, *4*, 5448. [CrossRef] [PubMed]
29. Nie, X.; Xiong, F.; Yang, L.; Li, C.; Zhou, G. Soil nitrogen storage, distribution, and associated controlling factors in the northeast Tibetan Plateau shrublands. *Forests* **2017**, *8*, 416. [CrossRef]
30. He, H. *Modeling Forest NPP Patterns on the Tibetan Plateau and Its Responses to Cliamte Change*; Chinese Academy of Forestry: Beijing, China, 2008.
31. Yang, Y. Carbon and Nitrogen in Alpine Grasslands on the Tibetan Plateau. Ph.D. Thesis, Peking University, Beijing, China, 2008.
32. Li, J. Biodiversity and Its Relationship to Ecosystem Production in Shrubland across Subtropical Region in China. Ph.D. Thesis, Institute of Botany, Chinese Academy of Sciences, Beijing, China, 2015.
33. Zhou, X.; Talley, M.; Luo, Y. Biomass, litter, and soil respiration along a precipitation gradient in southern Great Plains, USA. *Ecosystems* **2009**, *12*, 1369–1380. [CrossRef]
34. Chave, J.; Navarrete, D.; Almeida, S.; Alvarez, E.; Aragão, L.E.; Bonal, D.; Châtelet, P.; Silva-Espejo, J.; Goret, J.Y.; Hildebrand, P. Regional and seasonal patterns of litterfall in tropical South America. *Biogeosciences* **2010**, *1*, 43–55. [CrossRef]
35. Wynn, J.G.; Bird, M.I.; Vellen, L.; Grand-Clement, E.; Carter, J.; Berry, S.L. Continental-scale measurement of the soil organic carbon pool with climatic, edaphic, and biotic controls. *Glob. Biogeochem. Cycles* **2006**, *20*, GB1007. [CrossRef]
36. Wieder, W.R.; Cleveland, C.C.; Townsend, A.R. Controls over leaf litter decomposition in wet tropical forests. *Ecology* **2009**, *90*, 3333–3341. [CrossRef]
37. Hilli, S.; Stark, S.; Derome, J. Litter decomposition rates in relation to litter stocks in boreal coniferous forests along climatic and soil fertility gradients. *Appl. Soil Ecol.* **2010**, *46*, 200–208. [CrossRef]

38. Hobbie, S.E. Temperature and plant species control over litter decomposition in Alaskan tundra. *Ecol. Monogr.* **1996**, *66*, 503–522. [CrossRef]
39. Fierer, N.; Craine, J.M.; McLauchlan, K.; Schimel, J.P. Litter quality and the temprtature sensitivity of decomposition. *Ecology* **2005**, *86*, 320–326. [CrossRef]
40. Nie, X.; Yang, L.; Xiong, F.; Li, C.; Fan, L.; Zhou, G. Aboveground biomass of the alpine shrub ecosystems in Three-River Source Region of the Tibetan Plateau. *J. Mt. Sci.* **2018**, *15*, 357–363. [CrossRef]
41. Duan, K.; Yao, T.; Wang, K.; Tian, L.; Xu, B. The Difference in precipitation variabil ity between the north and south Tibetan Plateaus. *J. Glaciol. Geocryol.* **2008**, *30*, 726–732.
42. Ding, J.; Li, F.; Yang, G.; Chen, L.; Zhang, B.; Liu, L.; Fang, K.; Qin, S.; Chen, Y.; Peng, Y.; et al. The permafrost carbon inventory on the Tibetan Plateau: A new evaluation using deep sediment cores. *Glob. Chang. Biol.* **2016**, *22*, 2688–2701. [CrossRef] [PubMed]
43. Cleveland, C.C.; Wieder, W.R.; Reed, S.C.; Townsend, A.R. Experimental drought in a tropical rain forest increases soil carbon dioxide losses to the atmosphere. *Ecology* **2010**, *91*, 2313–2323. [CrossRef] [PubMed]
44. Vergutz, L.; Manzoni, S.; Porporato, A.; Novais, R.F.; Jackson, R.B. Global resorption efficiencies and concentrations of carbon and nutrients in leaves of terrestrial plants. *Ecol. Monogr.* **2012**, *82*, 205–220. [CrossRef]
45. Schreeg, L.A.; Mack, M.C.; Turner, B.L. Nutrient-specific solubility patterns of leaf litter across 41 lowland tropical woody species. *Ecology* **2013**, *94*, 94–105. [CrossRef]
46. Sun, X.; Kang, H.; Chen, H.; Berg, B.; Bartels, S.F.; Liu, C. Biogeographic patterns of nutrient resorption from Quercus variabilis Blume leaves across China. *Plant Biol.* **2016**, *18*, 505–513. [CrossRef]
47. Fang, J.; Chen, A.; Peng, C.; Zhao, S.; Ci, L. Changes in forest biomass carbon storage in China between 1949 and 1998. *Science* **2001**, *292*, 2320–2322. [CrossRef] [PubMed]
48. Guo, Z.; Hu, H.; Li, P.; Li, N.; Fang, J. Spatio-temporal changes in biomass carbon sinks in China's forests from 1977 to 2008. *Sci. China Life Sci.* **2013**, *56*, 661–671. [CrossRef]
49. Piao, S.; Fang, J.; Zhou, L.; Tan, K.; Tao, S. Changes in biomass carbon stocks in China's grasslands between 1982 and 1999. *Glob. Biogeochem. Cycles* **2007**, *21*, GB2002. [CrossRef]
50. Ma, W.; Fang, J.; Yang, Y.; Mohammat, A. Biomass carbon stocks and their changes in northern China's grasslands during 1982–2006. *Sci. China Life Sci.* **2010**, *53*, 841–850. [CrossRef] [PubMed]

4

Effect of Timber Harvest Intensities and Fertilizer Application on Stocks of Soil C, N, P and S

Marcella L.C. Menegale [1,*], Jose Henrique T. Rocha [2], Robert Harrison [1], Jose Leonardo de M. Goncalves [2], Rodrigo F. Almeida [3], Marisa de C. Piccolo [3], Ayeska Hubner [2], Jose Carlos Arthur Junior [4], Alexandre de Vicente Ferraz [5], Jason N. James [1] and Stephani Michelsen-Correa [1]

1 School of Environmental and Forest Sciences, University of Washington, Seattle, WA 98195-2100, USA; robh@uw.edu (R.H.); jajames@uw.edu (J.N.J.); smcorrea@uw.edu (S.M.-C.)
2 Forest Science Department, "Luiz de Queiroz" College of Agriculture, University of São Paulo, Piracicaba SP 13418-900, Brazil; rocha.jht@gmail.com (J.H.T.R.); jlmgonca@usp.br (J.L.d.M.G.); ayeskahubner@yahoo.com.br (A.H.);
3 Center of Nuclear Energy in Agriculture, University of São Paulo, Piracicaba, São Paulo 13416-000, Brazil; rodrigo.fa18@gmail.com (R.F.A.); mpiccolo@cena.usp.br (M.d.C.P.)
4 Federal Rural University of Rio de Janeiro, Seropédica, Rio de Janeiro 23890-000, Brazil; jcarthur@ufrrj.br
5 Forest Science and Research Institute (IPEF), Piracicaba, São Paulo 13400-970 Private Bag 530, Brazil; alexandre@ipef.br
* Correspondence: marcism@uw.edu

Academic Editors: Scott X. Chang and Xiangyang Sun

Abstract: The purpose of this study was to determine the stocks of available P and S, total N, and oxidizable C at depth in an Oxisol cultivated with Eucalyptus in Brazil following different timber harvest intensities and fertilizer application over 12 years. The harvest regimes considered were (i) conventional stem-only harvest (all forest residues were maintained on the soil); (ii) whole-tree harvest (only litter was maintained on the soil—all slash, stemwood, and bark were removed); and (iii) whole-tree harvest + litter layer removal. The site was planted in 2004 considering three timber harvest intensities, some with and some without N and P fertilization. In 2012 the experiment was reinstalled, and all the treatments were reapplied in the each plot. From 2004 to 2016, nutrient accumulation and soil N, P, and S stocks were assessed in the 0–20 cm layer. Also in 2016, soil N, P, S, and oxidizable C stocks were measured to 2 m depth. For each treatment, the net balance of N, P, and S were calculated from soil stocks and harvest outputs during two forest rotations. A reduction in all nutrient stocks was observed in the 0–20 cm layer for all treatments. For N, this reduction was 20% smaller in the stem-only harvest treatment and 40% higher when no N fertilizer was applied, when compared to other treatments. Stem-only harvest treatment was observed to reduce the loss of N, P, and S due to harvest by 300, 30, and 25 kg·ha^{-1}, respectively, when compared to the whole-tree harvest + litter layer removal treatment.

Keywords: nutrient cycling; forest residue management; Eucalyptus; nitrogen; carbon; phosphorus; sulfur; minimum tillage

---

1. Introduction

Forest soils form an important reservoir in ecosystem nutrient and carbon budgets, which are crucial for the sustained productivity of forests [1]. Nutrient fluxes and transformations in forest soils are a result of a complex interchange between the atmosphere, plants, and soil. Consequently, soils are a critical source of plant nutrition [2], and integral to the recovery of ecosystems following natural or human disturbances. As essential plant nutrients, the concentrations of plant available nitrogen (N), phosphorous (P), and sulfur (S) are crucial controls on the net primary productivity of a site. As the plants uptake these nutrients, their concentrations within the soil are reduced until they are replaced either by nutrient recycling or fresh inputs. In order to maintain the supply of these nutrients back into the soil, sustainable silviculture practices focus on high timber productivity while efficiently balancing the additional export of harvest residues with the cost of fertilizer to replace nutrients.

Highly weathered tropical soils are characteristically deficient in essential plant nutrients. This is especially true of the macronutrients essential for plant growth and development [3]. Organic matter (OM) plays a fundamental role in soil fertility, contributing to cation exchange capacity (CEC) and as a source of nutrients [4], while also stabilizing soil aggregates and increasing water holding capacity. Soil management practices can sustain adequate levels of OM while providing sufficient quantities of plant available P. The need for balancing high timber productivity while maintaining soil nutrient concentrations has led to the development of sustainable silvicultural practices [5,6].

Highly weathered soils may conserve more carbon (C) with increasing OM inputs because P frequently limits decomposition [7]. Such P limitations will also effect microbially driven processes, such as N mineralization. While most temperate ecosystems—which are usually N limited—may store added mineral N for long periods, a P limited tropical forest can rapidly loose substantial amounts of mineral N (NO_3^- and NH_4^+) following N fertilization [8]. The nutrient limitations of tropical soils not only decrease plant productivity but also impact decomposition rates and immobilization processes and therefore the soil's capacity to store and cycle C and N [9–11].

Many consider the proper management of forest residues to be a practice which maintains the harvest residues (i.e., slash, bark, and litter layer) on site as a source of nutrients. The practice of leaving forest residues on site can sustain soil quality with respect to the productivity of medium and long-term species such as Eucalyptus [12–16]. This practice helps to improve nutrient pools and OM content in the soil [3]. Harvest residues, when retained on site, deposit nutrients back to the soil following decomposition. Plants cultivated in the area will then readily absorb these nutrients. This reduces the need for synthetic fertilizer and other methods for soil preparation that can disrupt soil physicochemical properties such as aggregation [17,18]. In addition, the protective layer formed due to the presence of harvest residues on the soil helps to reduce extreme surface heating and water loss through evaporation, and decreases soil loss due to erosion.

Due to the coupling of soil nutrient status (i.e., C, N, P, and S) and forest productivity, elucidating the impacts of forest residue additions will aid in the sustainable management of high quality, rapidly growing timber crops. Under tropical conditions, it is still unclear how the presence of harvest residues in the soil can help to create a more efficient silviculture system with regards to productivity, improved preservation of soil characteristics, and reduced fertilizer applications. For this reason, the objective of this study was to determine the stocks of available P and S, total N, and oxidizable C at depth in an Oxisol which was cultivated with Eucalyptus in Brazil over 12 years, under different timber harvest intensities and fertilizer applications. During each forest rotation the net balance between soil N, P, and S stocks and harvest outputs was also calculated.

2. Materials and Methods

2.1. Study Site

The study was carried out at the Itatinga Forest Science Experimental Station of the University of São Paulo in Brazil (23°06′ S, 48°36′ W, and 857 m above sea level). The Köppen climate classification of the site is humid subtropical (Cfa) with an average annual temperature of 19.4 °C. In the coldest month (July), the temperatures average 15.6 °C, and in the hottest month (January), the temperatures average 22.3 °C. The mean annual rainfall is 1319 mm, with 75% of the rainfall concentrated between October and March [19].

The topography of the region is flat to undulating, and the soil is a very deep (>10 m) Ferralsol (International Union of Soil Sciences (IUSS) Working Group—World Reference Base for

Soil Resources (WRB), 2015; red-yellow Latosol—Brazilian Classification System, and Oxisols—United State Department of Agriculture (USDA) Soil Taxonomy) that developed on Cretaceous sandstone. The clay content ranges from 17% in the A1 horizon to 25% in deeper Bo horizons. The mineralogy was dominated by quartz, kaolinite, and oxyhydroxides of Al and Fe with a low pH and small amounts of exchangeable cations (Table 1).

Table 1. Physical and chemical attributes of experimental site *.

Depth	Sandy	Silt	Clay [1]	pH [2]	CEC_7 [3]	C [4]	N [5]	P [6]	Exchangeable Cations [4]			
									K	Ca	Mg	Al
cm		$g \cdot kg^{-1}$			$mmol_c\ kg^{-1}$	$g \cdot kg^{-1}$		$mg \cdot kg^{-1}$	$mmol_c\ kg^{-1}$			
0–10	802	22	175	3.8	63.98	9.61	1.44	4	0.25	4.28	2.81	7.50
10–20	811	12	176	3.9	51.42	10.05	1.67	3	0.27	2.80	2.17	8.43
20–30	790	34	176	3.9	39.98	6.77	1.53	1	0.20	1.32	1.00	6.09
30–40	777	23	200	3.9	40.18	5.33	1.29	1	0.15	0.88	0.81	7.03
40–60	747	14	239	3.9	38.46	5.42	1.14	1	0.15	0.99	0.72	7.50
60–100	712	12	276	3.9	32.72	5.04	0.99	1	0.15	0.66	0.54	6.56
100–150	712	11	277	4.0	30.09	3.44	1.04	1	0.08	0.71	0.54	2.34
150–200	704	20	276	4.2	22.05	0.87	1.04	1	0.05	0.55	0.54	2.81

* Samples were taken in 2014. The total area of the experiment was considered (average of all plots). [1] Pipette method; [2] Determined in $CaCl_2$ 0.01 $mol \cdot L^{-1}$ in soil/solution reason of 1:5; [3] Cation exchange capacity with soil at pH 7; [4] wet oxidation; [5] Determined using the micro Kjeldahl method after sulphuric digestion; [6] Extracted with exchange ion resin [20].

The original vegetation of the site was Cerrado *stricto sensu* (Brazilian savannah) [21]. The site has been planted with *Eucalyptus* species since 1940. From 1940 to 1992, it was cropped with *Eucalyptus saligna* and managed by coppicing with clearcutting every seven or eight years. In 1992, the plantation was harvested and replanted with *Eucalyptus grandis*, which was harvested (via clear-cutting) in 2004 when the study site was installed.

2.2. Experimental Design and Treatments

The study site was installed in 2004 (R1) and reinstalled in 2012 (R2) with three replicates of five treatments in a randomized block design. The plot sizes were 27 m × 18 m, with 81 trees per plot. The assessments were carried out in an inner plot of 15 m × 10 m (25 trees per plot). Five treatments were implemented with different management levels of forest residue removal and fertilizer applications (Table 2). The forest residues manipulated in this experiment include all of the organic residues remaining on the soil after wood harvesting of *E. grandis* plantations after 12 years of growth: the leaves and branches less than 3 cm in diameter (canopy), bark, and litter layer. The treatments tested were:

1. ReM + F—Only stemwood was harvested; all of the forest residues (bark, canopy, and litter layer from the previous rotation) were maintained on the soil after the clear-cutting, and all nutrients were applied as fertilizer and the soil was dressed with limestone;
2. ReR + F—All of the forest residues (bark, canopy, and litter layer from the previous rotation) were removed from the plot after the clear-cutting, and all nutrients were applied as fertilizer and the soil was dressed with limestone;
3. ACR + F—The canopy (leaves and branches) and bark were removed after clear-cutting, but the litter layer was maintained; all nutrients were applied as fertilizer and the soil was dressed with limestone;

4. ACR − N—The canopy (leaves and branches) and bark were removed after the clear-cutting, but the litter layer was maintained; all nutrients except N fertilizer were applied and the soil dressed with limestone. However, a small quantity of N was applied to ensure tree survival;
5. ACR − P—The canopy (leaves and branches) and bark were removed after the clear-cutting, but the litter layer was maintained; all nutrients except P fertilizer were applied and the soil was dressed with limestone;

Table 2. Forest residue management and nutrients applied in treatments.

Treatment [1]	**Forest Residue** [2]			**Nutrients** [3]					
	Canopy	**Bark**	**Litter Layer**	**N**	**P**	**K**	**Ca**	**Mg**	**S**
				$kg \cdot ha^{-1}$					
ReM + F	M	M	M	130	44	125	480	120	140
ReR + F	R	R	R	130	44	125	480	120	140
ACR + F	R	R	M	130	44	125	480	120	140
ACR − N	R	R	M	10	44	125	480	120	11
ACR − P	R	R	M	130	-	125	480	120	140

[1] Detailed description in Material and Methods; [2] M = maintained on the soil, R = removed from the area; [3] N, P, K, Ca and Mg sources were ammonium sulphate, triple superphosphate, potassium chloride, and limestone, respectively.

The ReR + F treatment was included to simulate a severe harvest condition where all forest residues were removed from the site, with fertilization being the only source of nutrients added to the soil. Although unusual, this harvest management system has already been implemented in some regions in Brazil, especially when biomass prices were high and all forest residues were removed with the objective of maximizing biomass production. Furthermore, by examining such an extreme treatment, this study brackets the range of harvest intensities actually utilized by different forest managers. Thus, the effects of more moderate treatments can be estimated by interpolating between two endmembers rather than extrapolating from a more limited range of harvest intensity.

2.3. Field Procedures

After clear-cutting the 12-year-old *Eucalyptus grandis* Hill Ex Maiden plantation, the treatments (Table 2) were applied and the soil was prepared by subsoiling to 0.4 m deep with a ripper. The plots were planted with a single progeny of *E. grandis* Hill ex Maiden seedlings in June 2004 (one month after harvesting the previous plantation). The fertilizer that was applied is shown in Table 2. Additionally, 3.4 $kg \cdot ha^{-1}$ of boron (B) and 30 $kg \cdot ha^{-1}$ of Fritted Trace Elements (FTE) (9% Zn, 1.8% B, 0.8% Cu, 2% Mn, 3.5% Fe, and 0.1% Mo) were applied in every treatment. The fertilizer was applied as one base fertilizer application and two topdressing applications. The base fertilizer application was made on the same day as the planting, the triple superphosphate, the FTE, and 10 $kg \cdot ha^{-1}$ of N and K were added in a small pit to the side of each seedling, whereas the lime (2 $Mg \cdot ha^{-1}$) was applied to the whole area. The ACR − N treatment received a small amount of N with the base fertilizer to ensure the survival and the initial development of plants. The topdressing application of N, K, S and B was applied around the seedling within the ground area covered by the canopy at three and eight months after planting. The plantation was harvested at eight years of age and replanted one month later (November 2012). The treatments were repeated at each plot (Table 2).

During clear-cutting, the canopy and bark from the trees of each plot were retained on the same plots. The clear-cut trees were felled with a chainsaw; the bark and branches were then removed manually. The seedlings in R2 were planted between the stumps of R1 without ripping. R2 was managed in the same way as R1 to evaluate the long-term effects of the residue removal and fertilizer application practices. Each experimental unit was maintained weed free throughout the two rotations.

2.4. Soil Sampling and Analysis

Soil samples were collected before the application of the treatments in the first (2004) and second (2012) rotations at 0–10 and 10–20 cm. In 2016, samples were taken from the soil down to two meters (0–10, 10–20, 20–40, 40–60, 60–80, 80–100, 100–150, and 150–200 cm) across three replications for each treatment. Each sample was composed of six subsamples taken from six points in the inner plot arranged in a diagonal design. For each plot, soil bulk density was calculated from soil samples taken from the walls of trenches opened with known volume rings.

Samples were then air-dried at 45 °C for three days and passed through a 2 mm sieve for chemical analysis. Soil analyses were carried out for all treatments. Total N was determined by dry combustion using an Elemental Analyzer CHNS/2400 (Perkin Elmer, Waltham, MA USA) [22]. Available P was determined by displacement using ion-exchange resins. Sulfur was displaced with $Ca(H_2PO_4)_2$ 0.01 mol·L^{-1} solution and determined by turbidimetry using $BaSO_4$ [20]. Oxidizable C was determined by wet oxidation [23].

2.5. Nutrient Accumulation in the Biomass

In order to assess the accumulation of nutrients in the biomass (canopy, bark, wood, and coarse roots), 10 trees were felled in 2004, 10 trees per treatment were felled in 2012, and 3 trees per treatment were felled in 2016, with all removed trees coming from the inner border of each plot. Felled trees were separated into the following compartments: leaves, branches, stemwood (diameter > 3 cm at the thinner end), and stem bark. Coarse root (diameter > 1 cm) was removed by excavation. Sub-samples were collected from all of the compartments and dried (65 °C) until reaching a constant weight, and then the dry biomass of the compartments in each tree was proportionally calculated. To estimate the wood, bark biomass, and stem volume of the plantation from the sampled trees, diameter at breast high (DBH) and total high (H) were used as independent variables to adjust a model following the form of Schumacher and Hall [24].

Samples were analyzed for N, P, and S. Total N was determined using the micro Kjeldahl method after sulphuric digestion. P was determined through colourimetry after nitric perchloric digestion, and S was determined through turbidimetry [25]. Nutrient accumulation per hectare was determined by the sum of the product of biomass accumulation of each compartment and the nutrient concentrations.

2.6. Statistical Analysis

Prior to statistical analysis, the data were tested for normality (Shapiro-Wilk) and homoscedasticity (Box-Cox). A two-way Analysis of Variance (ANOVA) test was performed to identify the differences between the factors considered in the experiment (soil depth and treatment). Equal sample sizes were used in order to obtain maximum power and robustness of the test. In the case of overall significant differences in the group means, Fisher's Least Significant Difference (LSD) post hoc testing was performed to determine the differences between groups [26]. Statistical tests were considered significant at $\alpha = 0.05$. Data was analyzed using SAS University Edition (University Edition, SAS Institute Inc., Cary, NC, USA).

3. Results

3.1. Nutrient Balance After Two Crop Rotations

Before the installation of the experiment in 2004, the site contained 370, 11, and 9.2 kg·ha^{-1} of N, P, and S, respectively, in the 0–20 cm layer (Figure 1). On that occasion, there was an accumulation of approximately 660, 90, and 100 kg·ha^{-1} of N, P, and S, respectively, in the biomass and litter layer (250 t·ha^{-1} in total). The nutrient losses following harvest within the ReR + F treatment were higher than that of the ReM + F treatment by 120%, 50%, and 40% of N, P, and S respectively. For the ACR treatments, the losses from harvest were 65%, 30%, and 20% higher, respectively, than the ReM + F treatment (Table 3).

In 2012 (the end of experiment's first rotation), the ReM + F treatment contained the largest stocks of N (206 kg·ha^{-1}) and P (13.8 kg·ha^{-1}) in the 0–20 cm layer, while the highest stocks of S (13.2 kg·ha^{-1}) for this layer were observed in the ACR − P treatment. The lowest stocks of N (151 kg·ha^{-1}) and P (9 kg·ha^{-1}) were observed in the ACR − N and ACR − P treatments, respectively (Figure 1).

At the end of the first rotation of the experiment (eight years) there were no observed differences in the volume of wood produced in the treatments ReM + F, ACR + F, ReR + F, and ACR − N, which were about 420 m^3·ha^{-1} on average. On the other hand, the absence of fertilization with P (ACR − P treatment) resulted in approximately a 10% reduction in the final wood volume [27]. Even with small differences in tree growth, the renewal of the experiment for the next crop rotation resulted in great differences in the quantity of nutrient losses among the treatments, due to the different parts of the plants removed at harvest (Table 3).

At 41 months into the second rotation, the ReM + F treatment contained 100 Mg·ha^{-1} of biomass (wood, bark, canopy, and coarse roots). Compared to the ReM + F treatment, the biomass accumulated in the ACR + F, ReR + F, ACR − N, and ACR − P treatments was lower by 14%, 15%, 16%, and 25%, respectively. The accumulated N, P, and S in the biomass of the ACR − P treatment was 33%, 40%, and 20% lower, respectively, compared to the ReM + F treatment.

Between 2004 and 2012, N contents in the soil from 0–20 cm depth were reduced by around 50% with larger losses in the ACR − N treatment and smaller losses in the ReM+F treatment. From 2012 to 2016, a smaller reduction in soil N was observed, with the exception of the ACR − P treatment. Only small losses of available P were observed between 2004 and 2012, but greater losses occurred after reinstallation of treatments, especially for repeated removal of harvest residues (ReR + F). In the ACR + F treatment the concentration of available P in surface soil was constant through the years. A larger reduction was observed for the ReR + F and ACR − P treatments with losses of approximately 30% from 2004 to 2016. In regards to available S, a reduction in the ReR + F treatment from 2004 to 2012 was observed, and an increase was observed in the ACR − P treatment. For the other treatments, S stocks were constant. From 2012 to 2016, available S stocks improved for all treatments (Figure 1).

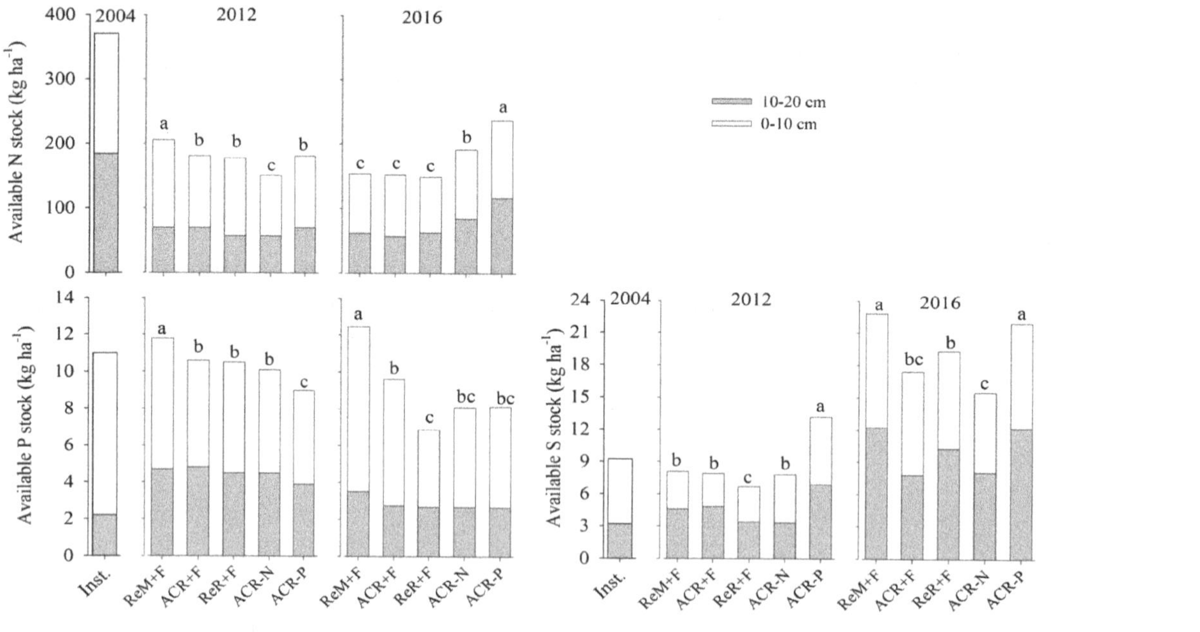

Figure 1. Available N, P, and S stocks in the 0–20 cm soil layer before the setup of treatments (2004), before the harvest of the first rotation (2012), and at four years into the second rotation (2016) in a Eucalyptus plantation under different harvest residue management and fertilizer application strategies. P was determined by resin extraction, S was determined by extraction with $Ca(H_2PO_4)_2$ 0.001 mol·L^{-1}, and available N was calculated assuming that 10% of total N is mineralizable [28]. Columns with the same letter are not significantly different from each other.

Table 3. Nutrient stocked in the forest biomass, inputs, and outputs of nutrients in two crop rotations.

Component	Biomass ($t \cdot ha^{-1}$)	N ($kg \cdot ha^{-1}$)	P ($kg \cdot ha^{-1}$)	S ($kg \cdot ha^{-1}$)
Nutrients Stocks Before 2004 Harvest				
Litter layer	24 (1) [1]	141 (17)	12 (1)	12 (1)
Stem wood	167 (15)	250 (20)	52 (3)	67 (5)
Stem bark	18 (1)	62 (3)	8 (1)	7 (1)
Canopy	9 (1)	100 (9)	7 (1)	6 (1)
Root	33 (5)	109 (18)	11 (2)	14 (2)
Total	251 (12)	662 (48)	90 (10)	106 (15)
Harvest Outputs in 2004				
Treatment ReM + F	167 (15)	250 (20)	52 (3)	67 (5)
Treatments ACR + F, ACR − N, and ACR − P	194 (18)	412 (46)	67 (4)	80 (7)
Treatment ReR + F	218 (18)	553 (49)	79 (15)	92 (8)
Inputs from 2004 to 2012				
Atmospheric deposition [2]		32	-	-
Fertilizer application [3]:				
Treatments ReM + F, ACR + F, and ReR + F		130	44	143
Treatment ACR − N		10	44	11
Treatment ACR − P		130	0	143
Harvest Outputs in 2012				
Treatment ReM + F	191 (10)	313 (17)	60 (3)	71 (4)
Treatment ACR + F	196 (20)	412 (37)	67 (7)	71 (7)
Treatment ReR + F	220 (16)	542 (39)	79 (6)	80 (6)
Treatment ACR − N	200 (32)	426 (60)	69 (11)	70 (11)
Treatment ACR − P	180 (23)	386 (48)	61 (8)	65 (8)
Inputs from 2012 to 2016				
Atmospheric deposition		16	-	-
Fertilizer application:				
Treatments ReM + F, ACR+F, and ReR + F		130	44	143
Treatment ACR − N		10	44	11
Treatment ACR − P		130	0	143
Accumulated in the Biomass in 2016				
Treatment ReM + F	100 (3)	255 (20)	25 (4)	40 (2)
Treatment ACR + F	86 (3)	207 (26)	21 (1)	37 (2)
Treatment ReR + F	85 (7)	202 (17)	18 (2)	37 (4)
Treatment ACR − N	84 (5)	195 (12)	18 (1)	36 (1)
Treatment ACR − P	75 (2)	172 (3)	15 (2)	32 (1)

[1] Standard deviation; [2] By Laclau et al. [29]; [3] It is assumed that 100% of fertilizer application was available.

3.2. Soil Nutrient Stocks in 2016

At four years into the second study rotation (2016), large differences between treatments were found in the stocks of Total N, Available P and S, and oxidizable C in the soil to 200 cm depth (Figure 2). The largest stock of oxidizable C in the soil was observed in the ReM + F treatment ($p > 0.001$), and this treatment contained the most oxidizable C at every individual sampling layer except between 60–100 cm. The ReR + F treatment contained the lowest stocks of oxidizable C ($p > 0.002$) up to 40 cm depth, and did not differ from the ACR treatments in the deeper layers. The supply of oxidizable carbon in the entire 0–200 cm profile of the ReM + F treatment was approximately 130 $Mg \cdot ha^{-1}$, which is 10% greater than the ACR + F treatment and 20% higher than the ReR + F treatment.

The ACR-P treatment contained the highest stock of N ($p > 0.001$) in all the layers considered. ReM + F, ACR + F, and ReR + F treatments presented the lowest stocks of N, not differing statistically from each other. The ReR + F and ACR − N treatments presented the lowest stocks of P, and were not significantly different. The stocks of N in the 0–200 cm layer of the soil were approximately 10 Mg·ha^{-1} in the ReM + F, ACR + F, and ReR + F treatments. Stocks of available P in the soil were approximately 40 kg·ha^{-1} in the ReM + F and ACR + F treatments, which is 60% higher than the stock of P observed in the ReR + F treatment.

The ACR + F and ReR + F treatments contained the largest stocks of S in the soil ($p > 0.001$). The ACR − N and ACR − P treatments contained the lowest stocks of S ($p > 0.001$). The stock of S available in the 0–200 cm layer was 280 kg·ha^{-1} in the ReM + F, ACR + F, and ReR + F treatments. The ACR − N treatment had a stock 30% lower than the previous treatments.

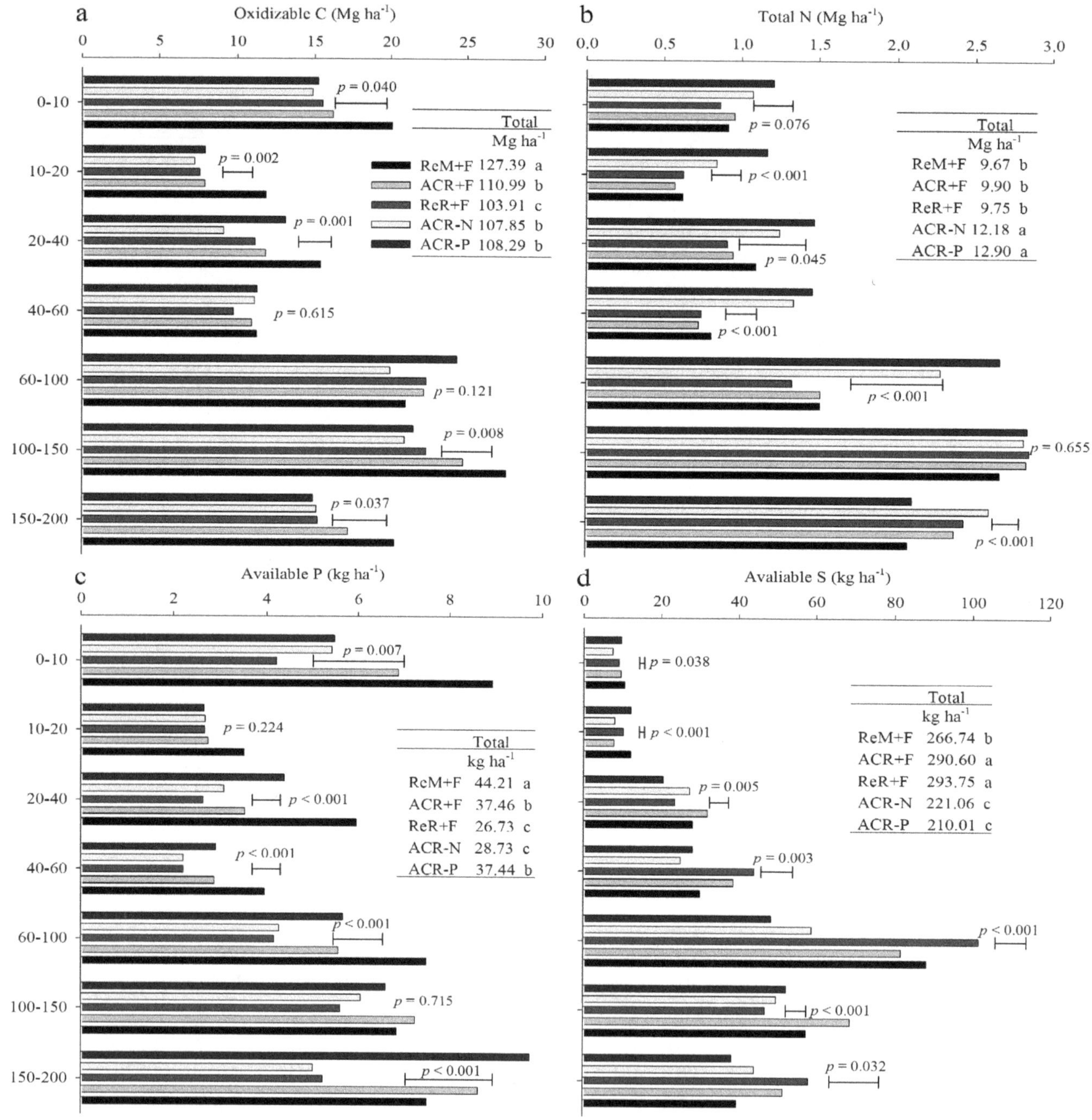

Figure 2. Stocks of oxidizable C (**a**), total N (**b**), available P (**c**) and S (**d**) to 2 m depth in a Eucalyptus plantation under different harvest residue management and fertilizer application strategies. The bars indicate the least significant difference based on the Fisher's Least Significant Difference (LSD) test at 5% probability, and the values next to the bars show the significance of the *F* test.

4. Discussion

4.1. Effects of Harvest Residue Strategies on Soil C and Available Nutrient Stocks

The effect of forest residue removal on soil nutrients stocks is inconsistent between different studies in the literature. In some cases, no losses in soil C or nutrient stocks have been reported, even with increases in nutrients exported during harvest [4,14,29–33]. This observation has been attributed to several mechanisms: high buffer capacity of the soil, slow decomposition of forest residues, a long harvesting return interval (more than seven years), and fast growth and litter deposition from new Eucalyptus plantations. However, reduction in soil C and nutrient stocks has been observed in wet tropical sites with sandy soils, high productivity forests, and successive harvest treatments [16,17,29,31,33]. Our study agrees with this latter set of studies and shows substantial changes in soil C and nutrients following removal of harvest residues.

In this study, the highest stocks of oxidizable C in the soil were observed in the 0–20 cm soil layer of the ReM + F treatment, followed by ACR + F and ReR + F treatments ($p > 0.001$). Similar results were observed by Moreno-Fernandez et al. [34] in Mediterranean mountain Scots pine forests, where the highest stocks of C were in the 0–20 cm layer of the soil under moderate intensity management.

Stocks of soil C decreased as soil depth increased, with the lowest stocks being observed in the ACR − P treatment ($p > 0.006$) at 150–200 cm depth. In a previous study developed in the same experimental site, Rocha [35] found that the removal of forest harvest residues from the soil reduces the oxidizable organic C from the surface layer of the soil by 50%, and 75% of this reduction happens in its labile fractions. Vanguelova et al. [36], on the other hand, concluded that there was no evidence that whole-tree harvest decreased soil organic C in a 28-year old second rotation stand of Sitka spruce in the UK. Slash removal has larger effects on soil C and nutrients when rotations are short, slash is removed repeatedly, clay content is low, temperatures are high, the site is relatively wet, and forest productivity is high [4,5]. It is well known that retaining harvest residues on the soil is extremely important to maintain high productivity of tropical forests [14–16]. However, more research is needed in order to reach a plausible conclusion about the subject due to the divergence in the data available, as observed by Nambiar and Harwood [14].

The forest residue removal for two rotations resulted in the cumulative loss of 23.5 $t \cdot ha^{-1}$ of C from 0–200 cm soil depth (ReM + F and ReR + F treatments). This reduction was more concentrated in the top 40 cm (Figure 2). This effect is a result of the additional removal of 80 $t \cdot ha^{-1}$ of biomass (slash and litter; 51 and 29 $t \cdot ha^{-1}$ in R1 and R2, respectively) in the ReR + F treatment. Furthermore, the reduced initial growth in the ReR + F treatment during both rotations resulted in less litter deposition [27]. In the second rotation, a 40% reduction in fine roots was observed in the 0–30 cm layer of the ReR + F treatment [37], which also contributed to low soil C content.

Forest residue removal resulted in a small reduction in the N stocks in the 0–20 cm layer in 2012 (Figure 1). However, no differences were observed between ReR + F and ReM + F treatments in 2016 (Figure 2) despite the higher harvest output in the ReR + F treatment. This can be attributed to the larger accumulation of N in the biomass in the ReM + F treatment, which was 33% higher when compared to the ReR + F treatment. This aligns with other long-term productivity studies that find that subsequent tree growth following stem only harvest tends generally to be higher than whole-tree harvest due to the maintenance of harvest residues on the soil [38].

The available P stock was larger in the ReM + F treatment when compared with ReR + F treatment. This result could be caused by two main factors: (i) less ReM + F nutrient outputs by harvest and (ii) larger quantities of organic matter in ReM + F which reduces the fixation of P on the soil colloids, thus improving its availability. No differences were observed between treatments after 100 cm in depth. In a long-term productivity experiment conducted in North Wales, UK, by Walmsley et al. [9], the alterations in soil characteristics and their effects were tested using different harvest treatments (whole tree harvest and bole-only harvest) on 23-year-old second rotation stands of Sitka spruce (*Picea sitchensis*). The authors found that the whole-tree harvest treatment is responsible for the depletion of three to four times greater quantities of N, P, and K than the conventional bole-only harvest in the first rotation.

ACR + F and ReR + F treatments contained the largest stocks of S in the 40–100 cm layer in the soil, being statistically different when compared to others. This can be attributed to the lower concentration of soil organic C (SOC) of the previously mentioned treatments. Low organic matter (OM) in highly weathered soils results in a higher quantity of positive charges, especially in the B horizon [39], with consequent retention of SO_4^{2-} in the soil.

Repeated removal of forest harvest residues is likely to reduce oxidizable C, available N, and total P stocks in the soil under *Eucalyptus* plantations. The use of harvest residues as soil coverage can protect the soil against erosion, improve or maintain the SOC, and improve both the quantity and availability of nutrients stored in the soil [5,11,15].

4.2. Soil Contribution to Nutrients Absorbed by Trees

The N and P losses as a result of the two harvest rotations (2004 and 2012) exceeded the inputs from fertilization and atmospheric deposition, resulting in a net loss of N and P from the site. In this period, the net N balance for ReM + F, ACR + F, ReR + F, and ACR − N treatments was −255, −516, −787, and −770 kg·ha^{-1}, respectively. With regards to P, there was also a negative net balance of −24, −48, −70, and −128 kg·ha^{-1} in the ReM + F, ACR + F, ReR + F, and ACR − P treatments, respectively. The net balance was positive only for S, with approximately +120 kg·ha^{-1} for all treatments with the exception of the ACR − N treatment, which contained a net balance of −128 kg·S·ha^{-1} (Table 4). N inputs through biological fixation are unlikely due to the absence of N-fixing weeds. In addition, one should not expect inputs of N from rock weathering as the soil was highly weathered (Ferralsol). The difference between nutrient losses due to harvesting and nutrients supplied via fertilization and atmospheric deposition must be made up for with nutrients absorbed from soil pools.

Table 4. Nutrient balance in two *Eucalyptus grandis* rotations under different timber harvest intensities and fertilizer applications.

Treatment	N (kg·ha^{-1})	P (kg·ha^{-1})	S (kg·ha^{-1})
Net Nutrient Balance [1]			
ReM + F	−255	−24	148
ACR + F	−516	−48	135
ReR + F	−787	−70	111
ACR − N	−770	−48	−128
ACR − P	−490	−128	141
Difference in Soil Nutrients Stocks from 2004 to 2012 (0–20 cm)			
ReM + F	−164	0.8	−1.1
ACR + F	−189	−0.4	−1.3
ReR + F	−192	−0.5	−2.5
ACR − N	−219	−0.9	−1.4
ACR − P	−189	−2.0	4.0

[1] Differences between the inputs by fertilizer application and atmospheric deposition with the harvest outputs (Table 3 and Figure 1).

From 2004 to 2012, soil N stocks declined in the 0–20 cm layer for all treatments (Figure 1). The ReR + F treatment was the most affected among the treatments due to the elevated nutrient outputs by harvest. For the ACR − N treatment, this was due to the small quantity of N applied via fertilization. In the ReM + F treatment, the 0–20 cm layer was responsible for providing 65% of the difference in the balance of soil N. The 0–20 cm layer was responsible for providing 35% of the difference in net soil N balance for the ACR treatments and 24% for the ReR + F treatment (Figure 1 and Table 4). The maintenance of harvest residues on the soil increases soil C and N pools—especially in labile fractions—thereby increasing the capacity of upper layers to provide N to plants [14–16,30].

The remaining difference possibly came from deeper layers of the soil profile and factors such as biological fixation by symbiotic associations which were not examined in this study [40]. Despite the large difference in the nutrient outputs between ReM + F and ReR + F treatments (more than 500 kg·ha^{-1} in the net balance), no differences in the N stocks between these treatments were found in the deeper layers of the soil (20–200 cm). More research is necessary in order to fully understand the additional sources of N in such highly weathered, tropical forests.

Little reduction was observed in the stocks of available P in the 0–20 layer (Figure 1). The ACR − P treatment presented the highest reduction (2 kg·ha^{-1}), while the ReM + F treatment presented a slight increase. This happened even with the negative balance of the nutrient for all treatments evaluated. No reduction in soil P below 40 cm depth was observed in 2016 (Figure 2). The small reduction in P content contradicts the highly negative net balance of this nutrient across treatments. P balance for all treatments was −60 kg·ha^{-1} on average. The ACR − P treatment had a deficit of −128 kg·ha^{-1} while the reduction was only 6 kg·ha^{-1} from 0–200 cm of soil depth. This discrepancy suggests that the Eucalyptus trees take up soil P hidden in fractions not extracted by the traditional resin method of analysis. Eucalyptus species can uptake organic and inorganic P fractions by phosphatase and exudation of low molecular mass acids [41]. More studies are necessary to clarify the complete origin of P absorbed by trees.

Even with a positive net balance of S stocks in the soil (Table 4), a small reduction was observed in the stocks of the nutrient when comparing the evaluations in 2004 and 2012. This is due to the elevated mobility of the element in the soil, especially in the upper layers [42], along with its consequent migration to and accumulation in the 40–100 cm layer (Figure 2) where it will adsorb to positive charges on soil surfaces exposed by losses of OM [39].

4.3. Management Considerations

Despite the small effect of whole-tree harvest on the soil nutrient stocks described by many authors [4,14,29–33], this and other studies emphasize the importance of the maintenance of forest residues on the soil, especially in tropical sites with low buffer capacity soils, high productivity, and short cycle plantations [14,16,29–31,33]. In these sites, removal of forest residues can result in loss of wood productivity of up to 40% due to the low nutrient pools remaining in the soil [16].

In the past several years, increasing demands for the use of forest residues for bioenergy purposes has grown in Brazil. Models for the utilization of these residues are based on research in temperate regions, where soils often have higher buffer capacities and higher organic C contents. This study shows that in tropical conditions the use of forest residues for bioenergy purposes should be carefully considered, taking into account the unique conditions of each site. On steep sites and/or those with low buffer capacity, all forest residues should be retained on the soil in order to avoid soil erosion [43] and depletion of soil nutrient pools [14,16,29,30,33,41]. In sites with favorable conditions for residue removal, preference should be given to the coarse residues due to their high caloric power and reduced nutrient concentration. With the removal of forest residues, the application of high rates of fertilizer is necessary in order to avoid productivity losses and to ensure the sustainability of the silviculture system.

The high negative net balance, especially for N and P, and the relative low reduction in the availability of these nutrients in the soil draws attention to two main points. First, more attention should be given to production sustainability, mainly in sites harvested in the whole-tree harvest system. Second, more studies should be implemented to better understand the contributions of organic and inorganic (mainly to P) fractions of low lability on the supply of nutrients to trees. Our results and others [29,41] suggest that Eucalyptus trees can access P fractions not identified by traditional soil analysis methods. Regarding N concentrations, even with high harvest outputs (more than 500 kg·ha^{-1} in each crop rotation) no response to N application by fertilization was found in terms of wood productivity in Brazilian conditions [28]. More research is essential to better understand the sources and the cycle of this nutrient within tropical soil-plant systems.

5. Conclusions

It was observed a reduction in organic C, total N, available P and S stocks in the whole tree harvest and the whole tree plus litter treatments, especially at 0–40cm. The balances of N and P in the two rotations of the experiment were negative in all treatments, with even lower values when whole tree and litter were removed or when no fertilizer was applied. The reduction in the total N and available P stocks in the soil explain around 50% of the negative balance observed for these nutrients; the source of the remaining 50% was not determined for this study. Our results indicate that N and P soil pools play an important role in forest nutrition; the reduction in these pools can possibly affect the sustainability of the production system in the long term. More research is necessary in order to better understand the balance of nutrients within the soil-plant system.

Acknowledgments: The authors would like to thank University of Washington/EUA, Experimental Station of Forest Sciences of Itatinga/Brazil, University of Sao Paulo/Brazil and FAPESP/Brazil (Processes: 2010/16623-9 and 2014/15876-1) for the financial support and for field collections.

Author Contributions: M.L.C.M. and J.H.T.R. wrote the manuscript as well as performed the soil sampling, and modeled and interpreted the results. A.R., J.C.A.J., A.V.F., J.N.J., and S.M.-C. helped on the interpretation of some of the results. R.R. and J.L.d.M.G. gave suggestions on the modeling and interpretation of the results. R.F.A. and M.d.C.P. performed the laboratory analysis.

References

1. Schlesinger, W.H.; Bernhardt, E.S. *Biogeochemistry: An Analysis of Global Change*, 3rd ed.; Elsevier: New York, NY, USA, 2013; pp. 1–664.
2. Jobbagy, E.; Jackson, R. The distribution of soil nutrients with depth: Global patterns and the imprint of plants. *Biogeochemistry* **2001**, *53*, 51–77. [CrossRef]
3. Vitousek, P.M.; Naylor, R.; Crews, T.; David, M.B.; Drinkwater, L.E.; Holland, E. Nutrient imbalances in agricultural development. *Science* **2009**, *324*, 1519–1520. [CrossRef] [PubMed]
4. Du Toit, B. Effects of site management on growth, biomass partitioning and light use efficiency in a young stand of *Eucalyptus grandis* in South Africa. *For. Ecol. Manag.* **2008**, *255*, 2324–2336. [CrossRef]
5. Gonçalves, J.L.N.; Stape, J.L.; Wichert, M.C.P.; Gava, J.L. *Conservação e Cultivo de Solos Para Plantações de Florestas (Soil Conservation and Cultivation for Forest Production)*; Instituto de Pesquisas e Estudos Florestais (IPEF): Piracicaba, (SP), Brazil, 2002; pp. 131–204.
6. Corbeels, M.; McMurtrie, R.E.; Pepper, D.A.; Mendham, D.S.; Grove, T.S.; O'Connell, A.M. Long-term changes in productivity of Eucalyptus plantations under different harvest residues and nitrogen and management practices: A modeling analysis. *For. Ecol. Manag.* **2005**, *217*, 1–18. [CrossRef]
7. Compton, J.E.; Cole, D.W. Phosphorus cycling and soil P fractions in Douglas-fir and red alder stands. *For. Ecol. Manag.* **1998**, *110*, 101–112. [CrossRef]
8. Hall, S.J.; Matson, P.A. Nutrient status of tropical rainforest influences soil N dynamics after N additions. *Ecol. Monogr.* **2003**, *73*, 107–129. [CrossRef]
9. Walmsley, J.D.; Jones, D.L.; Reynolds, B.; Price, M.H.; Healey, J.R. Whole tree harvesting can reduce second rotation forest productivity. *For. Ecol. Manag.* **2009**, *257*, 1104–1111. [CrossRef]
10. Jones, H.; Beets, P.; Kimberley, M.; Garrett, M. Harvest residue management and fertilization effect on soil carbon and nitrogen in a 15-year old *Pinus radiata* plantation forest. *For. Ecol. Manag.* **2011**, *262*, 339–347. [CrossRef]
11. Rocha, J.H.T.; Marques, E.R.G.; Goncalves, J.L.M.; Hubner, A.; Brandini, C.B.; Ferraz, A.V.; Moreira, R.M. Decomposition rates of forest residues and soil fertility after clear-cutting of *Eucalyptus grandis* stands in response to site management and fertilizer application. *Soil Use Manag.* **2016**. [CrossRef]
12. Xu, X.; Luo, Y.; Zhou, J. Carbon quality and the temperature sensitivity of soil organic carbon in a tallgrass prairie. *Soil Biol. Biochem.* **2012**, *50*, 142–148. [CrossRef]
13. Mendham, D.S.; White, D.A.; Battaglia, M.; McGrath, J.F.; Short, T.M.; Ogden, G.N.; Kinal, J. Soil water depletion and replenishment during first- and early second-rotation *Eucalyptus globulus* plantations with deep soil profiles. *Agric. For. Meteorol.* **2011**, *151*, 1568–1579. [CrossRef]

14. Nambiar, E.K.S.; Harwood, C.E. Productivity of acacia and eucalypt plantations in Southeast Asia 1. Biophysical determinants of production: Opportunities and challenges. *Intern. For. Rev.* **2014**, *16*, 225–248.
15. Achat, D.L.; Deleuze, C.; Landmann, G.; Pousse, N.; Ranger, J.; Augusto, L. Quantifying consequences of removing harvesting residues on forest soils and tree growth—A meta-analysis. *For. Ecol. Manag.* **2015**, *348*, 124–141. [CrossRef]
16. Rocha, J.H.T.; Goncalves, J.L.M.; Gava, J.L.; Godinho, T.O.; Melo, E.S.A.C.; Bazani, J.H.; Hubner, A.; Arthur Junior, J.C.; Wichert, M.P. Forest residue maintenance increased the wood productivity of a *Eucalyptus* plantation over two short rotations. *For. Ecol. Manag.* **2016**, *379*, 1–10. [CrossRef]
17. Peng, Y.; Thomas, S.C.; Tian, D. Forest management and soil respiration: Implications for carbon sequestration. *Environ. Rev.* **2008**, *16*, 93–111. [CrossRef]
18. Holub, S.M.; Terry, T.A.; Harrington, C.A.; Harrison, R.B.; Meade, R. Tree growth ten years after residual biomass removal, soil compaction, tillage, and competing vegetation control in a highly productive Douglas-fir plantation. *For. Ecol. Manag.* **2013**, *305*, 60–66. [CrossRef]
19. Alvares, C.A.; Stape, J.L.; Sentelhas, P.C.; Goncalves, J.L.M.; Sparovek, G. Koppen's climate classification map for Brazil. *Meteorol. Z.* **2013**, *22*, 711–728. [CrossRef]
20. Van Raij, B.; Andrade, J.C.; Cantarella, H.; Quaggio, J.A. *Análise Química Para Avaliação da Fertilidade de Solos Tropicais (Chemical Analysis for the Evaluation of Fertility in Tropical Soils)*; Instituto Agronomico de Campinas (IAC): Campinas, (SP), Brazil, 2001; pp. 1–58.
21. Ribeiro, J.F.; Walter, B.M.T. Fitofisionomias do Bioma Cerrado. Phytophisiognomy of Cerrado biome. In *Cerrado: Ambiente e Flora (Cerrado: Environment and Flora)*; Sano, S.M., Almeida, S.P., Eds.; Embrapa: Brasilia, Brazil, 1998; pp. 1–556.
22. Nelson, D.W.; Sommers, L.E. Total carbon, organic carbon, and organic matter. In *Methods of Soil Analysis. Part 3. Chemical Methods*; SSSA Book Series No. 5; Sparks, D.L, Ed.; Soil Science Society of America/SSSA and American Society of Agronomy (ASA): Madison, WI, USA, 1996; pp. 961–1010.
23. Walkley, A.; Black, I.A. An examination of the Degtjareff method for determining soil organic matter, and a proposed modification of the chromic acid titration method. *Soil Sci.* **1934**, *37*, 29–38. [CrossRef]
24. Schumacher, F.X.; Hall, F.D.S. Logarithmic expression of timber-tree volume. *J. Agric. Res.* **1933**, *47*, 719–734.
25. Malavolta, E.; Vitti, G.C.; Oliveira, A.S. *Avaliação do Estado Nutricional das Plantas: Princípios e Aplicações (Evaluation of Nutritional Condition of Plants: Principles and Application)*; Associação Brasileira para Pesquisa da Potassa e do Fosfato: Piracicaba, Brazil, 1989; pp. 1–201.
26. Zar, J.H. *Biostatistical Analysis*, 5th ed.; Pearson Education Ltd.: DeKalb, IL, USA, 2010.
27. Hubner, A. Ciclagem de Nutrientes e Produtividade de Madeira em Povoamento de Eucalyptus Grandis Sob Diferentes Manejos de Resíduos Florestais e Fertilização Mineral (Nutrient Cycling and Growth in *Eucalyptus Grandis* Plantation under Different Forest Residues Management and Fertilization). Ph.D. Thesis, ESALQ-USP, Piracicaba, Brazil, 2015.
28. Pulito, A.P.; Goncalves, J.L.M.; Smethurst, P.J.; Arthur Junior, J.C.; Alvares, C.A.; Rocha, J.H.T.; Hubner, A.; Moraes, L.F.; Miranda, A.C.; Kamogawa, M.Y.; et al. Available nitrogen and responses to nitrogen fertilizer in Brazilian Eucalypt plantations on soils of contrasting textures. *Forests* **2015**, *6*, 973–991. [CrossRef]
29. Laclau, J.-P.; Ranger, J.; Goncalves, J.L.M.; Maquere, V.; Krusche, A.V.; M'Bou, A.T.; Nouvellon, Y.; Saint-Andre, L.; Bouillet, J.-P.; Piccolo, M.D.C. Biogeochemical cycles of nutrients in tropical *Eucalyptus* plantations Main features shown by intensive monitoring in Congo and Brazil. *For. Ecol. Manag.* **2010**, *259*, 1771–1785. [CrossRef]
30. Mendham, D.S.; Ogden, G.N.; Short, T.; O'Connell, T.M.; Grove, T.S.; Rance, S.J. Repeated harvest residue removal reduces *E. globulus* productivity in the 3rd rotation in south-western Australia. *For. Ecol. Manag.* **2014**, *329*, 279–286. [CrossRef]
31. Mendham, D.S.; O'Connell, A.M.; Grove, T.S.; Rance, S.J. Residue management effects on soil carbon and nutrient contents and growth of second rotation eucalypts. *For. Ecol. Manag.* **2003**, *181*, 357–372. [CrossRef]
32. Huang, Z.; He, Z.; Wan, X.; Hu, Z.; Fan, S.; Yang, Y. Harvest residue management effects on tree growth and ecosystem carbon in a Chinese fir plantation in subtropical China. *Plant Soil* **2013**, *364*, 303–314. [CrossRef]
33. Kumaraswamy, S.; Mendham, D.S.; Grove, T.S.; O'Connell, A.M.; Sankaran, K.V.; Rance, S.J. Harvest residue effects on soil organic matter, nutrients and microbial biomass in eucalypt plantations in Kerala, India. *For. Ecol. Manag.* **2014**, *328*, 140–149. [CrossRef]

34. Moreno-Fernandez, D.; Dias-Pines, E.; Barbeito, I.; Sanchez-Gonzalez, M.; Montes, F.; Rubio, A.; Canellas, I. Temporal carbon dynamics over the rotation period of two alternative management systems in Mediterranean mountain Scots pine forests. *For. Ecol. Manag.* **2015**, *348*, 186–195. [CrossRef]
35. Rocha, J.H.T. Reflexos do Manejo de Resíduos Florestais na Produtividade, Nutrição e Fertilidade do Solo em Plantações de *Eucalyptus grandis* (Implications of Forest Harvest Residues Management on the Productivity, Nutrition and Soil Fertility in *Eucalyptus grandis* Plantation in Brazil). Master's Thesis, ESALQ-USP, Piracicaba, Brazil, 2014.
36. Vanguelova, E.; Pitman, R.; Luiro, J.; Helmisaari, H.-S. Long-term effects of whole tree harvesting on soil carbon and nutrient sustainability in the UK. *Biogeochemistry* **2010**, *101*, 43–59. [CrossRef]
37. Rocha, J.H.T.; Franci, A.F.; Gonçalves, J.L.M. Effect of forest residue management on fine roots density in an *Eucalyptus grandis* plantation. "Luiz de Queiroz" College of Agriculture, University of São Paulo: Piracicaba, Brazil, Unpublished data. 2016.
38. Smolander, A.; Kitunen, V.; Tamminen, P.; Kukkola, M. Removal of logging residue in Norway spruce thinning stands: Long-term changes in organic layer properties. *Soil Biol. Biochem.* **2010**, *42*, 1222–1228. [CrossRef]
39. Brady, N.C.; Weil, R.R. *The Nature and Properties of the Soil*, 14th ed.; Bookmam: DeKalb, IL, USA, 2013; pp. 1–716.
40. Silva, M.D.S.; Paula, T.D.; Moreira, B.C.; Carolino, M.; Cruz, C.; Bazzolli, D.M.S.; Silva, C.C.; Kasuya, M.C.M. Nitrogen-Fixing Bacteria in *Eucalyptus globulus* Plantations. *PLoS ONE* **2014**, *9*, e111313. [CrossRef] [PubMed]
41. Costa, M.G.; Gama-Rodrigues, A.C.; Gonçalves, J.L.M.; Gama-Rodrigues, E.F.; Sales, M.V.S.; Aleixo, S. Labile and non-labile fractions of phosphorus and its transformations in soil under *Eucalyptus* plantations, Brazil. *Forests* **2016**, *7*, 1–15. [CrossRef]
42. Bissani, C.A.; Tedesco, M.J. O enxofre no solo (Sulphur in the soil). In *Reunião Brasileira de Fertilidade do Solo*, 17nd ed.; Borkert, C.M., Lantmann, A.F., Eds.; EMBRAPA/IAPAR/SBCS: Londrina, Brazil, 1988; pp. 11–29.
43. Candido, B.M.; Silva, M.L.N.; Curi, N.; Batista, P.V.G. Water erosion post-planting in eucalyptus forests in the Parana river basin, in the eastern Mato Grosso do Sul. *Revista Brasileira De Ciencia Do Solo* **2014**, *38*, 1565–1575.

5

Mixed Broadleaved Tree Species Increases Soil Phosphorus Availability but Decreases the Coniferous Tree Nutrient Concentration in Subtropical China

Wen-Sheng Bu [1,2], Han-Jiao Gu [1,2], Can-can Zhang [1,2], Yang Zhang [1], Anand Narain Singh [3], Xiang-Min Fang [1,2], Jing Fan [1,2], Hui-Min Wang [4] and Fu-Sheng Chen [1,2,*,†]

1 Jiangxi Provincial Key Laboratory of Silviculture, College of Forestry, Jiangxi Agricultural University, Nanchang 330045, China; bws2007@163.com (W.-S.B.); guhanjiao@163.com (H.-J.G.); brightzcc@163.com (C.-c.Z.) zhangyang0558@163.com (Y.Z.); xmfang2013@126.com (X.-M.F.); fxh2022@163.com (J.F.)

2 Jiulianshan National Observation and Research Station of Chinese Forest Ecosystem, 2011 Collaborative Innovation Center of Jiangxi Typical Trees Cultivation and Utilization, Jiangxi Agricultural University, Nanchang 330045, China

3 Department of Botany, Panjab University, Chandigarh 160014, India; dranand1212@gmail.com

4 Qianyanzhou Ecological Station, Key Laboratory of Ecosystem Network Observation and Modeling, Institute of Geographic Sciences and Natural Resources Research, Chinese Academy of Sciences, Beijing 100101, China; wanghm@igsnrr.ac.cn

* Correspondence: chenfush@yahoo.com

† Current address: No. 1101 Zhiminda Road, Economic & Technological Development Zone, Nanchang 330045, China.

Abstract: Phosphorus (P) is a key limiting nutrient in subtropical forests and mixed forests with broadleaved species have been expected to stimulate P cycling, compared to pure conifer plantations. However, the mixture effect of Chinese fir (*Cunninghamia lanceolata* (Lamb.) Hook.) and broadleaved species on rhizosphere soil and coniferous tree P dynamics is unclear. In our study, eight plots of a single species of a Chinese fir plantation (pure plantation, PP) and eight mixed plantations (mixed plantation, MP) with broadleaved tree species (*Michelia macclurei* Dandy in Hunan Province or *Schima superba* Gardn. et Champ. in Fujian Province) were selected in subtropical China. Six P fractions in the rhizosphere and bulk soils were analyzed by a modified Hedley P fractionation method. Phosphorus fractions and nitrogen (N) concentrations in different root orders, different age fresh needles and twigs, and needle and twig litter of Chinese fir were measured. Our results showed that available P, slowly released P, occluded P, and the total extractable P in rhizosphere soil were significantly higher in MP than PP ($p < 0.05$). In contrast, P and N concentrations in the transportive roots and two-year old needles were generally higher in PP than MP. Meanwhile, the slowly released P, occluded P, total extractable P, and residual P in rhizosphere soil were negatively correlated with P concentrations in young (absorptive and transportive roots, one- and two-year old needles) but not old tissues (storative roots, three-year old needles and litters). In conclusion, mixture may increase soil P availability through the rhizosphere effect, but can decrease P and N concentration of Chinese fir tissues by competition between Chinese fir and broadleaved species. Clearly, the mixture effect may differ in soil and plant nutrients, and this issue needs be taken into consideration when converting a pure conifer plantation into a mixed-species forest.

Keywords: *Cunninghamia lanceolata*; mixture effect; nutrient cycling; rhizosphere effect; species competition

---

1. Introduction

Forest plantations in China have an area of 69 million hectares, occupy 31.8% of the total area of Chinese forests, and play important social, economic, and environmental roles [1]. However, most of these plantations were planted as monoculture conifers, such as Chinese fir (*Cunninghamia lanceolate* (Lamb.) Hook.) and *Pinus massoniana* Lamb., which might lead to nutrient imbalance, soil degradation, and the reduction of ecosystem stability [2]. In Southern China, Chinese fir has been widely planted and provides excellent commercial woods with easy processing because it is a native, fast-growing woody species with a 20–25 year rotation period and more than 1000 years of management practice history [3]. The removal of harvested wood and accumulated litter could alter the biogeochemistry in Chinese fir plantations, which may result in the decline of forest productivity due to multiple rotations [4]. Therefore, converting Chinese fir monoculture into mixed with broadleaved trees has become a common trend of forest management, since the mixed plantations seemed to have many advantages over pure stands, such as higher rates of litter decomposition and maintaining soil nutrient cycling [5], i.e., there were hardly any, or even negative effects of mixture on ecosystem functions, such as tree biomass accumulation and useful wood yield [6].

Phosphorus (P) is a key limiting nutrient in terrestrial ecosystems especially in the tropics and subtropics, where P is limited in the highly weathered soil and is found in pools with low amount available for plant [7]. Moreover, soil P utilization percentage is only 10%–25% in subtropical China [8] in comparison to P-rich soil from other tropics of the world. Shenoy et al. found that suboptimal levels of P could decrease crop growth and resulted in 5%–15% yield losses [9]. Many Chinese fir plantations have been established in these P-deficient sites, therefore, the availability of soil P is one of the most important limiting factors causing a decline in forest productivity because Chinese fir has a high use efficiency of which P is required for its optimal growth [8]. As mentioned above, the mixed-species forests have many advantages over pure plantations, and can improve root distribution patterns, plant residue compositions, and soil microbial diversity [10]. Theoretically, mixed forest plantations with Chinese fir and broadleaved trees would stimulate P cycling by mobilizing soil fixed P and increasing plant uptake, compared to pure stands of Chinese fir. However, the effect of mixed species on soil–tree P cycling in the Chinese fir forest ecosystem is little studied [8] and future P scenario cannot be effectively forecasted with the increasing conversion from pure plantations to mixed forests.

The nutrient availability is usually a major constraint to tree growth in most terrestrial ecosystems, while trees take up most of the mineral nutrients through the rhizosphere. The rhizosphere is a vital region of ecosystem biogeochemistry and governs the nutrient availability through plant roots to stimulate soil microbial activity [11]. Nutrients in the rhizosphere can rapidly transform from unavailable to available forms through the interaction of root exudates and microorganisms [12]. Plants generally secrete 10%–30% root exudates into rhizosphere soil through the root system. In order to increase the carbon availability in rhizosphere, microorganisms utilized those root exudates as food and energy [13]. Therefore, the microorganism population density might be much larger in the rhizosphere than in bulk soil. Plants absorb most mineral nutrients with effective nutrient transformation and mineralization dominated by rich microorganisms in rhizosphere soil [11]. Unfortunately, the difference of nutrient transformation and availability between the rhizosphere and bulk soil is less understood especially in forest ecosystems [14]. The modified P functional fractionation method in acidic soils provides an effective tool to explore the P supply process in the rhizosphere soil of subtropical forests [15].

On the other hand, the effects of species interactions on ecological processes such as nutrient cycling and tree growth in mixed stands have raised significant interest in recent years [2,16]. However, little is known about the influence of mixture on tree internal nutrient cycling [17], for example, the nutrient concentrations in various components of fresh tissues found in trees and their litters [18]. This result suggests that there is a change in plant physiology based on what is around the plant, or is it simply a function of competition? The understanding of internal nutrient cycling is of great importance because it not only depends on site nutrient availability but also reflects the competition among the different tree species in the mixed forest [19]. By now, mixed-species plantations containing a leguminous species have shown the potential to improve nutrient cycling [20]. However, there is a lack of evidence to validate the potential of mixed-species plantations without leguminous species, particularly in the tropics and subtopics [16]. There is no significant difference between mixtures and monocultures, when all of the empirical studies are combined [21]. The lack of a general trend, and the limited number of studies on the impact of mixed stands on tree nutrients, switches further studies. In addition, nutrient dynamics generally vary with the age of plant tissue. For example, more nutrients were resorbed to young leaves from old leaves when the acquisition of nutrients from the soil became more difficult [22]. Thus, the differentiation among the different ages of plant tissues might be helpful to explore nutrient internal cycles in response to the changing environment.

In the present study, eight plots of single-species Chinese fir plantation (pure plantation, PP) and eight plots of mixed plantation with a broadleaved tree species (mixed plantation, MP) with similar standing density in mature forest were selected to assess the effect of mixed species on soil and tree P cycling including belowground rhizosphere processes and nutrient concentrations of different functional root orders, and nutrient concentrations in aboveground components (fresh twigs, needles, and their litters) of different ages in Hunan and Fujian provinces of China. Both are the center distribution zones of Chinese fir production in China [23]. Our main hypotheses are: (1) the mixture with broadleaved trees improves soil P availability, and then increase tree nutrient levels; (2) the mixture effect (the difference between PP and MP) on rhizosphere soil P supply varies with P fraction, and depends on its rhizosphere effect (the increasing percentage in rhizosphere soil compared to bulk soils); and (3) the response of Chinese fir nutrients to the mixed with broadleaved tree varies with root orders, needle and twig ages due to combined effects of soil nutrient supplies, tree species competition, and plant physiological requirement. In addition, we explored the relationships among the nutrients of belowground component (root–soil), and aboveground nutrients (fresh twigs, needles, and their litters) in these ecosystems. Our results will help assess the potential effect of stand transformation from the monoculture plantation to mixed forest, and will provide theoretical support for forest management.

2. Materials and Methods

2.1. Study Sites

This study was conducted at two selected sites in subtropical China. One is located at Huitong Forest Ecological Station in Hunan Province (26°52′N, 109°42′E), and the other at Datian Experimental Forest Farm in Fujian Province (25°45′N, 117°33′E) (Figure S1). Both sites are approximately 1000 km away from each other but have a similar humid, mid-subtropical, monsoonal climate with the average annual temperatures of 15~18 °C and average annual precipitation of 1350~1450 mm. The zonal vegetation is evergreen broadleaved forest; but establishing forests by plantation is more pronounced, and this has occupied more than 90% due to logging. The altitudes of both sites are 500–600 m ASL, soil types belong to red soil, but are classified as Oxisol in Hunan Province and Humic Planosol in Fujian Province, respectively, according to the FAO classification system. Chinese fir has become the most important afforestation tree due to its fast-growing, easy processing, and wide utilization in both provinces [24].

2.2. Experimental Design and Investigation

At each selected site from both provinces, we randomly established total eight plots (circles with 200-m^2 size); At each site of both provinces, there are four plots in pure Chinese fir plantations and four plots in mixed plantations with a broadleaved tree species (*Michelia macclurei* in Hunan Province or *Schima superba* in Fujian Province), respectively, which are located on the mid-slope with the slopes ranging from 25° to 30°. Both *M. macclurei* and *S. superba* are native evergreen tree species, and widely afforested as pure or mixed forest due to rapid growth characteristic. In Hunan Province, both stand types were established in 1983 and common management practices were used in the early stages of the two stand types, including weeding and thinning. In Fujian Province, both stands were established in 1991 and similarly managed, such as weed-controlling during the first three years and thinning during the first 10 years (for more details see [23]). All sixteen selected plots were independently distributed in different hills, and generally far away, with more than 500 m between two plots of the same stand type.

In May 2013, the diameter at breast-height (DBH) and species of trees with a DBH > 2 cm were measured and recorded in all plots of both sites. The stand basal area was obtained at plot level. Meanwhile, general stand characteristics, including stem density, shrub density, and herb cover were investigated (See Table 1).

Table 1. Stand characteristics (means and standard errors, $n = 4$) of pure Chinese fir plantations (PP) and its mixed plantations with a broadleaved tree species (MP) at two sites located in Hunan and Fujian provinces of China.

Parameters	Hunan		Fujian	
	PP	MP	PP	MP
Basal area (m^2 ha^{-1})				
Chinese fir	35.31 ± 2.20	23.47 ± 2.26	69.75 ± 4.16	42.43 ± 2.02
Broadleaved tree	—	11.04 ± 1.21	—	11.97 ± 1.49
Total	35.31 ± 2.20	34.51 ± 2.23	69.75 ± 4.16	54.40 ± 2.96
Stem density (trees ha^{-1})				
Chinese fir	1410 ± 102	620 ± 86	2510 ± 129	2010 ± 76
Broadleaved tree	—	730 ± 68	—	710 ± 29
Total	1410 ± 102	1350 ± 47	2510 ± 129	2720 ± 85
Shrub density (stems ha^{-1})	8160 ± 1063	6320 ± 1038	14000 ± 2449	8500 ± 1000
Herb cover (%)	27.4 ± 12.4	14.1 ± 4.1	8.6 ± 4.0	2.6 ± 2.6
Understory plant biomass (g m^{-2})	154 ± 79	145 ± 42	214 ± 76	157 ± 63

Note: The broadleaved tree species are *Michelia macclurei* and *Schima superba* in Hunan and Fujian provinces, respectively.

2.3. Sampling

Within each circular plot in both PP and MP, three individuals of Chinese fir tree were chosen as reference plants and the rhizosphere and bulk soils, twig and needle litters under these tree canopies, roots of various orders, stems, fresh twigs, and needles varying with age in the trees were sampled. Rhizosphere soils were sampled using a hand shaking method and defined as the soils <4 mm away from the fine roots distributed 0–20 cm in the surface soil layer (>50% of the total Chinese fir fine roots [23]). Bulk soil was collected using soil cores (10 cm diameter) at 0–20 cm depth in the middle locations of forest gaps. Roots (living roots with <4 mm diameter) were extracted by shovel, the Chinese fir roots carefully collected by hand, and divided as three functional orders of absorptive roots (AR, the first three orders), transportive roots (TR, the 4th–5th orders), and storative roots (SR, >5th orders) in the laboratory [25]. Stems were sampled at breast height of the reference trees using an increment borer. Fresh needles and twigs from one first-order branch were collected from all three reference trees. Needles and twigs were divided into the first, second, and third orders of branching based on their ages (one-year old, two-year old, and three-year old needles or twigs) (see [18]). Meanwhile, twig and needle litters were directly cut from the trees using the combined method of people climbing and a tree pruner, since part of new branch litters (including the twig and needle litters) generally remain on the Chinese fir trunks. In our study, the concentrations of nutrients remaining in twig and needle litters were defined as nutrient resorption proficiencies in twig and needle organs, respectively [26]. All same samples were mixed as a sample within a plot. Soil and plant samples were air-dried at room temperature and dried at 60 °C for more than 72 h in an oven before nutrient measurement.

2.4. Chemical Analyses

Rhizosphere and bulk soils were cleared of roots and all organic debris, and ground to pass through a 0.25-mm sieve before analyses of P fractions, total P, organic carbon (C), and total N. In this study, total P concentrations of soil were determined by the molybdenum-antimony colorimetric method after samples through digestion with 1.84 M H_2SO_4 [27]. For the determination of extractable P, we employed improved Hedely P fractionation methods to quantify soil P functional fractions [28,29]. Air-dried soil samples were processed to follow the soil P fractionation sequential procedure [15]. The corresponding supernatants sequentially exacted with anion exchange resin (weak base), 0.5 M $NaHCO_3$, 0.1 M NaOH, 0.1 M NaOH with sonication, 1.0 M HCl were collected by centrifuge at 1.7×10^4 m·s^{-1} (3200 rpm) for five minutes, followed by filtering samples through a 0.45-μm micropore filter. Phosphorus concentration in each supernatant was determined by the phosphomolybdic acid blue color method. The extractable P, including Resin-P, $NaHCO_3$-P, NaOH-P, sonication-P, and HCl-P, are defined as available P, soluble P, slowly released P, occluded P, and weathered mineral P, respectively, based on their functions in soils [15]. The residual P is the difference between total P and extractable P.

Soil pH was determined using a pH Meter with a soil:water ratio of 1:2.5. Soil organic carbon (SOC) was determined by dichromate oxidation and titration with ferrous ammonium sulfate [29]. Total N was measured using the microkjeldahl method after digestion with 1.84 M H_2SO_4. Plant samples were washed with dematerialized water to remove dust, oven-dried, and ground and screened with a 0.25 mm sieve. Total N and total P concentrations in plant tissues were determined using the micro-Kjeldahl method and molybdenum-antimony colorimetric method, respectively [29].

2.5. Statistical Analyses

The data were tested for homogeneity of variances (Brown and Forsythe's variation of Levene's test) before statistical analysis. In order to assess the Chinese fir fine root function on soil P supply, the rhizosphere effect (%) of P availability was defined as the difference between rhizosphere and bulk soils. Multiple-way analysis of variance (ANOVA) was used to determine the effect of stand type, site location, soil sources (rhizosphere vs. bulk soil), or tissue ages, and the interactions between stand type and site location on nutrient variables. One-way ANOVA and least significant difference (LSD) was used to compare the differences of mean values among the three orders of roots, or three age classes of needles and twigs. Pearson's tests were used for comparing the significance of correlations among soil P fractions and tree nutrient concentrations. SPSS 16.0 software (SPSS, Inc., Chicago, IL, USA) [30] was used to perform all analyses. The standard 0.05 level was used throughout as a cutoff for statistical significance.

3. Results

3.1. Soil General Properties

Organic C and total N concentrations in rhizosphere and bulk soils were generally higher in MP than PP in both sites. However, the C/N ratio of rhizosphere soil in PP at the Hunan site was higher than that in MP, while it showed the opposite trend at the Fujian site. The C/N ratio in bulk soil was not significantly different between PP and MP in both sites. Organic C and total N were generally higher in rhizosphere than bulk soil under all the treatments. Meanwhile, organic C and C/N in rhizosphere and bulk soils were generally higher in the Fujian than in the Hunan sites, but the total N was not significantly different between both sites. In addition, there is not a significant difference in pH between rhizosphere and bulk soils and among the four treatments (Table 2).

3.2. Phosphorus Fractions in Rhizosphere and Bulk Soils

Average concentrations of soil available P, slowly released P, and occluded P (not soluble P), mineral P and residual P reflected significantly higher concentrations in MP than PP whether in rhizosphere or bulk soil (Figure 1 and Table S1). Average concentrations of soil soluble P, slowly released P, and residual P were higher in rhizosphere than bulk soil, and these three P fractions showed positive rhizosphere effects in all treatments. In contrast, the rhizosphere effects of available P, occluded P, and mineral P varied with forest type and site location (Figure 1 and Table S1). Soil P fractions including slowly released P, occluded P, mineral P, and residual P, but not available P, and soluble P was generally higher in the Fujian than the Hunan sites (Figure 1 and Table S1).

Table 2. General chemical properties of pure Chinese fir plantations (PP) and mixed plantations with a broadleaved tree species (MP) in Hunan and Fujian provinces of China. Values are means and standard errors of four replicates).

Parameter	Hunan		Fujian	
	PP	MP	PP	MP
pH				
Rhizosphere soil	4.36 ± 0.20 a	4.39 ± 0.09 a	4.35 ± 0.03 a	4.29 ± 0.03 a
Bulk soil	4.36 ± 0.21 a	4.32 ± 0.08 a	4.33 ± 0.01 a	4.20 ± 0.03 a
Organic carbon (g kg^{-1})				
Rhizosphere soil	16.0 ± 0.1 c	18.1 ± 0.8b c	21.7 ± 1.1 b	34.4 ± 2.7 a
Bulk soil	11.3 ± 2.0 d	15.6 ± 1.6 c	18.9 ± 2.0 b	23.3 ± 1.1 a
Total nitrogen (g kg^{-1})				
Rhizosphere soil	0.6 ± 0.0 b	1.0 ± 0.0 a	0.8 ± 0.1 ab	1.0 ± 0.1 a
Bulk soil	0.5 ± 0.00 b	0.8 ± 0.1 a	0.6 ± 0.1b	0.7 ± 0.1 ab
C/N				
Rhizosphere soil	27.6 ± 1.0 c	18.0 ± 0.9 d	30.7 ± 5.3 b	36.0 ± 1.2 a
Bulk soil	19.1 ± 1.5 b	20.9 ± 2.6 b	33.7 ± 2.0 a	35.3 ± 4.4 a

Note: Different letters indicate significant differences among four treatments at probability level of $p < 0.05$.

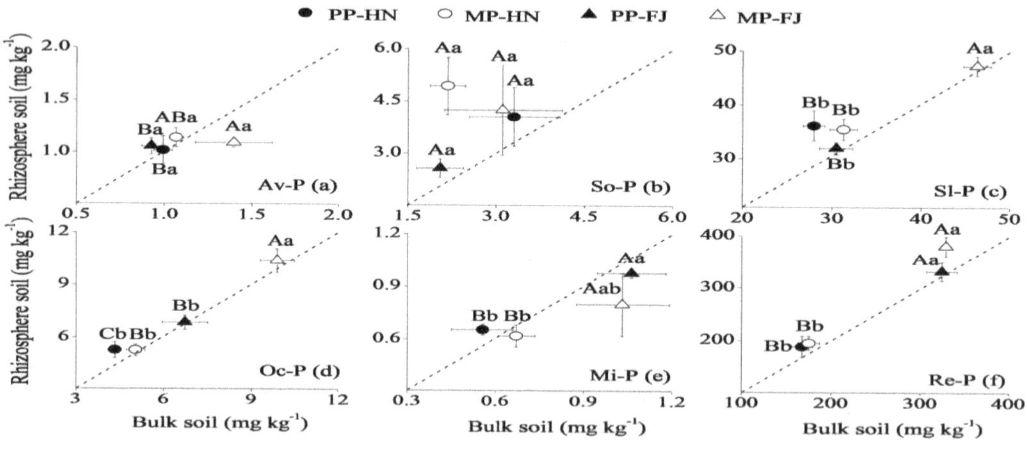

Figure 1. Variations in rhizosphere and bulk soil phosphorus fractions in pure Chinese fir plantations (PP) and mixed plantations with broadleaved tree species (MP) in Hunan and Fujian provinces of China. Note: The bar represents standard error and meaning of abbreviated words are as follow: PP-HN indicates the pure plantation in Hunan province, MP-HN indicates the mixed plantation in Hunan province, PP-FJ indicates the pure plantation in Fujian province, MP-FJ indicates the mixed plantation in Fujian province. Av-P indicates available P, So-P indicates soluble P, Sl-P indicates slowly released P, Oc-P indicates occluded P, Mi-P indicates mineral P, Re-P indicates residual P. The symbols located above and below the dotted line indicate the positive and negative rhizosphere effect, respectively. Among treatments, different small and capital letters indicate the significant differences in rhizosphere and bulk soils, respectively, probability level $p < 0.05$.

When the data from Hunan and Fujian were pooled, total P values were not significantly different between PP and MP either in rhizosphere or bulk soil, but the total extractable P was higher in MP than PP in both rhizosphere and bulk soils (Figure 2). The percentage of total extractable P to total P increased from 14.8% to 16.3% in bulk soil and from 13.8% to 16.8% in rhizosphere soil with the broadleaved tree species mixture. Among the five extractable fractions, slowly released P was the dominant form and its percentage to total extractable P ranged from 73.6% to 77.7%, followed by occluded P (11.2%~14.1%) in our study (Figure 2).

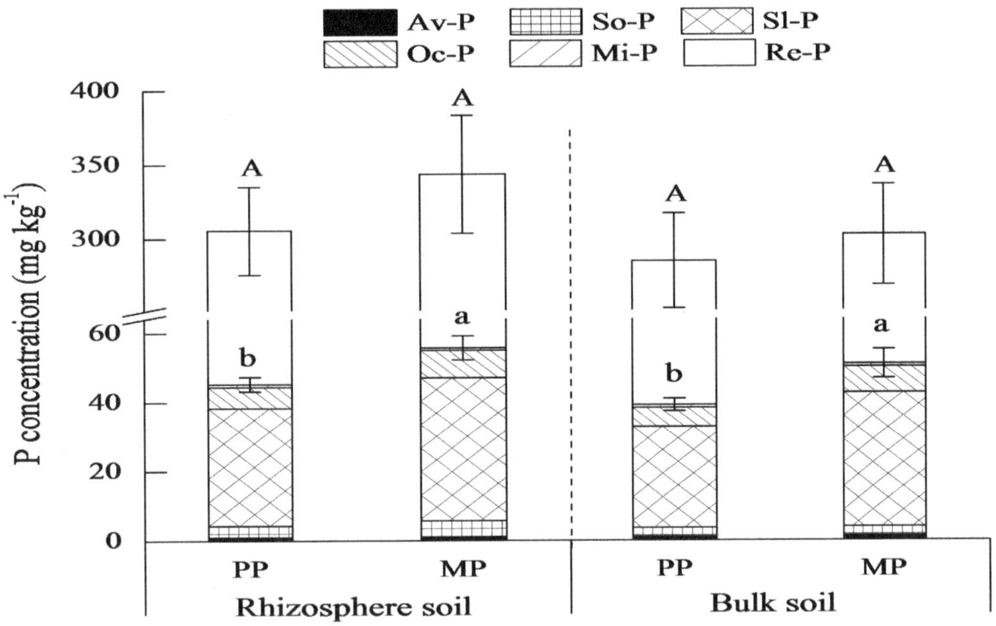

Figure 2. The components of soil phosphorus fractions in pure Chinese fir plantations (PP) and mixed plantations with broadleaved tree species (MP) in Hunan and Fujian provinces of China. Note: The bar represents standard error. Av-P indicates available P, So-P indicates soluble P, Sl-P indicates slowly released P, Oc-P indicates occluded P, Mi-P indicates mineral P, Re-P indicates residual P. Different small and capital letters indicate the significant differences of total extractable P (the sum of Av-P, So-P, Sl-P, Oc-P, and Mi-P) and total P, respectively, among the four soil sources (two forest types in rhizosphere or bulk soils, probability level $p < 0.05$.

3.3. Nutrient Levels in Different Root Orders

Total P concentration in absorptive and transportive roots is higher in PP than MP in the Fujian site, whereas it was not significantly different in roots between PP and MP in the Hunan site. Total N concentrations in absorptive and transportive roots were higher in PP than MP in each site of Hunan and Fujian provinces. Thus, the average N concentration in roots was also higher in PP than MP, although there were not statistical differences between both forest types in storative roots. The N/P ratio in all orders of roots seemed to be higher in PP than MP in the Hunan site (Figure 3 and Table S2). Meanwhile, total P and N concentrations were generally higher in absorptive than transportive and storative roots, and the N/P ratio was highest in storative roots, followed by transportive roots, and lowest in absorptive roots. In addition, the average P concentration was significantly higher in the Fujian than the Hunan site, while the average N concentration and N/P ratio showed the opposite trends (Figure 3).

3.4. Nutrient Distributions in Aboveground Tissues

Average P concentrations in the two-year old fresh needles were generally higher in PP than MP, although P concentrations in other aged needles were not significantly different between the two forest types (Figure 4a,b). In contrast, average N concentrations of all ages' needles and in each ages' needles were not significantly different between two forest types except higher N concentration in PP than MP in the two-year old fresh needles (Figure 4a,b and Table S3). Average P concentration in needles across all plots was highest in fresh needle of one-year old, followed by two- and three-year old needles and lowest in needle litter, while average N concentration was higher in fresh needles of one- and two-year old than three-year old needles and needle litter (Figure 4a,b). The ratio of N/P varied slightly between the two forest types and between the two site locations, but was higher in needle litter than fresh needles of various ages (Figure 4c).

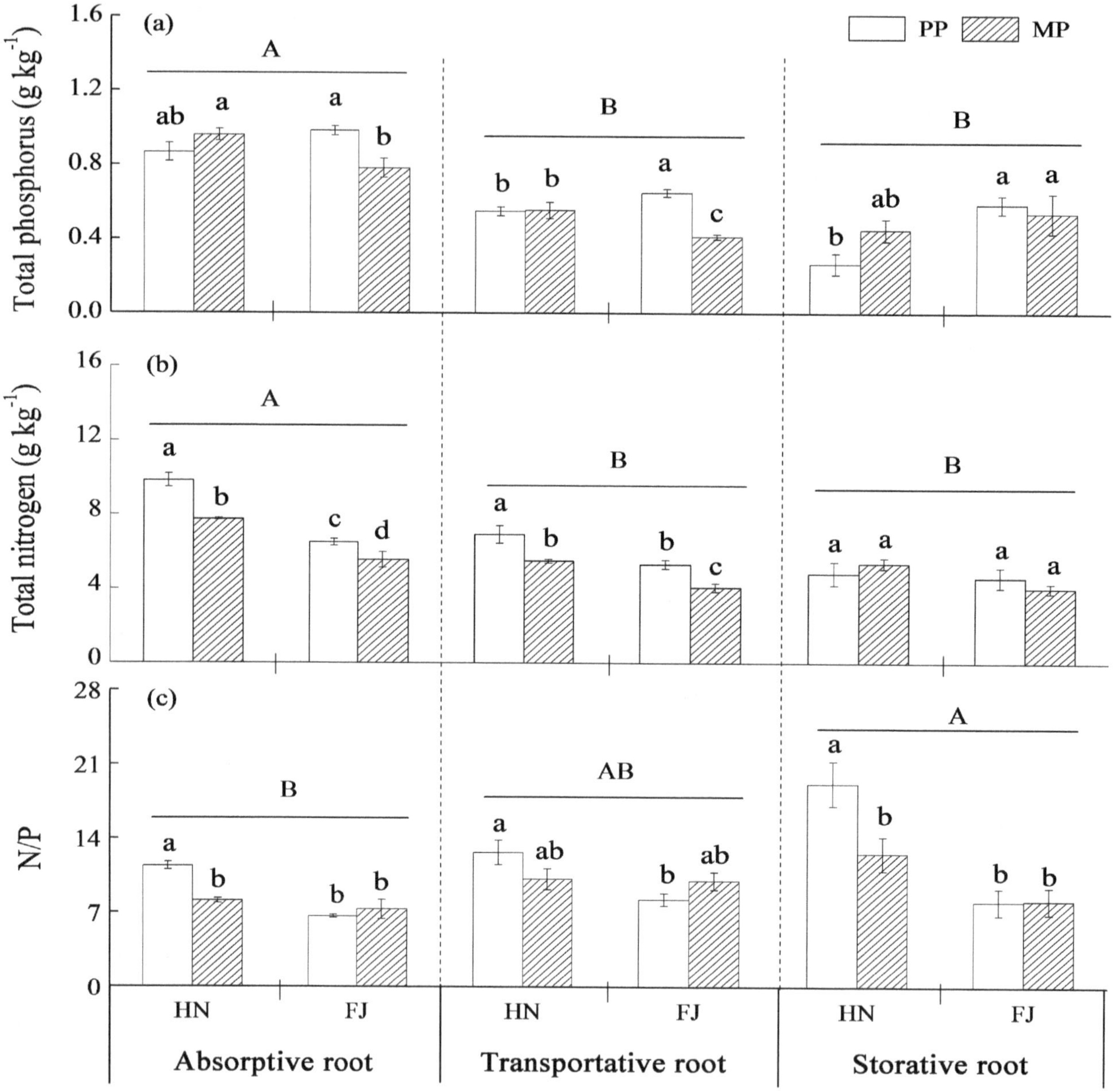

Figure 3. Phosphorus and nitrogen concentrations and N/P ratios of Chinese fir various functional roots in pure Chinese fir plantations (PP) and mixed plantations with broadleaved tree species (MP) in Hunan and Fujian provinces of China. (**a**) Total Phosphorus (g kg^{-1}), (**b**) Total nitrogen (g kg^{-1}), (**c**) N/P. Note: The bar represents standard error. Different small and capital letters indicate the significant differences among the four treatments within the root order, and among the three root orders, respectively, probability level $p < 0.05$.

Both P and N concentrations of twig were not significantly different between PP and MP, although N concentrations of the two- and three-year old twigs were significantly larger in PP than those in MP at the Hunan site. Additionally, those nutrient concentrations were highest in fresh twigs of one-year old, and lowest in twig litters, and significantly higher in Hunan than Fujian sites when just considered the effect of site location (Figure 5a,b and Table S4). The average N/P ratio was not significantly different between two forest types and between two site locations, but highest in twig litters and lowest in fresh twig of one-year old (Figure 5c and Table S4). In addition, the stem N and P concentrations and N/P ratio were not significantly different between both forest types except for a higher N/P ratio in PP than MP in the Fujian site (Figure 5).

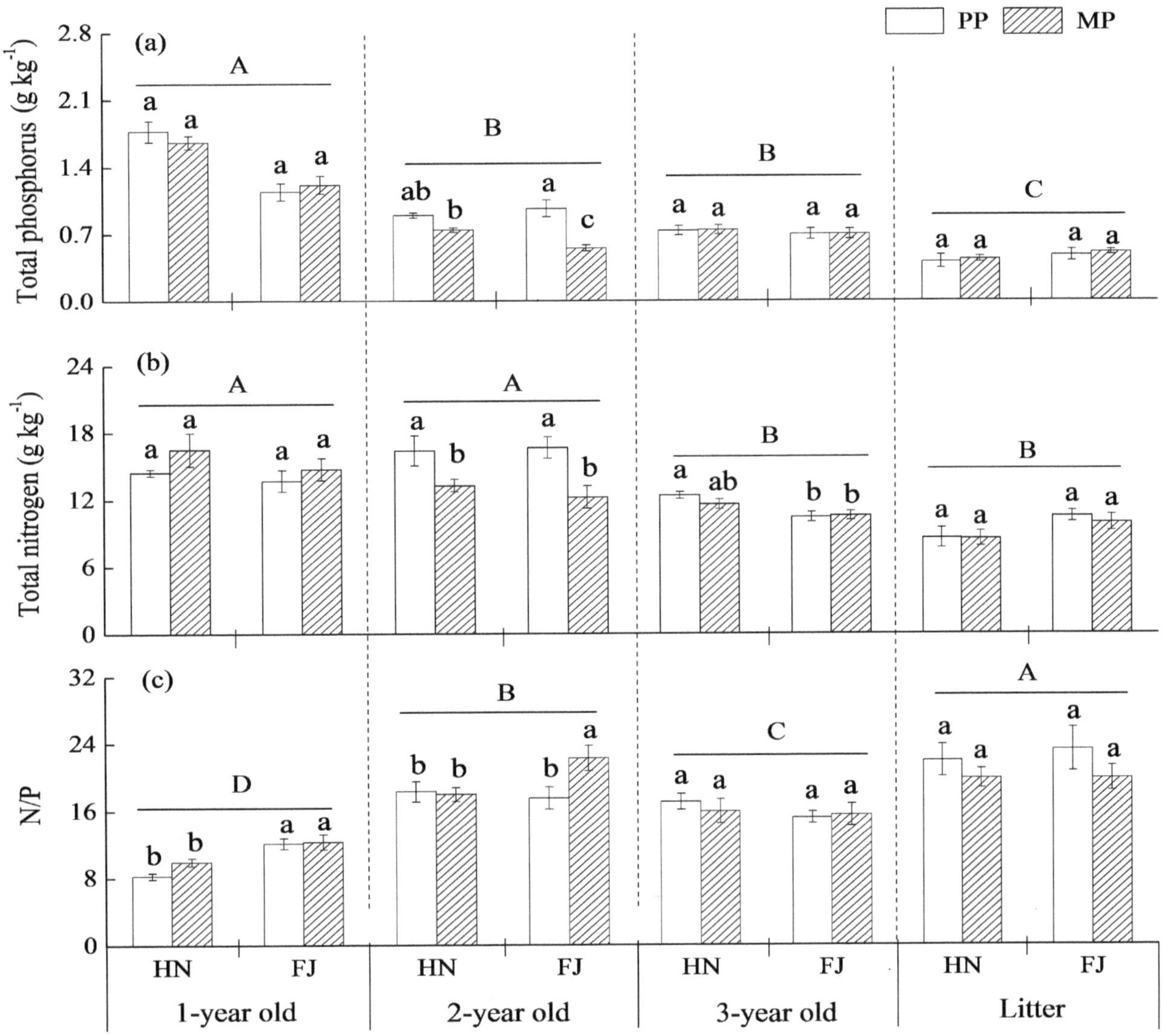

Figure 4. Phosphorus and nitrogen concentrations and N/P ratios of Chinese fir different-aged fresh needles and needle litters between pure Chinese fir plantations (PP) and mixed plantations with broadleaved tree species (MP) in Hunan and Fujian provinces of China. (**a**) Total Phosphorus (g kg^{-1}), (**b**) Total nitrogen (g kg^{-1}), (**c**) N/P. Note: The bar represents standard error. Different small and capital letters indicate the significant differences among the four treatments within a same age needle, and among these four age needles, respectively, probability level $p < 0.05$.

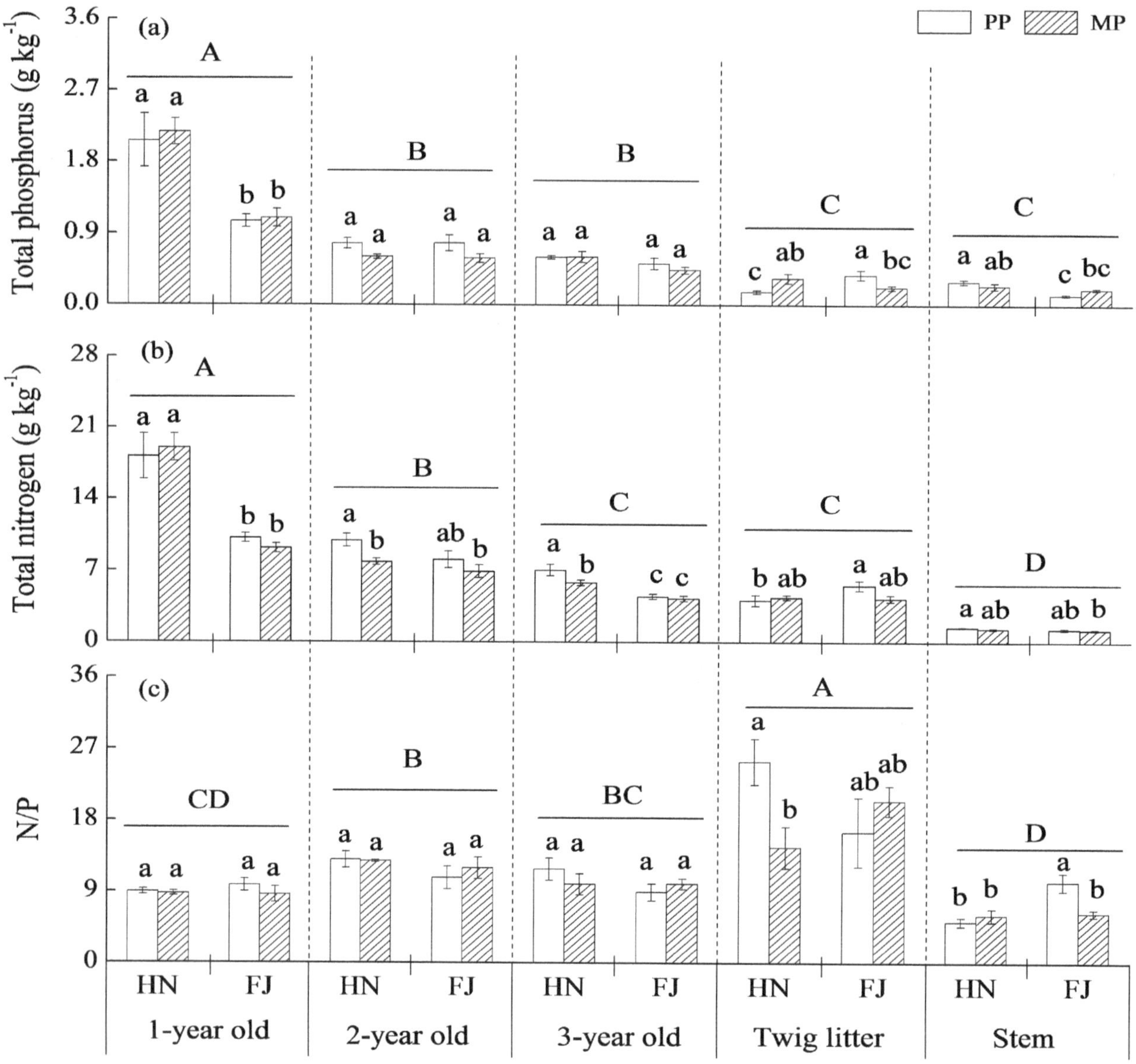

Figure 5. Phosphorus and nitrogen concentrations and N/P ratio of Chinese fir different-aged fresh twigs, twig litters, and stems between pure Chinese fir plantations (PP) and mixed plantations with broadleaved tree species (MP) in Hunan and Fujian provinces of China. (**a**) Total Phosphorus (g kg^{-1}), (**b**) Total nitrogen (g kg^{-1}), (**c**) N/P. Note: The bar represents standard error. Different small and capital letters indicate the significant differences among the four treatments within a same-aged twig, and among the four ages of twigs and stems, respectively, probability level $p < 0.05$.

3.5. Linkages Among Soil, Root and Needle Nutrients

Unlike our expectations, we did not find any positive relationships between all P fractions in rhizosphere soils and P concentrations in roots of different functions and needles of various ages. In contrast, slowly released P, occluded P, and residual P negatively correlated with the P concentrations in some functional roots and fresh needles of some ages, and the extractable P negatively correlated with P concentration in one-year old needles (Table 3). Meanwhile, the rhizosphere effect of some P fractions negatively correlated with P concentration in some roots, fresh needles, and needle litters except for a positive correlation between soluble P and P concentration in absorptive roots (Table 3).

Likewise, slowly released P, occluded P, mineral P, extractable P, and residual P in rhizosphere soils negatively correlated with N concentrations in transportive or storative roots, and in two- or three-year old needles. However, rhizosphere soil soluble P concentration was positively correlated with absorptive root N concentration, and both mineral P and extractable P were positively correlated with N concentration in one-year old needle (Table 4). Meanwhile, the rhizosphere effects of slowly released P and residual P were positively correlated with N concentrations in two- and three-year old needles, but negatively correlated with N concentration in one-year old needle. In addition, N concentration in needle litter was not significantly correlated with any P fractions and their rhizosphere effects (Table 4).

Table 3. Correlation efficiencies (n = 16) between rhizosphere soil phosphorus supplies (rhizosphere soil P fractions and their rhizosphere effects) and tree tissue phosphorus concentrations (total P of various Chinese fir tissues) under all studied stands in Hunan and Fujian provinces, China.

Variables	Total P of Roots			Total P of Fresh Needles			Total P of Needle Litter
	Absorp-tive	Transpor-tive	Storative	1-yr old	2-yr old	3-yr old	
			P fractions in rhizosphere soil				
Av-P	−0.04 ns	−0.16 ns	−0.11 ns	−0.34 ns	−0.12 ns	0.05 ns	−0.21 ns
So-P	0.08 ns	−0.26 ns	−0.50 ns	0.19 ns	−0.26 ns	−0.16 ns	−0.35 ns
Sl-P	**−0.58 ***	**−0.77 ****	−0.11 ns	−0.21 ns	**−0.78 ****	−0.02 ns	−0.04 ns
Oc-P	−0.39 ns	**−0.63 ****	0.25 ns	**−0.64 ****	**−0.62***	−0.10 ns	0.21 ns
Mi-P	0.34 ns	0.25 ns	0.25 ns	−0.38 ns	0.16 ns	0.17 ns	0.24 ns
Ex-P	−0.18 ns	−0.31 ns	0.47 ns	**−0.86 ****	−0.40 ns	−0.06 ns	0.39 ns
Re-P	−0.49 ns	**−0.76 ****	−0.11 ns	−0.29 ns	**−0.76 ****	−0.06 ns	−0.04 ns
			Rhizosphere effect of phosphorus fractions				
Av-P	0.43 ns	0.36 ns	−0.27 ns	−0.12 ns	0.43 ns	0.21 ns	−0.34 ns
So-P	**0.51 ***	0.00 ns	−0.17 ns	0.08 ns	−0.08 ns	0.10 ns	−0.27 ns
Sl-P	−0.02 ns	0.02 ns	**−0.51 ***	0.42 ns	0.21 ns	−0.03 ns	**−0.60 ***
Oc-P	0.07 ns	−0.12 ns	−0.39 ns	−0.04 ns	0.05 ns	−0.18 ns	**−0.54 ***
Mi-P	−0.02 ns	0.27 ns	0.01 ns	0.48 ns	0.35 ns	−0.08 ns	0.03 ns
Ex-P	0.15 ns	**−0.5 2 ***	−0.31 ns	0.06 ns	−0.39 ns	0.38 ns	−0.03 ns
Re-P	0.18 ns	0.03 ns	**−0.50 ***	0.34 ns	0.19 ns	−0.06 ns	**−0.69 ****

Note: Av-P indicates available P, So-P indicates soluble P, Sl-P indicates slowly released P, Oc-P indicates occluded P, Mi-P indicates mineral P, Re-P indicates residual P. ns indicates not significant, * $p < 0.05$, ** $p < 0.01$.

Table 4. Correlation efficiencies (n = 16) between rhizosphere soil phosphorus supplies (rhizosphere soil P fractions and their rhizosphere effects) and tree tissue nitrogen concentrations (total N of various Chinese fir tissues) under all studied stands in Hunan and Fujian provinces of China.

Variables	Total N of Roots			Total N of Fresh Needles			Total N of Needle Litter
	Absorp-tive	Transpor-tive	Storative	one-year old	two-year old	3-year old	
			Phosphorus fractions in rhizosphere soil				
Av−P	−0.18 ns	−0.31 ns	−0.14 ns	−0.24 ns	−0.09 ns	−0.12 ns	−0.14 ns
So−P	**0.51 ***	−0.33 ns	0.03 ns	−0.49 ns	0.11 ns	−0.10 ns	0.01 ns
Sl−P	−0.05 ns	**−0.55 ***	−0.13 ns	−0.02 ns	−0.38 ns	−0.46 ns	−0.32 ns
Oc−P	−0.11 ns	−0.46 ns	**−0.60 ***	0.34 ns	**−0.71 ****	**−0.73 ****	−0.39 ns
Mi−P	−0.07 ns	0.16 ns	−0.33 ns	**0.69 ****	**−0.51 ***	−0.28 ns	−0.33 ns
Ex−P	−0.31 ns	−0.36 ns	**−0.70 ****	**0.50 ***	**−0.76 ****	**−0.72 ****	−0.43 ns
Re−P	0.03 ns	**−0.57 ***	−0.25 ns	−0.01 ns	−0.44 ns	**−0.54 ***	−0.34 ns
			Rhizosphere effect of phosphorus fractions				
Av−P	−0.12 ns	0.27 ns	0.11 ns	−0.12 ns	0.12 ns	0.17 ns	−0.14 ns
So−P	0.46 ns	−0.14 ns	−0.03 ns	−0.35 ns	−0.02 ns	−0.03 ns	0.11 ns
Sl−P	−0.09 ns	0.27 ns	**0.55 ***	**−0.58***	**0.61 ***	**0.72 ****	0.24 ns
Oc−P	−0.06 ns	0.02 ns	−0.10 ns	−0.49 ns	0.31 ns	0.28 ns	0.01 ns
Mi−P	0.04 ns	0.47 ns	0.25 ns	0.07 ns	0.30 ns	**0.56 ***	0.26 ns
Ex−P	0.38 ns	−0.47 ns	0.14 ns	−0.21 ns	0.12 ns	−0.24 ns	−0.28 ns
Re−P	0.05 ns	0.20 ns	0.40 ns	**−0.68 ****	**0.54 ***	**0.62 ***	0.23 ns

Note: Av−P indicates available P, So−P indicates soluble P, Sl−P indicates slowly released P, Oc−P indicates occluded P, Mi−P indicates mineral P, Re−P indicates residual P. ns indicates not significant, * $p < 0.05$, ** $p < 0.01$.

4. Discussion

Overall, there is a trend of converting pure plantations into mixed plantations for increasing stability and sustainability due to, in theory, complementary resource use, environmental benefits, and soil improvement in mixed forests [5]. However, the practical effect does not always match with our expectation. Until now, there have been few reports on the negative effect of mixture with broadleaved trees and coniferous trees [31].

4.1. Mixture Effect on Nutrient Concentrations in Rhizosphere and Bulk Soils

Our results showed that organic C, total N, and total extractable P in both rhizosphere and bulk soils were higher in MP than PP. In other words, mixed Chinese fir plantations with broadleaved species increased nutrient concentrations in both rhizosphere and bulk soils. These results were in line with the hypothesis that the mixture with broadleaved trees improves soil P availability in Chinese fir plantations. Greater litter production or changes in the timing of litter inputs or differences of leaf nutrient concentration in mixed-species plantations could increase the soil nutrient supply, relative to monocultures, if decomposition rates were constant or faster [30]. Generally, the decomposition rate of litter and fine roots for coniferous species are slower than broadleaf species [32]. Additionally, some studies support that the decomposition rate of leaves and roost for *M. macclurei* and *S. superba* were faster than those of Chinese fir [33,34]. Moreover, compared with a pure Chinese fir plantation, the soil fertility and nutrient return were raised significantly in the mixed Chinese fir plantation with *M. macclurei*, due to substantial amount of litter and higher turnover rate of fine root [24]. Therefore, broadleaved trees with higher nutrient concentrations in the mixture could increase soil nutrient availability by altering the amount and quality of litter input as well as the amount and chemical composition of the root mass and exudates through fine root decomposition.

4.2. Mixture Effect Varies with Soil P Fraction

The average concentrations of soil available P, slowly released P, and occluded P, but not soluble P, mineral P, and residual P, were significantly higher in MP than PP and three of these showed positive rhizosphere effects. Since P is an element of depositional cycle, P absorption by plants depends on its concentration gradient and diffusivity in the soil near the roots. Both available P and soluble P are labile and are considered as the available fractions for plant growth [35]. Some studies have found that amending phosphates increases the immediate phosphorus availability and the rate of available P dissolution can be enhanced by the rhizosphere effect [36]. Moreover, the litter decomposition experiment of *S. superba* on a Chinese fir plantation shows that P concentration in the *S. superba* litter dropped rapidly, which could release 32% of the initial P into the soil during the first three months in Hunan Province [37].

Furthermore, the total extractable P was higher in MP than PP in both rhizosphere and bulk soils through the introduction of a broadleaf species in the mixed plantation. Recent studies indicated that the inactive P fractions could be converted into plant available forms with the help of the necessary manipulation of the rhizosphere environment [38]. The release of root exudates, such as organic ligands, is an activity of the root that can alter the concentration of P in the soil solution [12]. Some studies showed that cyclic dipeptides, which caused autoinhibition of Chinese fir, may be released into the soil through litter decomposition and root exudation. Moreover, root exudates provided more contributions to soil cyclic dipeptide levels than litter in Chinese fir plantations [39]. Thus, the introduction of broadleaf species to a pure Chinese fir plantation may relieve the autoinhibition of Chinese fir and alter the concentration of P in the soil solution. Furthermore, *S. superba* is a Mn-accumulating subtropical tree species and Mn hyperaccumulation is associated with Mn mobilization in the rhizosphere, most likely due to the release of protons. The carboxylates generated to produce the protons released into the rhizosphere are used internally in the plant to mobilize soil inorganic and organic P [40]. Therefore, the overall combination of changes in the P fractions in soil demonstrates that a mixture effect may induce an increase in soil P availability through the solubility of the unavailable form of soil P and the rhizosphere effect. After all, the mixture effect on P availability in rhizosphere soil varies with the P fraction, and may depend on the litter decomposition and root exudates of introduced broadleaf species.

4.3. The Response of Chinese Fir to Mixture Effect Varies with Root Orders, Needle, and Twig Ages

Our results showed that P and N concentrations in one-year old fresh needles and one-year old twigs were not significantly different between PP and MP in both study sites, although the mixture effect may induce an increase in soil P availability. Generally, evergreen trees tend to maintain a relatively favorable nutrient status in active young leaves for positive carbon (C) gain and high N use efficiency [41]. Trees can transport nutrients from old and senescing leaves to new organs to support new growth, which has been verified as a key mechanism of nutrient conservation and reuse in plants [42]. Thus, old leaves might be more sensitive to environmental variation, such as those mixed with a broadleaf species than young leaves. Generally, P and N concentrations in absorptive and transportive roots, two-year old fresh needles and two-year old twigs were significantly higher in PP than MP, although these concentrations in the other tissues were not different between both stand types. These results suggest that the introduction of a broadleaf species to pure Chinese fir plantation might decrease the nutrient concentration of Chinese fir. In contrast, most of studies suggested that mixed species stands have higher nutrient availability than monoculture, even in the absence of N-fixing species [43], though few studies have contrasted nutrient concentrations for species in mixtures [21]. Therefore, foliar nutrient concentrations did not show a general trend in the mixtures studied so far. Moreover, since the Chinese fir biomass in the mixed plantation was lower than that in the pure Chinese fir stand in our study, we deduced that lower P concentration in Chinese fir tissue may be attributing to the stronger competition with the broadleaved tree. Obviously, our result did not support the hypothesis that the mixture can increase tree tissue nutrient availability but, rather, supported the hypothesis that the response of Chinese fir nutrients to the mixed with broadleaved tree varies with root orders, and needle and twig ages. These results may be caused by the lower competition of Chinese fir than broadleaved species for nutrient uptake.

4.4. Linkages between Soil, Root, and Needle Nutrients

For the linkages among soil, root, and needle nutrients, we did not find any positive relationships between all P fractions in rhizosphere soil and P concentration in roots of different orders and in leaves of various ages. In contrast, slowly released P, occluded P, and residual P negatively correlated with the P concentrations in some root orders and fresh leaves of some ages, and the extractable P negatively correlated with P concentration in one-year old needle. Many studies suggest that any shift in plant species composition, which is able to alter soil nutrient stoichiometry, can influence rhizosphere microbial and soil enzyme activities, which can further induce plant community species shifts and alter ecosystem function [44]. Two adjacent plants can simultaneously participate in competition and facilitation processes, and the direction and intensity of plant interactions are determined by the sum of the co-occurring negative and positive effects of one to another. These negative linkages may cause by the negative plant–plant interactions between Chinese fir and the broadleaf species. In general, plants take up most mineral nutrients through the rhizosphere where microorganisms interact with root exudates. A study reported that the seedling survivorship of *S. superba* was significantly inhibited by *eucalyptus* robusta Smith litter addition alone, meanwhile the seedling height of *S. superba* and *M. macclurei* was significantly suppressed when eucalyptus roots were present [45]. This negative nutrient feedback (homeostasis) supports plant species coexistence with lower proportional changes in consumer stoichiometry compared with resource stoichiometry [46]. Evaluating homeostatic relationships can provide valuable insight into assessing plant competition or plant coexistence [47]. Therefore, negative relationships between P fractions in rhizosphere soil and tree tissue nutrients of Chinese fir may be caused by plant competition between Chinese fir and the broadleaved species.

5. Conclusions

We found that, compared with the Chinese fir plantation, the mixed Chinese fir plantation with a broadleaved species increased nutrient availability in both rhizosphere and bulk soils. Meanwhile, the introduction of broadleaf species to pure Chinese fir plantation generally decreased the tree nutrient availability of the transportive root and two-year old needles of Chinese fir, possibly due to the lower competition of Chinese fir than broadleaved species for nutrient uptake. Therefore, the effect of mixture may bring about a negative nutrient feedback between P availability in rhizosphere soil and P concentration in plant tissues attributed to root exudates and plant competition between Chinese fir and broadleaf species. In conclusion, replacing monoculture plantations of Chinese fir into mixed-species forest promotes soil nutrient availability, and it might be an option for multi-purpose forest management in China, from where the largest areas of Chinese fir plantations alter the functioning of the soil ecosystem. However, the mixed forest may affect P and N stocks of Chinese fir tissues. Thus, the mixture effect may differ in soil and plant nutrients, and when soil nutrient availability is expected to be improved, converting a pure conifer plantation into a mixed-species forest can be considered.

Supplementary Materials

Figure S1: The study sites in Huitong Forest Ecological Station of Hunan Province and Datian Experimental Forest Farm of Fujian Province, Table S1: Summary of ANOVA about effects of forest type, soil type, site location and their interaction on phosphorus fractions under pure Chinese fir plantation (PP) and its mixed plantation with broadleaved tree species (MP) at Hunan and Fujian provinces, China, Table S2: ANOVA of effect of forest type, root functional order, site location and their interaction on root nitrogen, phosphorus concentrations and N/P between pure Chinese fir plantation (PP) and mixed plantation with broadleaved tree species (MP) at Hunan and Fujian province, China.. Table S3: ANOVA of effect of forest type, leaf age, site location and their interaction on leaf nitrogen, phosphorus concentrations and N/P ratio between pure Chinese fir plantation (PP) and mixed plantation with broadleaved tree species (MP) at Hunan and Fujian province, China. Table S4: ANOVA of effect of forest type, twig age, site location and their interaction on twig nitrogen, phosphorus concentrations and N/P ratio between pure Chinese fir plantation (PP) and mixed plantation with broadleaved tree species (MP) at Hunan and Fujian province, China.

Author Contributions: F.-S.C. contributed to the study conception and design. Material preparation, data collection and analysis were performed by W.-S.B., H.-J.G., C.-c.Z., Y.Z., A.N.S., X.-M.F., J.F., H.-M.W. and F.-S.C. The first draft of the manuscript was written by W.-S.B. and all authors commented on previous versions of the manuscript. All authors have read and agreed to the published version of the manuscript

References

1. Xu, J. The 8th forest resources inventory results and analysis in China. *For. Econ.* **2004**, *24*, 6–8.
2. Wen, L.; Lei, P.; Xiang, W.; Yan, W.; Liu, S. Soil microbial biomass carbon and nitrogen in pure and mixed stands of Pinus massoniana and Cinnamomum camphora differing in stand age. *Ecol. Manag.* **2014**, *328*, 150–158. [CrossRef]
3. Chen, H. Phosphatase activity and P fractions in soils of an 18−year−old Chinese fir (*Cunninghamia lanceolata*) plantation. *For. Ecol. Manag.* **2003**, *178*, 301–310. [CrossRef]
4. Ma, X.; Heal, K.V.; Liu, A.; Jarvis, P.G. Nutrient cycling and distribution in different−aged plantations of Chinese fir in southern China. *For. Ecol. Manag.* **2007**, *243*, 61–74. [CrossRef]
5. Meng, J.; Lu, Y.; Zeng, J. Transformation of a degraded *Pinus massoniana* plantation into a mixed-species irregular forest: Impacts on stand structure and growth in Southern China. *Forests* **2014**, *5*, 3199–3221. [CrossRef]
6. Nunes, L.; Coutinho, J.; Nunes, L.F.; Rego, F.C.; Lopes, D. Growth, soil properties and foliage chemical analysis comparison between pure and mixed stands of *Castanea saliva* Mill.; *Pseudotsuga menziesii* (Mirb.) Franco, in Northern Portugal. *System* **2011**, *20*, 496–507.
7. Yang, X.; Thornton, P.E.; Ricciuto, D.M.; Post, W.M. The role of phosphorus dynamics in tropical forests−A modeling study using CLM−CNP. *Biogeosciences* **2014**, *11*, 1667–1681. [CrossRef]
8. Zou, X.; Wu, P.; Chen, N.; Wang, P.; Ma, X. Chinese fir root response to spatial and temporal heterogeneity of phosphorus availability in the soil. *Can. J. For. Res.* **2015**, *45*, 402–410. [CrossRef]

9. Shenoy, V.; Kalagudi, G. Enhancing plant phosphorus use efficiency for sustainable cropping. *Biotechnol. Adv.* **2005**, *23*, 501–513. [CrossRef]
10. Hasegawa, M.; Ota, A.T.; Kabeya, D.; Okamoto, T.; Saitoh, T.; Nishiyama, Y. The effects of mixed broad–leaved trees on the collembolan community in larch plantations of central Japan. *Appl. Soil. Ecol.* **2014**, *83*, 125–132. [CrossRef]
11. Pii, Y.; Mimmo, T.; Tomasi, N.; Terzano, R.; Cesco, S.; Crecchio, C. Microbial interactions in the rhizosphere: Beneficial influences of plant growth–promoting rhizobacteria on nutrient acquisition process. A review. *Biol. Fert. Soils* **2015**, *51*, 403–415. [CrossRef]
12. Hinsinger, P. Bioavailability of soil inorganic P in the rhizosphere as affected by root–induced chemical changes: A review. *Plant Soil* **2001**, *237*, 173–195. [CrossRef]
13. Brzostek, E.R.; Greco, A.; Drake, J.E.; Finzi, A.C. Root carbon inputs to the rhizosphere stimulate extracellular enzyme activity and increase nitrogen availability in temperate forest soils. *Biogeochemistry* **2013**, *115*, 65–76. [CrossRef]
14. Dotaniya, M.L.; Meena, V.D. Rhizosphere effect on nutrient availability in soil and Its uptake by plants: A Review. *Proc. Nat. Acad. Sci. India Sect. B Biol. Sci.* **2015**, *85*, 1–12. [CrossRef]
15. Hu, X.F.; Chen, F.S.; Nagle, G.; Fang, Y.T.; Yu, M.Q. Soil phosphorus fractions and tree phosphorus resorption in pine forests along an urban–to–rural gradient in Nanchang, China. *Plant Soil* **2011**, *346*, 97–106. [CrossRef]
16. Redondo-Brenes, A.; Montagnini, F. Growth, productivity, aboveground biomass, and carbon sequestration of pure and mixed native tree plantations in the Caribbean lowlands of Costa Rica. *For. Ecol. Manag.* **2006**, *232*, 168–178.
17. Lemma, B. Soil chemical properties and nutritional status of trees in pure and mixed-species stands in south Ethiopia. *J. Plant Nutr. Soil Sci.* **2012**, *175*, 769–774. [CrossRef]
18. Chen, F.S.; Niklas, K.J.; Liu, Y.; Fang, X.M.; Wan, S.Z.; Wang, H. Nitrogen and phosphorus additions alter nutrient dynamics but not resorption efficiencies of Chinese fir leaves and twigs differing in age. *Tree Physiol.* **2015**, *35*, 1106–1117. [CrossRef]
19. Pretzsch, H. Canopy space filling and tree crown morphology in mixed-species stands compared with monocultures. *For. Ecol. Manag.* **2014**, *327*, 251–264. [CrossRef]
20. Montagnini, F. Accumulation in above–ground biomass and soil storage of mineral nutrients in pure and mixed plantations in a humid tropical lowland. *For. Ecol. Manag.* **2000**, *134*, 257–270. [CrossRef]
21. Rothe, A.; Binkley, D. Nutritional interactions in mixed species forests: A synthesis. *Can. J. For. Res.* **2001**, *31*, 1855–1870. [CrossRef]
22. Estiarte, M.; Peñuelas, J. Alteration of the phenology of leaf senescence and fall in winter deciduous species by climate change: effects on nutrient proficiency. *Glob. Chang. Biol.* **2015**, *21*, 1005–1017. [CrossRef] [PubMed]
23. Fu, X.; Wang, J.; Di, Y.; Wang, H. Differences in Fine–Root Biomass of Trees and Understory Vegetation among Stand Types in Subtropical Forests. *PLoS ONE* **2015**, *10*, e0128894. [CrossRef] [PubMed]
24. Wang, Q.; Wang, S. Soil microbial properties and nutrients in pure and mixed Chinese fir plantations. *J. Res. (Harbin)* **2008**, *19*, 131–135. [CrossRef]
25. Guo, D.; Xia, M.; Wei, X.; Chang, W.; Liu, Y.; Wang, Z. Anatomical traits associated with absorption and mycorrhizal colonization are linked to root branch order in twenty–three Chinese temperate tree species. *New Phytol.* **2008**, *180*, 673–683. [CrossRef]
26. Killingbeck, K.T. Nutrients in senesced leaves: Keys to the search for potential resorption and resorption proficiency. *Ecology* **1996**, *77*, 1716–1727. [CrossRef]
27. Allen, S.E.; Grimshaw, H.M.; Parkinson, J.A.; Quarmby, C. *Chemical Analysis of Ecological Materials*; Blackwell Scientific Publications: Oxford, UK, 1974.
28. Chen, F.S.; Li, X.; Nagle, G.; Zhan, S.X. Topsoil phosphorus signature in five forest types along an urban–suburban–rural gradient in Nanchang, southern China. *J. Res.* **2010**, *21*, 39–44. [CrossRef]
29. Hedley, M.J.; Stewart, J.W.B.; Chauhan, B.S. Changes in inorganic and organic soil phosphorus fractions induced by cultivation practices and by laboratory incubations. *Soil Sci. Soc. Am. J.* **1982**, *46*, 970–976. [CrossRef]
30. Spss, I. *SPSS for Windows (16.0)*; SPSS Inc: Chicago, IL, USA, 2007.
31. Richards, A.E.; Forrester, D.I.; Bauhus, J.; Scherer-Lorenzen, M. The influence of mixed tree plantations on the nutrition of individual species: a review. *Tree Physiol.* **2010**, *30*, 1192–1208. [CrossRef]
32. Eshel, A.; Beeckman, T. *Plant Roots: The Hidden Half*; CRC Press: Bocaraton, FL, USA, 2013.

33. Wang, Q.; Wang, S.; Fan, B.; Yu, X. Litter production, leaf litter decomposition and nutrient return in *Cunninghamia lanceolata* plantations in south China: effect of planting conifers with broadleaved species. *Plant Soil* **2007**, *297*, 201–211. [CrossRef]
34. Wang, Q.; Wang, S.; Huang, Y. Leaf litter decomposition in the pure and mixed plantations of *Cunninghamia lanceolata* and *Michelia macclurei* in subtropical China. *Biol. Fert. Soils* **2009**, *45*, 371–377. [CrossRef]
35. Li, G.; Li, H.; Leffelaar, P.A.; Shen, J.; Zhang, F. Dynamics of phosphorus fractions in the rhizosphere of fababean (*Vicia faba* L.) and maize (*Zea mays* L.) grown in calcareous and acid soils. *Crop Pasture Sci.* **2017**, *66*, 1151. [CrossRef]
36. Phillips, R.P.; Fahey, T.J. Tree species and mycorrhizal associations influence the magnitude of rhizosphere effects. *Ecology* **2006**, *87*, 1302–1313. [CrossRef]
37. Zhong, M.; Wang, Q.; Gao, H.; Yu, X. Decomposition and nitrogen and phosphorus release of leaf litters from main tree species in a mid−subtropical forest. *Chin. J. Ecol.* **2013**, *32*, 1653–1659.
38. Gerke, J. The acquisition of phosphate by higher plants: Effect of carboxylate release by the roots. A critical review. *J. Plant Nutr. Soil Sci.* **2015**, *178*, 351–364. [CrossRef]
39. Chen, L.; Kong, C. Autoinhibition and soil allelochemical (cyclic dipeptide) levels in replanted Chinese fir (*Cunninghamia lanceolata*) plantations. *Plant Soil* **2014**, *374*, 793–801. [CrossRef]
40. Lambers, H.; Hayes, P.E.; Laliberte, E.; Oliveira, R.S.; Turner, B.L. Leaf manganese accumulation and phosphorus−acquisition efficiency. *Trends Plant Sci.* **2015**, *20*, 83–90. [CrossRef]
41. Di, Y.; Fu, X.; Wang, H.; Li, W.; Wang, S. N concentration of old leaves and twigs is more sensitive to stand density than that of young ones in Chinese fir plantations: A case study in subtropical China. *J. For. Res.* **2018**, *29*, 163–169. [CrossRef]
42. Wang, M.; Murphy, M.T.; Moore, T.R. Nutrient resorption of two evergreen shrubs in response to long−term fertilization in a bog. *Oecologia* **2014**, *174*, 365–377. [CrossRef]
43. Zeugin, F.; Potvin, C.; Jansa, J.; Scherer-Lorenzen, M. Is tree diversity an important driver for phosphorus and nitrogen acquisition of a young tropical plantation? *For. Ecol. Manag.* **2010**, *260*, 1424–1433. [CrossRef]
44. De Graaff, M.A.; Classen, A.T.; Castro, H.F.; Schadt, C.W. Labile soil carbon inputs mediate the soil microbial community composition and plant residue decomposition rates. *New Phytol.* **2010**, *188*, 1055–1064. [CrossRef] [PubMed]
45. Zhang, C.; Fu, S. Allelopathic effects of eucalyptus and the establishment of mixed stands of eucalyptus and native species. *For. Ecol. Manag.* **2009**, *258*, 1391–1396. [CrossRef]
46. Bell, C.; Carrillo, Y.; Boot, C.M.; Rocca, J.D.; Pendall, E.; Wallenstein, M.D. Rhizosphere stoichiometry: are C: N: P ratios of plants, soils, and enzymes conserved at the plant species-level? *New Phytol.* **2014**, *201*, 505–517. [CrossRef] [PubMed]
47. Elgersma, K.J.; Yu, S.; Vor, T.; Ehrenfeld, J.G. Microbial−mediated feedbacks of leaf litter on invasive plant growth and interspecific competition. *Plant Soil* **2012**, *35*, 341–355. [CrossRef]

6

Grand Fir Nutrient Management in the Inland Northwestern USA

Dennis R. Parent [1] and Mark D. Coleman [2,*]

1 DRP Forestry, Hayden, ID 83835, USA; drparent@roadrunner.com
2 Forest, Rangeland and Fire Sciences, University of Idaho; Moscow, ID 83844, USA
* Correspondence: mcoleman@uidaho.edu

Academic Editors: Scott X. Chang and Xiangyang Sun

Abstract: Grand fir (*Abies grandis* (Douglas ex D. Don) Lindley) is widely distributed in the moist forests of the Inland Northwest. It has high potential productivity, its growth being nearly equal to western white pine, the most productive species in the region. There are large standing volumes of grand fir in the region. Nutritionally, the species has higher foliage cation concentrations than associated conifers, especially potassium (K) and calcium (Ca). In contrast, it has lower nitrogen (N) foliage concentrations, which creates favorable nutrient balance on N-limited sites. Despite concentration differences, grand fir stores proportionally more nutrients per tree than associated species because of greater crown biomass. Although few fertilization trials have examined grand fir specifically, its response is inferred from its occurrence in many monitored mixed conifer stands. Fertilization trials including grand fir either as a major or minor component show that it has a strong diameter and height growth response ranging from 15% to 50% depending in part on site moisture availability and soil geology. Grand fir tends to have a longer response duration than other inland conifers. When executed concurrently with thinning, fertilization often increases the total response. Late rotation application of N provides solid investment returns in carefully selected stands. Although there are still challenges with the post-fertilization effects on tree mortality, grand fir will continue to be an important species with good economic values and beneficial responses to fertilization and nutrient management.

Keywords: forest nutrition; grand fir; fertilization; Inland Northwest

---

1. Introduction

Grand fir is widely distributed in the moist forests of the Pacific Northwest (Figure 1). Its range of 39–51 degrees latitude and 114–125 degrees longitude includes southern Canada in the Okanogan and Kootenay Lakes region south through northeastern Washington, northern Idaho, western Montana, and northeastern Oregon [1]. In the Inland Northwest, it occurs on moister sites from 450 to 1800 m elevation on a wide variety of soils derived from basalts, granites, gneiss, and sandstone as long as moisture is adequate [1]. Grand fir does best throughout the region where a deep layer of volcanic ash occurs [2,3]. There are similar occurrences of grand fir throughout western Montana habitat types and northeastern Washington and Oregon plant associations [4,5]. Grand fir is shade tolerant and often takes a subordinate position in the stand in mixture with other inland species.

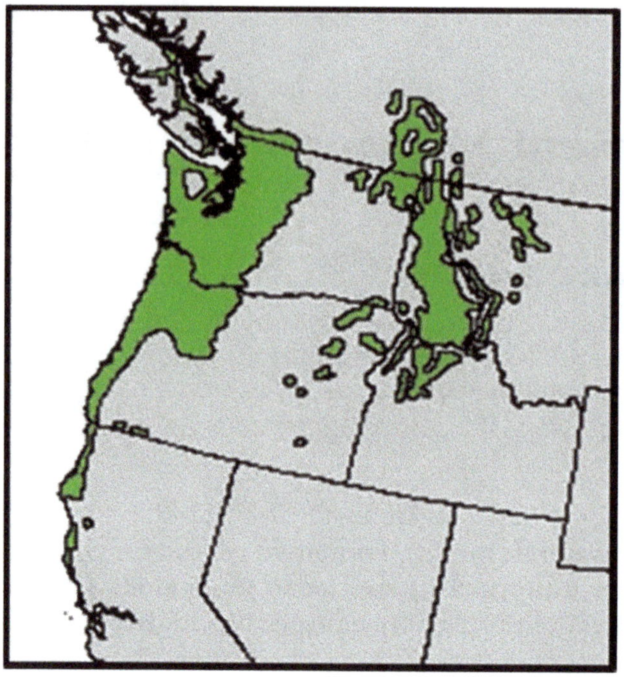

Figure 1. Range map of grand fir in the Pacific Northwest [6].

Grand fir has high potential productivity. Once established, it grows rapidly nearly equal to Douglas-fir (*Pseudotsuga mensiesii* (Mirb.) Franco) on the Pacific Coast and western white pine (*Pinus monticola* Dougl. ex D. Don) in Idaho [1]. As it approaches 50 years old, height is comparable to other associated species in the Inland Northwest including western larch (*Larix occidentalis* Nutt.). At 100-years-old, grand fir may approach or surpass other species in total height and volume yield. Table 1 is a general comparison of grand fir and its associates in the Inland Region using data from various historical authors [7–9]. The last two columns of Table 1 include data for coastal Douglas-fir for an approximate comparison with inland Douglas-fir [10].

Table 1. Grand fir height and total volume yield compared to associated species in the US Inland Northwest. All yields are from yield tables of fully stocked stands.

Stand Age	Grand Fir		Western White Pine		Western Larch		Inland Douglas-Fir		Coastal Douglas-Fir [1]	
	Height (m) [2]	Volume (m^3) [3]	Height (m) [2]	Volume (m^3) [4]	Height (m) [2]	Volume (m^3) [5]	Height (m) [2]	Volume (m^3) [6]	Height (m) [7]	Volume (m^3) [8]
50	20.6	74.1	21.5	140.3	22.8	37.9	20.0	73.4	23.7	110.0
100	36.3	367.1	42.4	393.3	37.8	141.2	35.4	206.6	33.8	260.1

[1] From McArdle [10], Site Class IV, 100-year site index = 110; [2] From Deitschman [7], Table 2, p. 14; [3] From Stage [9], Table 7, p. 25; [4] From Haig [8], Table 6, p. 20; [5] Stage, op. cit, Table 12, p. 41; [6] Stage, op. cit., p. 9, Table 2; [7] McArdle, op. cit., Table 1, p. 12; [8] Ibid., Table 3, p. 23.

There are large standing volumes of grand fir in the Inland Northwest. It ranks first in total growing stock volume at 26 percent [11]. Douglas-fir ranks second at 23 percent. In Idaho, grand fir also makes up the largest percentage of species harvested, ranking first at 36 percent with Douglas-fir second at 26 percent [12].

Grand fir log and lumber values are lower than Douglas-fir, which is the framing lumber standard in the Inland Region. Grand fir has compensating wood quality advantages over Douglas-fir, such as higher log scale, lighter weight, less shrinkage during drying, and excellent fiber characteristics. Alternative forest products that add value are Christmas trees, floral boughs, and oils obtained from the foliage.

2. Nutrient Characteristics

Grand fir has some unique nutritional characteristics compared to other inland conifers. When grand fir was compared to other conifers in over 160 stands, it had higher foliar base cation levels and also higher variation [13]. Grand fir N foliar concentrations are slightly less than associated conifers (Figure 2a). However, foliar K is higher (Figure 2b) and Ca levels are even higher (Figure 2c). Grand fir also has higher foliar concentrations of sulfur (S) and boron (B) than the other species (not shown). Foliar nutrient data from several studies confirm that N concentrations were similar for all species but other cations were higher in grand fir [14].

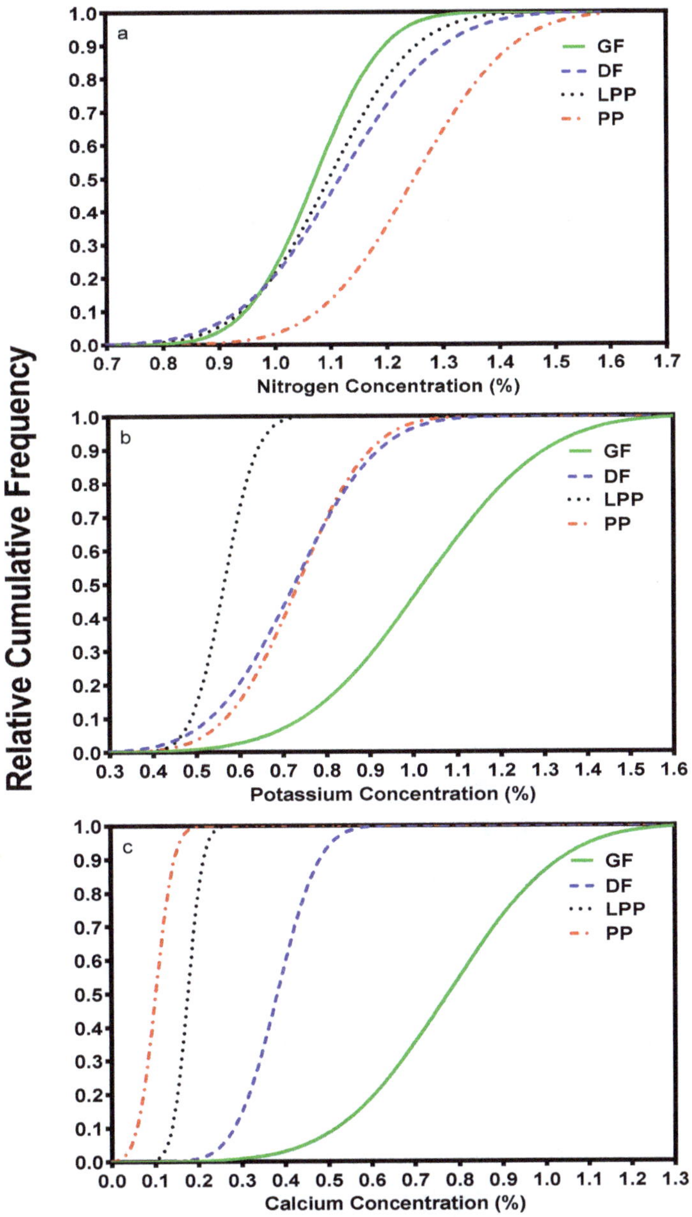

Figure 2. Relative cumulative frequency graphs of foliar N (**a**); K (**b**); and Ca (**c**) concentrations for grand fir (GF), inland Douglas-fir (DF), lodgepole pine (LP), and ponderosa pine (PP). *Y*-axis shows the proportion of trees sampled that are below the nutrient concentration on the horizontal axis. Foliage was collected from over 160 mixed conifer stands that were randomly selected within categories representing geographic region, habitat type, stand density, and rock type to assure that differences in foliar nutrient characteristics reflect species and not sample site differences [13].

Canopy nutrient content shows that grand fir acquires more nutrients than other conifer species (Figure 3). Grand fir holds twice the canopy foliar biomass as ponderosa pine and lodgepole pine and 1.4 times the foliar mass as Douglas-fir for a given diameter class [15], which is in part due to greater needle weight in comparison to other inland species (Figure 4).

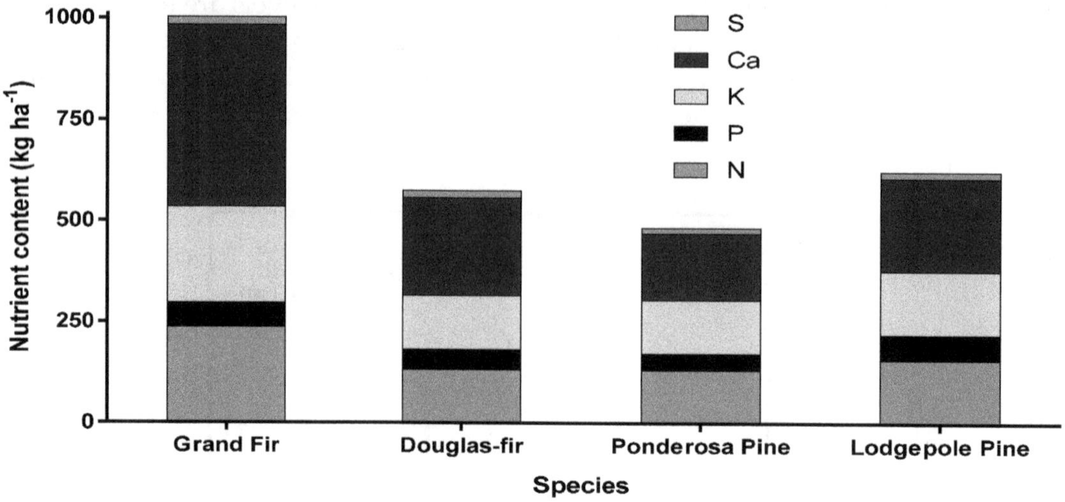

Figure 3. Total nitrogen (N), phosphorus (P), potassium (K), calcium (Ca), and sulfur (S) content of aboveground portion of four inland tree species: grand fir, Douglas-fir, ponderosa pine, and lodgepole pine [16].

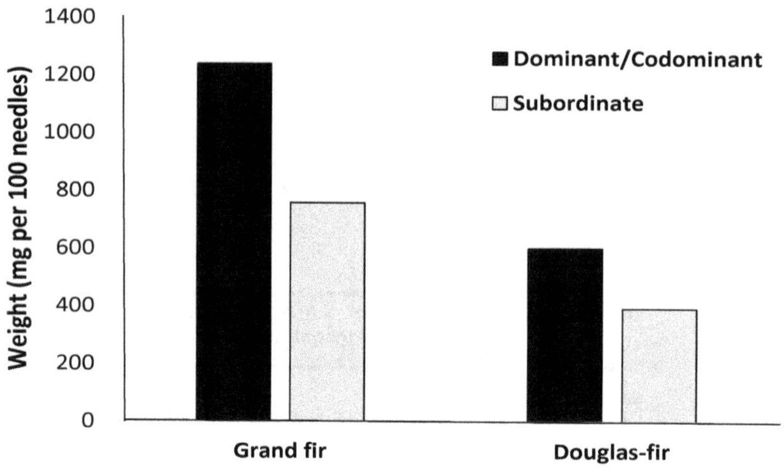

Figure 4. Foliage weight comparison between crown classes of grand fir and Douglas-fir [17].

Shade tolerant tree species (such as grand fir) incorporate cations at higher rates than intolerants (Figure 2b,c). Consequently, in mixed-conifer stands, grand fir may be more competitive for nutrients than less tolerant conifers, particularly when nutrient supply is limited [14]. This indicates that grand fir may have greater nutrient demand than associated species, which would affect its competitive ability in mixed stands and its response to nutrient additions.

We evaluate nutrient status by comparing foliage nutrient concentrations of field-collected test samples to critical foliar nutrient levels determined from greenhouse-grown seedlings. The critical nutrient level is the concentration above which growth does not respond to nutrient additions [14,18]. Table 2 shows the critical foliar nutrient levels for four Inland Northwest tree species. Seedling-based critical levels are used as reference values in mature stands. The critical level of foliar N in grand fir is 11.5 mg·g^{-1}, 18 percent less than Douglas fir. Critical levels of other selected macro- and micro-nutrients are about the same as for other species, suggesting that grand fir may be able to cope with lower N concentrations than other species due to its larger crown biomass, and relatively low N values (Figure 2a). Although critical levels of N are lower, only 20 percent of unfertilized grand fir has adequate levels of N (Figure 2a) indicating that there still may be considerable opportunity for growth increases from N additions. However, there are important limitations in our knowledge of seedling critical levels relate to fertilizer responses in mature trees.

Table 2. Critical nutrient concentrations for inland conifer species: grand fir (GF), Douglas-fir (DF), lodgepole pine (LP), and ponderosa pine (PP) [14].

Foliar Concentration	GF	DF	LP	PP	Ingestad Ratios *
N ($mg \cdot g^{-1}$)	11.5	14.0	12.0	11.0	100
P ($mg \cdot g^{-1}$)	1.5	1.2	1.2	0.8	16
K ($mg \cdot g^{-1}$)	5.8	6.0	5.0	4.8	50
S ($mg \cdot g^{-1}$))	0.8	1.1	0.9	0.8	7
Cu ($\mu g \cdot g^{-1}$)	3	2	3	3	0.03
B ($\mu g \cdot g^{-1}$)	10	10	4.3	20	0.2

* Optimum nutrient ratios, $R = 100$ (M/N), where M is the concentration of the element of interest and N is the concentration of N ([19] except S/N).

Grand fir's shade tolerant ecological status may help it survive and grow in a subordinate position in mixed stands. The lower critical N levels compared to Douglas-fir and lodgepole pine and foliar N levels in untreated stands indicate that grand fir is able to use N conservatively when in low supply. Conversely, when N is available, grand fir may utilize N more aggressively. Higher foliar concentrations of other nutrients such as K and B in untreated grand fir stands suggest that it is commonly in good nutritional balance [14].

Nutrient balance is an important consideration in tree nutrition. Plants require nutrients in specific proportions, which Ingestad [19,20] expressed as optimal nutrient ratios calculated relative to N. Nutrient ratios are consistent among plant taxa due to metabolic requirements, and therefore, are important indicators of nutrient balance (Table 2). As with critical nutrient concentrations, optimal nutrient ratios are based on seedling studies. We assume they are consistent in mature trees because there is important stoichiometric balance among all taxa [21], but we are not aware of studies that consider nutrient balance in mature trees. Grand fir nutrient ratios across three studies found more than 80% of trees measured had K, S, B, and Cu ratios above the Ingestad ratios listed in Table 2. Phosphorous ratios of grand fir averaged near Ingestad's values. Grand fir has relatively high nutrient ratios compared with inland Douglas-fir [22] and perhaps other inland northwest conifers because of the relatively low N concentrations and high concentrations of other nutrients in tree foliage (Figure 2). The resulting good nutritional balance of grand fir further suggests that it may have a competitive advantage.

The previous data show that grand fir has lower N concentrations and N critical levels than Douglas-fir and other inland conifer species. However, the concentrations and ratios of other nutrients are relativity high in grand fir. Grand fir may be better able to utilize available site N than other inland conifer species because of this favorable nutrient balance. Furthermore, grand fir maintains relatively high foliage biomass and therefore stores site nutrients for its own use and out of reach from competitors. These unique features may be partially explained by grand fir's physiology as a shade tolerant species [23]. Perhaps grand fir can tolerate low levels of soil N by sequestering site N in foliage while maintaining a favorable nutrient balance. It is then well-prepared for aggressive growth when opportunity arises.

3. Response to Fertilization

Fertilization is an important tool in nutrient management of conifer stands containing grand fir. The commonly deficient N levels along with subordinate canopy position and shade-tolerance of grand fir suggest that it may have strong foliar and growth response to N additions. However, several complicating factors must be accounted for when considering foliar nutrient and stand growth response to fertilizer. These factors include stand density, the relationships with site quality, and the fact that N fertilizer growth response can be offset by an increase in mortality. We consider each of these factors in the following sections.

Foliar data for grand fir reveals a strong response to N additions, both in young stands at crown closure and mature stands at the end of the rotation (Figure 5). There was also good foliar response to additions of S and B (data not shown). Foliage did not respond positively to K additions (Figure 6) perhaps because all the trees were above critical K levels before fertilization. It is also possible that the amount of added K was insufficient relative to added N, or that the results simply reflect the dilution effect of expanding foliage in response to the N. Limited trials with phosphorous (P) in the Inland Northwest show no response in either foliar nutrient or growth responses [24]. Consequently, P has receive little attention as a limiting nutrient in the Inland Northwest. There is some evidence that mature trees respond to P amendments in more alkaline, N-rich soils where additional research is warranted [25].

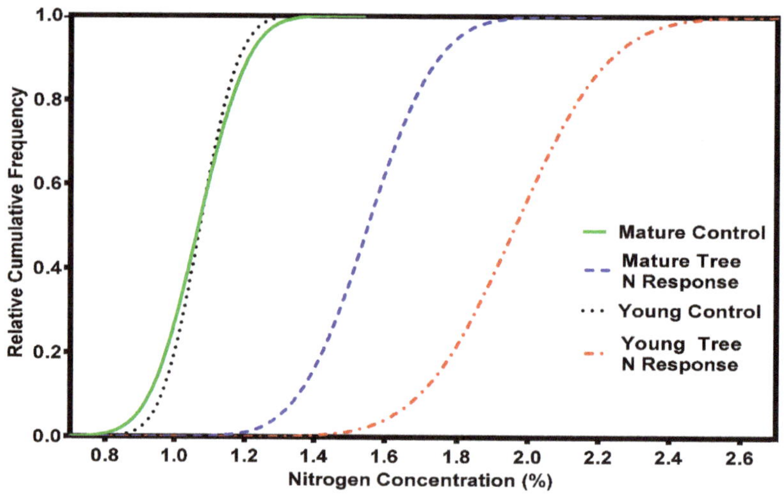

Figure 5. Grand fir foliar N response additions for young (~15 year-old) and mature (~40 year-old) stand ages [16]. (Foliage samples were collected one-year after fertilization with 224–336 kg N ha^{-1} as urea).

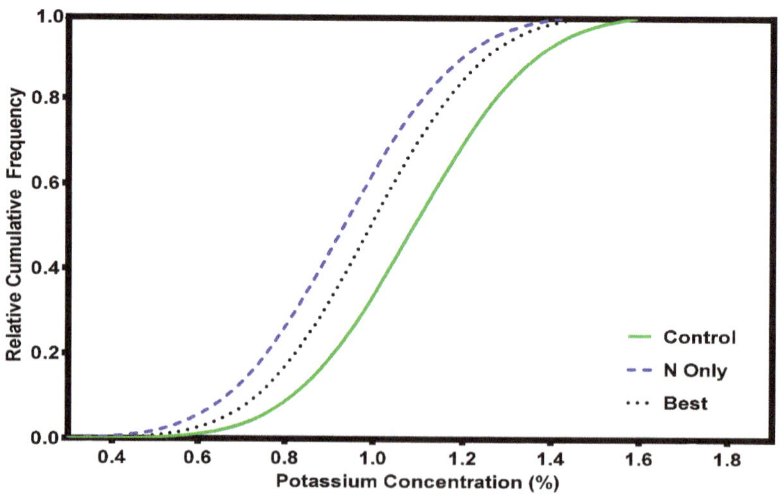

Figure 6. Grand fir foliar K response after fertilization with 224 kg N ha^{-1} (N-only) or a blend of 224 kg N ha^{-1}, 170 kg K ha^{-1} + 73 kg S ha^{-1} + 2.7 kg B ha^{-1} (Best) [16].

Reflecting foliar responses, N fertilization trials generally show a strong response in both tree diameter, and height [26]. Various trials show consistent growth responses to N rates or combinations of N with other nutrients [16,27] (Figure 7). Growth responses over 20 percent are within the realm of economic feasibility. Additional macro- and micronutrients have resulted in a response in some, but not all instances. For example, multi-nutrient trials showed that grand fir responds well to S and B applications but not necessarily K [16,27,28]. Limited responses to K are also observed in Douglas-fir, even though K is suspected of protecting trees from mortality following N-only fertilization [29].

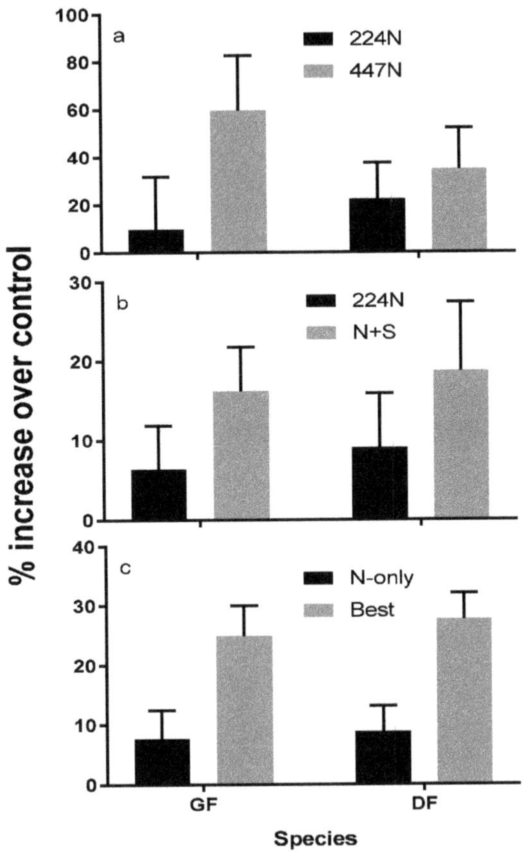

Figure 7. Grand fir (GF) and Douglas-fir (DF) volume increment response in three different fertilization trials at various locations. (**a**) Average annual ten-year response to two levels of N-only fertilization at 224 and 447 kg N ha^{-1} in 30 Douglas-fir dominated (80% basal area) stands in Idaho, Montana, Oregon and Washington including five parent material classes and three vegetation series representing the range of regional productivity [30]; (**b**) Average annual eight-year response to either 224 kg N ha^{-1} or 224 kg N ha^{-1} + 73 kg S ha^{-1}) in seven mixed conifer stands growing on basalt parent materials at 1370 to 1680 m in *Abies grandis* and *Abies lasiocarpa* vegetation series [31]; (**c**) Average annual two-year response to individual-tree treatments of 224 kg N ha^{-1} (N-only) or a blend of 224 kg N ha^{-1}, 190 kg K ha^{-1} + 73 kg S ha^{-1} + 2.7 kg B ha^{-1} (Best) in eight young (5–10 cm average diameter) stands all on metasedimentary parent material [16].

Response trials have shown that the length of the growth response is 4–6 years for unthinned stands [23,32] and up to 14 years in properly thinned stands [32,33]. Thinning concentrates the growth on fewer trees allowing them to take advantage of other site resources and sustain the response for a longer period.

Repeating a fertilization application can have differing effects depending on the site [27]. For example, grand fir showed a strong repeat response with an additional 90 kg N ha^{-1} treatment on the drier sites, but did not respond to higher rates. Grand fir also showed a good repeat response to 89 and 180 kg N ha^{-1} on the wetter sites. However, Douglas fir did not respond to these repeat treatments. Repeat fertilizations of grand fir on moist sites at rates of 90 to 180 kg N ha^{-1} are expected to result in a significant basal area growth response compared to a single fertilization. However, a significant response of grand fir is not likely on drier sites.

Differences in site quality can partly explain the range in fertilizer response. Important factors affecting fertilizer response to site quality include soil parent material, and soil moisture as indicated by climax vegetation series. Soil types modified by bedrock geology can influence growth and response to nutrient additions. One of the earliest investigations involving geologic rock types found no differentiation between responses on three different rock types [34]. More precise experimental designs have since detected parent material effects by using individual tree screening trials. Screening trials include different treatments of replicated trees across a uniform site. Sites with various parent materials then contain replicates of that design.

Rock types divide into three categories when unfertilized controls are compared using N-only fertilization, and a multi-nutrient treatment that included N, K, S, and B (Figure 8). Grand fir does not respond to fertilizer additions on soils derived from some parent material types (schist and basalt), indicating they are not deficient in any of the applied nutrients. Sites with soils from other parent materials respond only to N additions, but not multi-nutrients (granite, carbonate bearing metasedimentary, indicating they are N deficient, but other applied nutrients are not limiting. Heavily weathered quartz metasedimentary rocks have the greatest response to multiple nutrients to achieve optimal growth, indicating more than N is deficient. Productivity of grand fir on quartz-meta parent material is equal to the highest growth observed among the screening trials when treated with multi-nutrient fertilizer.

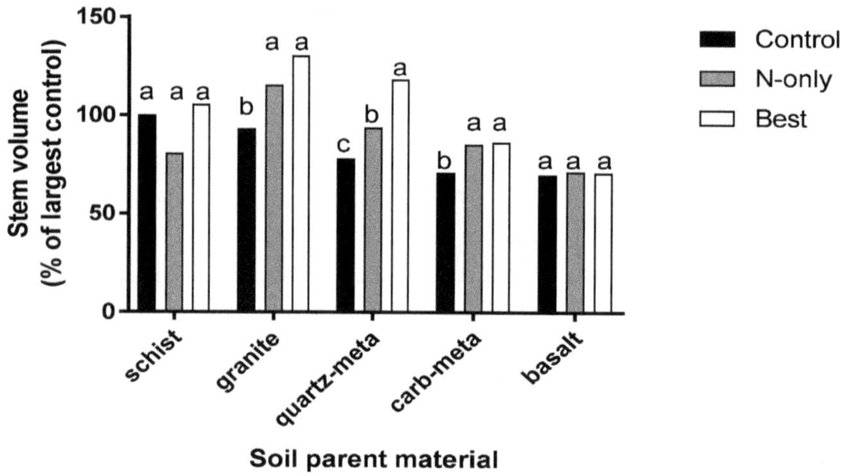

Figure 8. Fertilizer treatment effects on 2-year stem volume by soil parent material. Stem volume is expressed as a percentage of untreated trees growing on schist soils. Fertilizer treatments included 224 kg N ha^{-1} (N-only), or a blend of 224 kg N ha^{-1}, 190 kg K ha^{-1} + 73 kg S ha^{-1} + 2.7 kg B ha^{-1} (Best). Two metasedimentary parent materials include a quartzite (quartz-meta) and carbonate (carb-meta) [24].

Vegetation series as described by habitat type [2] also affect growth response due to soil moisture variations. Habitat types containing more available soil moisture or higher amounts of organic matter would tend to have better response. Soils containing volcanic ash display increased productivity and N response for all conifer species due to the increase in water-holding capacity [3]. Yet, only half of the sites where grand fir is the climax vegetation contain significant ash deposits. Higher elevation vegetation series are associated with greater moisture availability and ash depth, and grand fir is capable of higher productivity on these sites due to increased moisture availability during the growing season. The limited ash deposits and dryer site conditions for the grand fir habitat type suggest that grand fir is tolerant of dryer conditions. However, when grown on sites with abundant moisture grand fir is capable of better growth than on drier sites. The above results demonstrate that there are differences in nutrient availability and fertilization response among soils, bedrock types, and vegetation series but these differences interact with moisture availability and will require focused research effort to discover mechanisms responsible for distinctions among bedrock geology, surficial deposits, and moisture availability.

The development stage of a forest stand explains some of the response variability observed in previous experiments with soils and site types. A good understanding of these stages and their effect on nutrient requirements helps determine when nutrient additions will provide a growth response. Stands of trees have differing nutritional needs depending on where they are in the successional pathway of development [35].

At the seedling and sapling stage, the trees are unrestricted in their quest for nutrients, especially if artificially planted with competition control. Consequently, during the seedling stage of development, sites meet seedling nutrient requirements, and seedlings can explore the soil for available nutrients reducing the need for additions. Furthermore, they can take advantage of the nutrient flush that often occurs after timber harvest. Grand fir plantations are rare because the species abundantly regenerates in association with planted seedlings of higher-value species. Consequently, we are not aware of seedling fertilizer tests specifically for grand fir. Nutrition of natural grand fir seedling development is a potential area of research. Tests using planted seedling of other inland conifer species on multiple sites show few consistent responses to broadcast applied nutrients [36]. During one seedling trial where fertilizer was applied into the planting hole, Douglas-fir and ponderosa pine exhibited increased height and diameter growth after outplanting, but also had higher mortality, especially at greater N application rates [37].

As stands develop through intermediate ages, nutrient demands increase until they reach the self-thinning stage when nutrient demands are greatest [35]. At this stage, the ability to respond to treatments is high because of nutrient limitations and high stand vigor. This stage is not necessarily age-related because stand age at self-thinning can vary depending on the size-density relationship, species, site quality, and stand establishment practices [38,39]. One would expect during self-thinning that nutrient additions would increase growth. Unfortunately, many of the recorded grand fir experiments do not specifically indicate the stage of stand development or measures of density.

Experiments in relatively young (10–30 years old), high-density stands showed strong response to fertilization [26,40–42]. As stands mature, nutrient demands continue and stand density becomes a controlling influence. If artificial fertilization occurs during this time, grand fir should have a strong response in comparison with associated species. Tree spacing control during this period is important to allow sufficient non-nutrient resources to support fertilizer response. As stands reach mature and old growth conditions, nutrient demands depend more on site factors. Grand fir's physiology as a late-successional, shade-tolerant tree may affect its response by allowing it to conserve N when necessary but be more demanding during competitive growth phases such as that which would occur during crown closure [23]. Focused study is required on the ability of grand fir to compete for nutrients during between crown closure and self-thinning stages of stand development.

During the crown closure through self-thinning stages of development, additional growth response is possible when thinning and fertilization are combined as demonstrated from numerous experiments [32,34,43–45]. After thinning a crowded stand, light, water, and temperature become more favorable for growth [46]. Fertilization within a decade after thinning, or in properly spaced stands allows the trees to take advantage of the additional resources [30,47]. Thinning and then nutrient additions are beneficial, especially if relative densities prior to fertilization allow adequate space for increased growth. Research trials show that combined thinning and fertilization are additive compared to the individual treatments [47,48]. In some instances the total growth response of the crop trees is greater than the sum of either fertilization or thinning effects alone [32].

Grand fir exhibits higher mortality after fertilization, especially with N-only amendments. Its mortality is naturally greater than that of Douglas-fir, and N fertilization only increases that trend (Figure 9) [49]. Of the known reasons for mortality, N fertilization increases insect-caused tree mortality more than other causes. This is commonly observed in trees and is attributed to improved foliage nutritional quality [50]. Addition of other macro- and micronutrients tended to reduce mortality but not significantly [16]. There is critical need to learn more about the causes of increase mortality to N-only fertilizer applications.

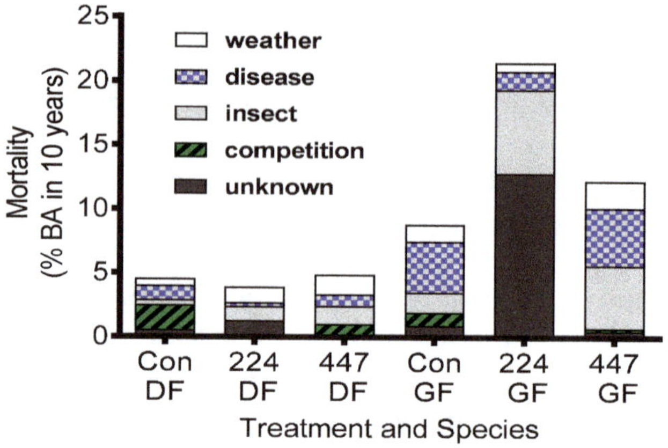

Figure 9. The causes of ten-year basal area mortality of Douglas-fir and grand fir treated with either 224 or 447 kg N ha^{-1} and compared with unfertilized controls. Data are from 30 mature (average 65 year-old) Douglas-fir dominated (80% basal area) stands in Idaho, Montana, Oregon and Washington including five parent material classes and three vegetation series representing the range of regional productivity. All stands were recently thinned or were naturally well spaced. Initial basal area averaged 36 ± 10 m^2 ha^{-1} [30].

Nitrogen fertilization can potentially stimulate pests and disease. Insects such as spruce budworm (*Choristoneura occidentalis* Freeman) reduce current growth, and can diminish fertilizer response [48]. Grand fir is more susceptible to Douglas-fir tussock moth (*Orgyia pseudotsugata* McDunnough) than other inland conifers, but experience confirms that susceptibility tends to decreases with soil volcanic ash content [51] which, as described above, is expected to interact with soil moisture, tree nutrition and fertilizer application. The effects of Armillaria root disease (*Armillaria ostoyae* Romagnesi) can substantially reduce net response to fertilization. Therefore, stand-level responses to fertilization may be lower than expected if root disease is present in the stand. Although there were initial experiments implying that fertilization (especially with K additions) [52] may reduce incidence of root disease, a 10-year study in a diseased stand found no treatment differences due to fertilization [27]. Moore et al. [29] suggested that tree nutrition (K/N ratio) controls production of phenolic defense compounds and resistance to the root rot disease Armillaria in Douglas-fir. If true, grand fir should exhibit a similar relationship because it too is susceptible to Armillaria infection. While the relationship between favorable K/N ratios and phenolic concentrations was confirmed in seedlings [53,54] they could not demonstrate an effect of stand-level N and K fertilization on root disease or mortality after ten years. The potential negative effects of pests and diseases following fertilization are important to recognize [55] and managers should avoid fertilization in stands that display these properties. Increased effort is needed to elucidate the relationship between nutrient balance, defense compounds, and root disease.

Vegetation control using herbicides is expanding in the Inland Northwest. Herbicide application is cheaper, more effective, and protects the soil surface from compaction and displacement by heavy site preparation machinery. Grand fir is generally more tolerant to herbicides than associated species [56] so prudent use presents few difficulties. Specific effects of herbicide application in combination with fertilization are not well understood in the Inland Northwest Region. However, in the southeastern U.S., research confirms the positive combined effects of herbicide treatments and nutrient additions [57].

There is evidence that vegetation control alone may provide additional nutrients to trees by eliminating competition and causing faster breakdown and mineralization of organic matter [55].

In summary, these studies indicate that grand fir growing under the right conditions can respond to N fertilization with volume growth ranging from 15% to 50%. Grand fir generally has a longer response duration than other inland conifers, the greatest response occurring during 3–5 years, tapering off during 5–8 years, with some response lasting through 10–12 years under optimum conditions. The response can be greater after well-timed thinning. Repeated fertilizer applications have the potential to maintain growth response depending on the site and the stage of stand development. It is important to recognize site types and their interaction on fertilizer response. Grand fir is sensitive to higher mortality rates after fertilization so when seeking locations for effective treatments, it is critical to avoid sites with heavily weathered parent material where nutrient imbalance may lead to high pest and disease susceptibility.

4. Economics of Grand Fir Fertilization

Large land-holding companies in the Inland Northwest are interested in the economics of fertilization. As a financial investment, forest fertilization must consider the time value of money. Success is very sensitive to the time between the initial outlay and the final return on investment; thus, fertilization in older stands is considered to be more profitable because the volume gains can be harvested sooner [58] and investment returns increase [59]. Late rotation fertilization has a short payback period with relatively low investment costs relative to other forestry investments [60]. Repeated fertilization can extend the response duration, however a second amendment doubles the cost and the response is not as strong as the first application. These economic principles have led foresters in the region to believe that a one-time fertilization late in the rotation can be a sound forestry investment. A recent Internal Revenue Service (IRS) tax law change that allowed expensing of fertilization costs further increased the investment potential [61].

Since most grand fir in the Inland Northwest grows in mixture with Douglas-fir and other species, fertilization experiments require additional analysis and assumptions to confine the results to just grand fir. Because of this, grand fir response, by itself, is usually inferred using data from mixed stands.

Properly spaced, 60–90 years old mature stands containing grand fir may respond well to N fertilization with responses similar to those obtained for Douglas-fir stands in western Washington [58]. The treatment generally involves application of 150–300 kg ha^{-1} of urea N followed by timber harvest in 8–12 years. This results in a basal area growth response of 20%–80% with comparable volume responses. Addition of multi-nutrients may improve protection against mortality but does not significantly improve individual tree growth response [29], so the added cost does not increase the economic benefit.

Recent operational fertilization treatments in mixed stands including grand fir confirm that investment returns are very good, often surpassing those of other forestry investments, if stands are carefully selected [62]. It is essential to identify stand conditions that will provide an acceptable economic response [63]. Incorporation of a carefully selected fertilization treatment into a general forest management plan can be economically warranted. These growth responses can be used for more sophisticated economic analysis for more accurate projections depending on management requirements. However, such growth responses may require adjustments in coming decades with increased atmospheric carbon dioxide concentrations and accompanying changes in climate.

5. Management Recommendations

We recommend that a grand fir nutrient management and fertilization program consider the following concepts.

1. The greatest response from added nutrients seems to occur just after crown closure when competitive stresses are high.

2. Grand fir is expected to best respond to nutrient additions when it is maintained in a healthy and vigorous condition through stand density management.

3. To conserve site nutrients, soil disturbance should be minimized during timber harvest and stand regeneration activities. Heavy-handed site preparation and slash cleanup may have a negative impact on the site's nutrient resources, especially low-quality sites [14]. Some alternate possibilities to consider include light broadcast burning; leaving small limbs, foliage, and larger material scattered across the site; and the use of herbicides.

4. Carefully analyze timber stands to determine tree and site conditions, species mix, stand density, soil parent material, and harvest plans before making fertilization decisions to assure effective results. Consult the latest fertilization research results to guide grand fir nutrient management.

6. Conclusions

Grand fir is ubiquitous in Inland Northwest forests growing in mixture with other species, mainly Douglas-fir. It often contains higher crown nutrient content than associate species [14]. Grand fir has good growth response to N fertilizer and possibly K. The effect of nutrition, especially K, on root disease potential remains unconfirmed. Good investment returns are possible from fertilization treatments after careful site evaluation. With good management, and a solid understanding of nutrient dynamics, grand fir will remain an important species with solid economic values and a positive response to fertilizer under the right conditions in Inland Northwest forests.

Acknowledgments: Funding for analysis of data included in this report and preparation of the manuscript was supplied by the Intermountain Forestry Cooperative, formerly Intermountain Forest Tree Nutrition Cooperative (IFTNC). We wish to thank the following people who assisted with the preparation and review of this manuscript: Mark J. Kimsey, Research Assistant Professor, IFTNC/University of Idaho, Terry M. Shaw Research Scientist, IFTNC/University of Idaho.

Author Contributions: Dennis Parent and Mark D. Coleman preformed equally in drafting, editing and revising the manuscript.

References

1. Foiles, M.W.; Graham, R.T.; Olson, D.F. Grand fir. In *Silvics of Forest Trees of the United States, Conifers*; Burns, R.M., Honkala, B.H., Eds.; USFS: Washington, DC, USA, 1990; Volume 1, pp. 52–59.
2. Cooper, S.V.; Neiman, K.E.; Roberts, D.W. *Forest Habitat Types of Northern Idaho: A Second Approximation*; USFS, Intermountain Research Station: Ogden, UT, USA, 1991.
3. Garrison-Johnston, M.T.; Mika, P.G.; Miller, D.L.; Cannon, P.; Johnson, L.R. Ash cap influences on site productivity and fertilizer response in forests of the Inland Northwest. In *Volcanic-ash-derived Forest Soils of the Inland Northwest: Properties and implications for management and restoration*; USFS, RMRS: Coeur d'Alene, ID, USA, 2007; pp. 137–163.
4. Pfister, R.D.; Kovalchik, B.L.; Arno, S.F.; Presby, R.C. *Forest Habitat Types of Montana*; USFS, INT Station: Ogden, UT, USA, 1977.
5. Williams, C.K.; Kelley, B.F.; Smith, B.; Lillybridge, T.R. *Forested Plant Associations of the Colville National Forest*; Pacific Northwest Research Station: Portland, OR, USA, 1995.
6. USGS; Little, E.L., Jr. Digital representations of "Atlas of United States Trees". Available online: http://esp.cr.usgs.gov/data/atlas/little/abiegran.pdf (accessed on 31 August 2016).
7. Deitschman, G.H.; Green, A.W. *Relations between Western White Pine Site Index and Tree Height of Several Associated Species*; USFS: Ogden, UT, USA, 1965.
8. Haig, I.T. *Second-growth Yield, Stand, and Volume Tables for the Western White Pine Type*; USFS, Northern Rocky Mountain Forest Experiment Station: Washington, DC, USA, 1932.
9. Stage, A.R.; Renner, D.L.; Chapman, R.C. *Selected Yield Tables for Plantations and Natural Stands in the Inland Northwest forests*; USFS, INT Research Station: Ogden, UT, USA, 1988.
10. McArdle, R.E.; Meyer, W.H.; Bruce, D. *The Yield of Douglas-Fir in the Pacific Northwest*; USFS: Washington, DC, USA, 1961.

11. USFS. Forest Inventory Data Online (FIDO). Available online: http://apps.fs.fed.us/fia/fido/index.html (accessed on 2 Jaunary 2016).
12. Simmons, E.A.; Hayes, S.W.; Morgan, T.A.; Keegan, C.E., III; Witt, C. *Idaho's Forest Products Industry and Timber Harvest 2011 with Trends through 2013*; USFS: Fort Collins, CO, USA, 2013.
13. Moore, J.A.; Mika, P.G.; Shaw, T.M.; Garrison-Johnston, M. Foliar nutrient characteristics of four conifer species in the interior Northwest United States. *West. J. Appl. For.* **2004**, *19*, 13–24.
14. Garrison, M.T.; Moore, J.A. Nutrient Management: A Summary and Review. Available online: http://digital.lib.uidaho.edu/cdm/singleitem/collection/iftnc/id/3357/rec/40 (accessed on 29 September 2016).
15. Brown, J.K. *Weight and Density of Crowns of Rocky Mountain Conifers*; USFS, INT Forest & Range Experiment Station: Ogden, UT, USA, 1978.
16. Mika, P.G. Grand Fir: Nutrient Ecology and Response to Fertilization. Available online: http://digital.lib.uidaho.edu/cdm/singleitem/collection/iftnc/id/3446/rec/6 (accessed on 29 September 2016).
17. Garrison-Johnston, M. Nutrient Cycling in a North Idaho Conifer Stand. Ph.D. Thesis, University of Idaho, Moscow, ID, USA, 2003.
18. Walker, R.B.; Gessel, S.P. *Mineral Deficiencies of Coastal Northwest Conifers*; College of Forest Resources, University of Washington: Seattle, WA, USA, 1991.
19. Ingestad, T. Mineral nutrient requirements of *Pinus silvestris* and *Picea abies* seedlings. *Physiol. Plant.* **1979**, *45*, 373–380. [CrossRef]
20. Ingestad, T. A definition of optimum nutrient requirements in birch seedlings. II. *Physiol. Plant.* **1971**, *24*, 118–125. [CrossRef]
21. Elser, J.J.; Sterner, R.W.; Gorokhova, E.; Fagan, W.F.; Markow, T.A.; Cotner, J.B.; Harrison, J.F.; Hobbie, S.E.; Odell, G.M.; Weider, L.W. Biological stoichiometry from genes to ecosystems. *Ecol. Lett.* **2000**, *3*, 540–550. [CrossRef]
22. Coleman, M.D.; Shaw, T.M.; Kimsey, M.J.; Moore, J.A. Nutrition of Douglas-fir in the Inland Northwest. *Soil Sci. Soc. Am. J.* **2014**, *78*, 11–22. [CrossRef]
23. Graham, R.T.; Tonn, J.R. *Ten-Year Results of Fertilizing Grand Fir, Western Hemlock, Western Larch, and Douglas-Fir with Nitrogen in Northern Idaho*; USFS, INT Station: Ogden, UT, USA, 1985.
24. Shaw, T.M.; Kimsey, M. Screening trial data. Intermountain Forestry Cooperative, University of Idaho: Moscow, ID, USA, Unpublished data. 2016.
25. Mainwaring, D.B.; Maguire, D.A.; Perakis, S.S. Three-year growth response of young Douglas-fir to nitrogen, calcium, phosphorus, and blended fertilizers in Oregon and Washington. *For. Ecol. Manag.* **2014**, *327*, 178–188. [CrossRef]
26. Loewenstein, H.; Pitkin, F.H. *Growth Responses and Nutrient Relations of Fertilized and Unfertilized Grand Fir*; University of Idaho, Forestry, Wildlife, and Range Experiment Station: Moscow, ID, USA, 1971.
27. Shaw, T.; Coleman, M.; Kimsey, M.; Mika, P. *Forest Health and Nutrition Ten-year Growth, Mortality and Foliar Nutrition*; Intermountain Forest Tree Nutrition Cooperative: Moscow, ID, USA, 2014.
28. Garrison, M.T.; Moore, J.A.; Shaw, T.M.; Mika, P.G. Foliar nutrient and tree growth response of mixed-conifer stands to three fertilization treatments in northeast Oregon and north central Washington. *For. Ecol. Manag.* **2000**, *132*, 183–198. [CrossRef]
29. Moore, J.A.; Mika, P.G.; Schwandt, J.W.; Shaw, T.M. Nutrition and forest health. In *Interior Cedar-Hemlock-White Pine Forests: Ecology and Management*; Washington State University: Spokane, WA, USA, 1994; pp. 173–176.
30. Moore, J.A.; Mika, P.G.; Vanderploeg, J.L. Nitrogen fertilizer response of Rocky Mountain Douglas-fir by geographic area across the Inland Northwest. *West. J. Appl. For.* **1991**, *6*, 94–98.
31. Garrison, M.T.; Moore, J.A.; Mika, P.G.; Shaw, T.M. Six-Year Growth Response of the Umatilla and Four-Year Growth Response of the Okanogan Mixed Conifer Stands to N, N + K and N + S Fertilization. Available online: http://digital.lib.uidaho.edu/cdm/ref/collection/iftnc/id/3358 (accessed on 3 October 2016).
32. Shafii, B.; Moore, J.A.; Olson, J.R. Effects of nitrogen fertilization on growth of grand fir and Douglas-fir stands in northern Idaho. *West. J. Appl. For.* **1989**, *4*, 54–57.
33. Scanlin, D.C.; Lowenstein, H. *Forest Fertilization in the Intermountain Region-six Year Results*; University of Idaho, Forestry, Wildlife and Range Experiment Station: Moscow, ID, USA, 1980.
34. Scanlin, D.C.; Loewenstein, H.; Pitkin, F.H. *Two-year Response of North Idaho Stands of Douglas-Fir and Grand Fir to Urea Fertilizer and Thinning*; University of Idaho: Moscow, ID, USA, 1976.

35. Kimmins, J.P. *Forest Ecology: A Foundation for Sustainable Forest Management and Environmental Ethics in Forestry*, 3rd ed.; Prentice Hall: Upper Saddle River, NJ, USA, 2004.
36. Shaw, T.M.; Johnson, L.R. Two-Year Growth Response of Douglas-fir and Ponderosa Pine Seedlings to Boron and Multi-Nutrient Fertilization in Northeast Oregon. Available online: http://digital.lib.uidaho.edu/cdm/ref/collection/iftnc/id/4860 (accessed on 4 October 2016).
37. Fan, Z.; Moore, J.A.; Osborne, H.L. Three-year response of ponderosa pine seedlings to controlled-release fertilizer applied at planting. *West. J. Appl. For.* **2002**, *17*, 154–164.
38. Oliver, C.D.; O'Hara, K.L.; McFadden, G.; Nagame, L. Concepts of thinning regimes. In *Douglas-Fir: Stand Management for the Future*; Oliver, C.D., Hanley, D.P., Johnson, J.A., Eds.; Institute of Forest Resources: Seattle, WA, USA, 1986; pp. 246–257.
39. Weiskittel, A.; Gould, P.; Temesgen, H. Sources of variation in the self-thinning boundary line for three species with varying levels of shade tolerance. *For. Sci.* **2009**, *55*, 84–93.
40. Loewenstein, H.; Pitkin, F.H. Response of grand fir and western white pine to fertilizer applications. *Northwest Sci.* **1963**, *37*, 23–30.
41. Graham, R.T.; Tonn, J.R. *Response of Grand Fir, Western Hemlock, Western White Pine, Western Larch, and Douglas-Fir to Nitrogen Fertilizer in Northern Idaho*; USFS, INT: Ogden, UT, USA, 1979.
42. Powers, R.F. Response of California true fir to fertilization. In *Forest Fertilization Conference*; Gessel, S.P., Ed.; University of Washington: Union, WA, USA, 1979; pp. 95–101.
43. Olson, J. *Four-Year Fertilization and Thinning Response Analysis of Douglas-fir and Grand Fir Stands in Northern Idaho*; Potlatch Corporation: Lewiston, ID, USA, 1979.
44. Olson, J. *Response of Intermountain Grand Fir and Douglas-Fir Stand Types to Nitrogen Fertilization and Thinning*; Potlatch Corporation: Lewiston, ID, USA, 1981.
45. Olson, J.; Hatch, C.R. Volume response of intermountain grand fir stand types to nitrogen fertilizer and thinning treatments. In *The Biology and Management of True Fir in the Pacific Northwest*; University of Washington: Seattle, WA, USA, 1981; pp. 247–250.
46. Chase, C.W.; Kimsey, M.J.; Shaw, T.M.; Coleman, M.D. The response of light, water, and nutrient availability to pre-commercial thinning in dry inland Douglas-fir forests. *For. Ecol. Manag.* **2016**, *363*, 98–109. [CrossRef]
47. Scanlin, D.C.; Loewenstein, H. Response of inland Douglas-fir and grand-fir to thinning and nitrogen fertilization in northern Idaho. In *Forest Fertilization*; Gessel, S.P., Kenady, R.M., Atkinson, W.A., Eds.; University of Washington: Alderbrook Inn, Union, WA, USA, 1979; pp. 82–88.
48. Cochran, P.H. *Response of Thinned White Fir Stands to Fertilization with Nitrogen Plus Sulfur*; USFS, PNW Research Station: Portland, OR, USA, 1991.
49. Shen, G.; Moore, J.A.; Hatch, C.R. The effect of nitrogen fertilization, rock type, and habitat type on individual tree mortality. *For. Sci.* **2001**, *47*, 203–213.
50. Herms, D.A. Effects of fertilization on insect resistance of woody ornamental plants: Reassessing an entrenched paradigm. *Environ. Entomol.* **2002**, *31*, 923–933. [CrossRef]
51. Stoszek, K.J.; Mika, P.J. Douglas-fir tussock moth outbreaks related to stand conditions and pest management. In *1980 SAF Convention: Land Use Allocation*; Society of American Foresters: Bethesda, MD, USA, 1981; pp. 134–137.
52. Mika, P.G.; Moore, J.A. Foliar potassium status explains Douglas-fir response to nitrogen fertilization in the inland Northwest, USA. *Water Air Soil Pollut.* **1990**, *54*, 477–491. [CrossRef]
53. Schwandt, J.W. Fertilization effects on root disease in two Douglas-fir stands in northern Idaho. In *Proceedings of the Fiftieth Western International Forest Disease Work Conference*; USFS: Powell River, BC, Canada, 2002; pp. 1–6.
54. Shaw, T.M.; Moore, J.A.; Marshall, J.D. Root chemistry of Douglas-fir seedlings grown under different nitrogen and potassium regimes. *Can. Jour. For. Res.* **1998**, *28*, 1566–1573. [CrossRef]
55. Mandzak, J.M.; Moore, J.A. The role of nutrition in the health of the inland western forest. *J. Sustain. For.* **1994**, *2*, 191–210. [CrossRef]
56. Miller, D.L.; Gravelle, P.J. *Species Selection Guidelines for Planting, Natural Regeneration and Crop Tree Selection on Potlatch Land in Northern Idaho*; Potlatch Corporation: Lewiston, ID, USA, 2009.
57. Will, R.E.; Munger, G.T.; Zhang, Y.; Borders, B.E. Effects of annual fertilization and complete competition control on current annual increment, foliar development, and growth efficiency of different aged *Pinus taeda* stands. *Can. J. For. Res.* **2002**, *32*, 1728–1740. [CrossRef]

58. Miller, R.E.; Webster, S.R. Fertilizer response in mature stands of Douglas-fir. In *Forest Fertilization;* University of Washington: Union, WA, USA, 1979; pp. 126–132.
59. Miller, R.E.; Fight, R.D. *Fertilizing Douglas-fir Forests;* USFS, PNW Research Station: Portland, OR, USA, 1979.
60. Olson, J.R. Experience with nitrogen fertilization in northern Idaho. In *Forest Fertilization: Sustaining and Improving Nutrition and Growth in Western Forests;* Chappell, H.N., Weetman, G.F., Miller, R.E., Eds.; University of Washington: Seattle, WA, USA, 1992; pp. 232–236.
61. National Timber Tax Website. Timber Fertilization. Available online: http://www.timbertax.org/research/revenuerulings/bda/04-62/ (accessed on 14 January 2015).
62. Shaw, T.M.; Kimsey, M.; Parent, D.R. *Growth Response of an Operationally Fertilized Stand at Brickel Creek, North Idaho;* IFTNC, University of Idaho: Moscow, ID, USA, 2010.
63. Olson, J. The fertilization decision: Using what we have learned to evaluate a nitrogen fertilization program. In *IFTNC Fertilization Workshop;* IFTNC: Moscow, ID, USA, 1988; pp. 1–15.

7

Effects of Experimental Nitrogen Addition on Nutrients and Nonstructural Carbohydrates of Dominant Understory Plants in a Chinese Fir Plantation

Fangchao Wang [1], Fusheng Chen [1,2,*], G. Geoff Wang [3], Rong Mao [1,2], Xiangmin Fang [1], Huimin Wang [4] and Wensheng Bu [1,2]

1. Jiangxi Provincial Key Laboratory of Silviculture, Jiangxi Agricultural University, Nanchang 330045, China; wfangch@163.com (F.W.); maorong23@163.com (R.M.); xmin007@163.com (X.F.); bws2007@163.com (W.B.)
2. Jiulianshan National Observation and Research Station of Chinese Forest Ecosystem, Key Laboratory of National Forestry and Grassland Administration on Forest Ecosystem Protection and Restoration of Poyang Lake Watershed, Jiangxi Agricultural University, Nanchang 330045, China
3. Department Forestry and Environmental Conservation, Clemson University, Clemson, SC 29634, USA; gwang@g.clemson.edu
4. Qianyanzhou Ecological Station, Key Laboratory of Ecosystem Network Observation and Modeling, Institute of Geographic Sciences and Natural Resources Research, Chinese Academy of Sciences, Beijing 100101, China; wanghm@igsnrr.ac.cn

* Correspondence: chenfusheng@jxau.edu.cn

Abstract: *Research Highlights:* This study identifies the nitrogen (N) deposition effect on understory plants by altering directly soil nutrients or indirectly altering environmental factors in subtropical plantation. *Background and Objectives:* N deposition is a major environmental issue and has altered forest ecosystem components and their functions. The response of understory vegetation to N deposition is often neglected due to a small proportion of stand productivity. However, compared to overstory trees, understory species usually have a higher nutrient cycle rate and are more sensitive to environmental change, so should be of greater concern. *Materials and Methods:* The changes in plant biomass, N, phosphorus (P), and nonstructural carbohydrates (NSCs) of three dominant understory species, namely *Dicranopteris dichotoma*, *Lophatherum gracile*, and *Melastoma dodecandrum*, were determined following four years of experimental N addition (100 kg hm^{-2} $year^{-1}$ of N) in a Chinese fir plantation. *Results:* N addition increased the tissue N concentrations of all the understory plants by increasing soil mineral N, while N addition decreased the aboveground biomass of *D. dichotoma* and *L. gracile* significantly—by 82.1% and 67.2%, respectively. The biomass of *M. dodecandrum* did not respond to N addition. In contrast, N addition significantly increased the average girth growth rates and litterfall productivity of overstory trees—by 18.28% and 36.71%, respectively. NSCs, especially soluble sugar, representing immediate products of photosynthesis and main energy sources for plant growth, decreased after N addition in two of the three species. The plant NSC/N and NSC/P ratios showed decreasing tendencies, but the N/P ratio in aboveground tissue did not change with N addition. *Conclusions:* N addition might inhibit the growth of understory plants by decreasing the nonstructural carbohydrates and light availability indirectly rather than by changing nutrients and N/P stoichiometry directly, although species-specific responses to N deposition occurred in the Chinese fir plantation.

Keywords: experimental nitrogen addition; understory plant growth; plant nutrient; nonstructural carbohydrates

1. Introduction

Fossil fuel combustion and chemical fertilization are increasing atmospheric nitrogen (N) deposition throughout the world and altering regional and global N cycles [1]. The annual input of reactive N into the Earth's land surface has approximately doubled since 1970, and this trend will continue in the rapidly developing regions of the world [2]. Considering that N is an essential nutrient element limiting plant photosynthetic capacity and productivity, increased N deposition would produce a cascading effect on forest ecosystem structure, process, and function [3–5]. The understory plants' responses to N deposition in forest ecosystems need to be of greater concern in the future.

The vegetation of forest ecosystems includes both overstory (or canopy) and understory species. Because of the dominance of the forest canopy, understory species and their responses to environmental changes are often neglected, especially in plantation forests [6]. Understory vegetation, despite often accounting for a small proportion of stand productivity, is an important component of forest ecosystems and plays a key role in regulating ecosystem processes and functions. Moreover, compared with tree species in the canopy, understory species usually have faster nutrient turnover rates and thus are more sensitive to environmental change. Previous studies have found that N deposition exerted a substantial influence on soil nutrient supply and decreased understory plant biomass due to increasing soil acidification and phosphorus (P) limitation [7–9]. Furthermore, some studies reported that increased N availability exaggerated asymmetrical competition for other resources (e.g., light and water), favoring the growth of overstory species at the expense of understory species [6,10,11]. For example, Strengbom suggested that understory vegetation was mainly limited by light, since N addition increased shading from the canopy that decreased the light available to understory plants in a boreal forest [12]. However, these studies concentrated on boreal and temperate natural forests, the results from which may not be applicable to subtropical plantations given that species coexistence and ecosystem biogeochemical cycles vary across forest types and climatic regions [6,13,14].

Nitrogen and P are often recognized as limiting nutrients in forest ecosystems; thus, N and P concentrations and their stoichiometry have been widely used to determine plant adaptation and feedback in response to resource alteration [15]. Nonstructural carbohydrates (NSCs) are the immediate products of photosynthesis and are often used to indicate plant light environment and growth rate [16]. The proportional relationships among N, P, and NSCs to a large extent reflect the available C and energy utilization efficiency of plant growth [17]. For example, N deposition increases N supply and causes an imbalance between N and P in plant tissues [18]. Furthermore, N deposition alters plant photosynthetic processes and the associated accumulation and consumption of NSCs including starch (ST) and soluble sugar (SS) [19]. Therefore, identifying NSCs and their interactions with N and P in aboveground (i.e., leaves) and belowground tissues (i.e., roots) might provide a new perspective on the effect of N deposition on understory vegetation growth.

In this study, a chronic N addition experiment was conducted to simulate N deposition in a Chinese fir (*Cunninghamia lanceolata*) plantation of subtropical China. We measured soil available N, P and plant biomass and N, P, and NSC concentrations in major understory plants to assess the responses of understory vegetation to N deposition in subtropical plantations. We hypothesized that (1) N deposition leads to an imbalance between N and P in soils and understory plants due to elevated N availability; (2) N deposition causes a decline in understory plant biomass due to elevated P limitation or increased shading by stimulated overstory growth; and (3) the changes in NSCs and NSCs/nutrients in aboveground and belowground tissues help identify the potential resource (such as light) competition mechanisms of understory plants in response to N deposition [14]. Our results may have some implications for understory plant management in plantation forests of subtropical areas experiencing N deposition.

2. Materials and Methods

2.1. Study Region

This study was conducted in a 12-year-old Chinese fir planation in the Jian-Taihe Basin of Jiangxi Province, China (26°42′ N, 115°04′ E, 100 m asl). The area is a typical red soil hilly region with a subtropical moist monsoon climate. The soil belongs to the typical Hapludult Ultisols with 68% sand and 15% clay [20]. The month average temperatures range from 6.5 °C in January to 29.7 °C in July, and a mean annual temperature is 18 °C. The annual precipitation ranges from 945 to 2144 mm, with an average of 1500 mm [21]. The area belongs to the core distribution of Chinese fir and the center of N deposition in China with 49 kg N ha^{-1} $year^{-1}$ [3,22].

2.2. Experimental Treatments and Sample Collection

The simulated N deposition experiment followed a paired design and was established in November 2011. Within each paired plot, two 20 m × 20 m plots were treated with four years of in situ N addition (100 kg hm^{-2} $year^{-1}$ of N) or no N addition (control, CK), with a buffer zone of more than 20 m between the plots. Four replicates were established on four separate hilly slopes. Nitrogen mixed with sand was added four times each year (March, June, September, and December) in the form of NH_4NO_3. In order to evenly spread NH_4NO_3 in the N addition plots, we added N together with a small amount of sand (8 kg $plot^{-1}$) and also added sand in control plots. For fertilization we generally chose a date without rain in the two days before or two days after, combined with a local weather forecast. General properties in the Chinese fir forest plantation before the experiment were not significantly different between the experimental units that received the two treatments (Table A1).

In August 2015, eight 1 m × 1 m sample subplots were randomly established in each plot. The understory species in these subplots were measured and recorded to obtain their richness. We used the number of each understory species in subplot as understory richness. Furthermore, the harvested understory plants were divided into three representative species, namely *Dicranopteris dichotoma* (a fern belonging to Gleicheniaceae; a sun plant), *Lophatherum gracile* (a perennial grass belonging to Gramineae; a neutral plant), and *Melastoma dodecandrum* (a creeping small shrub belonging to Melastomataceae; a shade plant) (see Figure A1), and other understory plants. The average proportion of biomass contributed by the three major plants to the total understory vegetation biomass in our study plots was more than 90%. All three species are perennial understory plants. We collected samples of all three understory species in maturation stage. The aboveground tissues (leaves and stems) of these species in each subplot were brought back to the laboratory to measure their dry biomass. We did not harvest the belowground tissue (roots) in order to avoid damage to these permanent study plots.

In addition, we collected samples of aboveground and belowground tissues and from rhizosphere and bulk soils for each of the three understory species within a plot. Soil strongly adhering within 4 mm of roots was considered rhizosphere soil, and the remaining soil was considered bulk soil [23]. Rhizosphere soil samples were collected for each species by separating soil from roots through hand shaking, while the bulk soil samples were collected using a soil auger under/near the plant crown. Each soil sample was divided into two replicates: one used for available nutrient measurement within five days and another stored in a refrigerator at 4 °C and then used for determination of soil enzyme activities. All the plant samples were immediately microwaved for 90 s to stop all enzymatic activity [24], washed with distilled water and oven-dried to a constant mass at 60 °C. After being finely ground using a mixer mill and through a sieve (<0.2 mm), these plant samples were used for determining nutrient and NSC concentrations.

2.3. Soil Nutrient and Enzyme Measurement

Soil NH_4^+-N and NO_3^--N were extracted with 2 M KCl for 30 min and then measured by spectrophotometry using the indophenol blue and cadmium reduction methods, respectively. Soil available P was extracted with 0.5 M $NaHCO_3$ for 30 min and determined using the molybdenum-antimony colorimetric method [25]. The activities of N-acetyl-β-D-glucosaminidase (NAG; EC 3.2.1.14) and acid phosphatase (AP; EC 3.1.3.2) in soils were determined by the fluorogenic microplate method [26].

2.4. Understory Plant Nutrient Measurement

Plant N and P concentrations were determined by the Kjeldahl method and the molybdenum blue spectrophotometric procedure, respectively, after the samples were digested with H_2SO_4 [25]. NSCs were measured by the anthrone colorimetry method [27]. Briefly, the powdered plant sample (0.5 g) was put into a 15 mL centrifuge tube, where 10 mL of 80% alcohol was added. The mixture was incubated in a 100 °C water bath for 20 min and then centrifuged at 4000 rpm for 10 min. The supernatants were retained for SS determination, and the residue was extracted two more times as described above. ST was extracted from the ethanol-insoluble pellet until ethanol was first removed by evaporation. The residue remaining after the SS extraction was extracted with 5 mL 1 M H_2SO_4, and the mixture was shaken for 15 min. The mixture was incubated in an 80 °C water bath for 40 min and then centrifuged at 4000 rpm for 10 min. The pellets were extracted two more times with 1 M H_2SO_4. SS and ST determinations were performed based on absorbance at 625 nm using the same anthrone reagent in a spectrophotometer [28,29]. NSC concentration was obtained by the sum of the total SS and ST.

2.5. Overstory Tree Growth and Litterfall Production Measurement

In each plot, 20 trees have been randomly selected to measure the girth growth rate at breast height using a self-made dendrometer (including a sheet steel, a wire spring two steel nails and a digital caliper) [3] in June and December since the establishment of the experiment. Meanwhile, five 75 cm × 75 cm litter traps were uniformly distributed under the stand canopies to measure litterfall biomass from November 2015 to April 2016. The growth rate and litterfall productivity of Chinese fir were calculated at a plot level in this study in order to assess the potential effect of overstory trees on light conditions of understory plants.

2.6. Data Analysis

The data were tested for homogeneity of variances (Brown and Forsythe's variation of Levene's test) before statistical analysis. A paired t-test was used to compare the differences in soil and plant variables between the control and N addition treatment. Multi-way analysis of variance (ANOVA) was used to determine the interactive effects of N addition, plant species and plant tissue on plant nutrient and NSC concentrations. Pearson's correlation analysis was performed to determine the relationship among soil available nutrients, plant nutrients and NSC parameters. All ANOVA and correlation analyses were conducted with a significance criterion of $p < 0.05$ using IBM SPSS 19 statistical software (SPSS, Chicago, IL, USA).

3. Results

3.1. Plant Growth and Soil Nutrients

After 4 years of N addition, the aboveground biomass of *D. dichotoma* and *L. gracile* significantly decreased by 82.1% and 67.2%, respectively ($p < 0.05$), while the biomass of *M. dodecandrum*, which was much lower than that of the other two species, did not respond to N addition (Table 1). Moreover, N addition did not alter the richness of understory plants (Table 1). In contrast, the average girth growth rates of overstory trees (Chinese fir) within four years after N addition treatment and litterfall productivity in the fifth year significantly increased by 18.28% and 36.71%, respectively (Figure 1, Table A2).

Table 1. The aboveground biomass and richness of the major understory plants in the Chinese fir forest plantation treated by nitrogen addition or in the control in 2015.

Variables	Control	Nitrogen Addition	*T*-test
Dicranopteris dichotoma			
Biomass (kg ha^{-1})	652.51 ± 83.71	117.37 ± 38.49	$p < 0.05$
Richness	4.00 ± 1.01	2.25 ± 0.63	ns
Lophatherum gracile			
Biomass (kg ha^{-1})	205.73 ± 16.15	67.84 ± 33.96	$p < 0.05$
Richness	7.67 ± 1.15	4.00 ± 2.45	ns
Melastoma dodecandrum			
Biomass (kg ha^{-1})	34.33 ± 16.40	14.80 ± 8.22	ns
Richness	3.57 ± 0.81	3.67 ± 1.20	ns

The values are the means ± SE (n = 4). ns = not significant at $p > 0.05$ level.

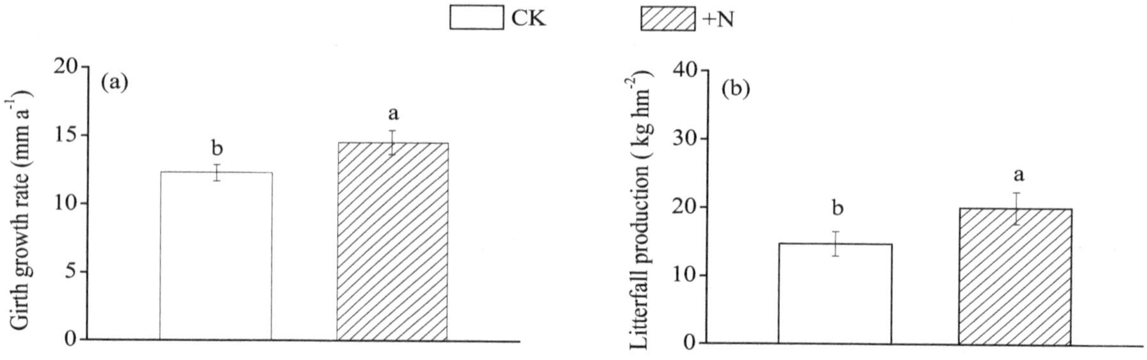

Figure 1. Girth growth from 2011 to 2015 (**a**) and litterfall production in 2016 (**b**) of the Chinese fir tree plantation treated by nitrogen addition (+N) or in the control (CK). Note: Values shown are the means ± SE (n = 4). Lowercase letters indicate significant differences at the $p < 0.05$ level between the control and N-treatment plots.

As expected, the rhizosphere soil NH_4^+-N and NO_3^--N concentrations of all three species were generally higher in the N addition treatment than in the control, while the available P concentration and AP and NAG activities were unaffected by N addition in both rhizosphere and bulk soils (except for NAG activity, which significantly increased in the rhizosphere soil of *L. gracile* in response to the N addition treatment) (Table 2).

Table 2. Available nutrients and enzyme activities in rhizosphere and bulk soils of the three understory plant species in the Chinese fir plantation treated by nitrogen addition or in the control.

Variables	Control	Nitrogen Addition	*T*-test
	Rhizosphere Soil		
	Dicranopteris dichotoma		
NH_4^+-N (mg kg^{-1})	16.03 ± 0.52	34.22 ± 0.89	$p < 0.05$
NO_3^--N (mg kg^{-1})	1.68 ± 0.11	2.93 ± 0.35	$p < 0.05$
Available P (mg kg^{-1})	3.81 ± 0.19	3.57 ± 0.57	ns
NAG activity (mmol g^{-1} h^{-1})	30.5 ± 8.47	14.3 ± 3.45	ns
AP activity (mmol g^{-1} h^{-1})	358.3 ± 57.33	419.8 ± 39.29	ns

Table 2. *Cont.*

Variables	Control	Nitrogen Addition	*T*-test
	Lophatherum gracile		
NH_4^+-N (mg kg^{-1})	9.95 ± 0.66	19.56 ± 0.23	$p < 0.05$
NO_3^--N (mg kg^{-1})	1.85 ± 0.31	3.04 ± 0.85	ns
Available P (mg kg^{-1})	3.82 ± 0.36	3.40 ± 0.15	ns
NAG activity (mmol g^{-1} h^{-1})	3 ± 2.38	20.8 ± 5.15	$p < 0.05$
AP activity (mmol g^{-1} h^{-1})	231.5 ± 52.34	303.5 ± 32.75	ns
	Melastoma dodecandrum		
NH_4^+-N (mg kg^{-1})	7.83 ± 0.71	12.49 ± 0.74	$p < 0.05$
NO_3^--N (mg kg^{-1})	1.16 ± 0.06	2.21 ± 0.37	$p < 0.05$
Available P (mg kg^{-1})	3.02 ± 0.42	3.77 ± 0.27	ns
NAG activity (mmol g^{-1} h^{-1})	22.5 ± 4.575	28.5 ± 7.89	ns
AP activity (mmol g^{-1} h^{-1})	272 ± 15.44	356.3 ± 56.79	ns
	Bulk Soil		
	Dicranopteris dichotoma		
NH_4^+-N (mg kg^{-1})	12.68 ± 3.09	31.99 ± 0.76	$p < 0.05$
NO_3^--N (mg kg^{-1})	1.55 ± 0.17	2.74 ± 0.84	ns
Available P (mg kg^{-1})	3.16 ± 0.14	3.46 ± 0.36	ns
NAG activity (mmol g^{-1} h^{-1})	26.01 ± 12.45	33.81 ± 3.04	ns
AP activity (mmol g^{-1} h^{-1})	344.82 ± 88.11	372.31 ± 29.19	ns
	Lophatherum gracile		
NH_4^+-N (mg kg^{-1})	9.66 ± 0.81	18.66 ± 1.20	$p < 0.05$
NO_3^--N (mg kg^{-1})	1.39 ± 0.09	3.23 ± 0.26	$p < 0.05$
Available P (mg kg^{-1})	4.06 ± 0.45	3.06 ± 0.41	ns
NAG activity (mmol g^{-1} h^{-1})	12.80 ± 7.98	19.55 ± 6.88	ns
AP activity (mmol g^{-1} h^{-1})	237.8 ± 21.17	317.3 ± 20.07	ns
	Melastoma dodecandrum		
NH_4^+-N (mg kg^{-1})	8.18 ± 1.47	14.88 ± 0.52	$p < 0.05$
NO_3^--N (mg kg^{-1})	0.92 ± 0.08	2.48 ± 0.37	$p < 0.05$
Available P (mg kg^{-1})	3.77 ± 0.27	3.54 ± 0.24	ns
NAG activity (mmol g^{-1} h^{-1})	13.34 ± 7.78	29.51 ± 10.51	ns
AP activity (mmol g^{-1} h^{-1})	361.3 ± 36.82	314.3 ± 19.26	ns

The values are the means ± SE ($n = 4$). Available P = Available phosphorus, NAG = N-acetyl-β-D-glucosaminidase, AP = acid phosphatase; ns = not significant at $p > 0.05$ level.

3.2. Nutrients and NSCs in Plant Tissues

Nitrogen addition, plant species, and plant tissue had significant interactive effects on the nutrients and NSC concentrations of the three understory plants (Figure 2). The average tissue N concentration increased, P and ST concentrations did not change, and average SS and NSC concentrations decreased in response to the N addition treatment (Figure 2). When analyzed by species and tissue type, N addition significantly increased the N concentration in both tissue types of all three species (Figure 2a); significantly increased the P concentration in the aboveground tissue of *M. dodecandrum* (Figure 2b); significantly decreased the ST concentration in the aboveground tissue of *M. dodecandrum* and belowground tissue of *D. dichotoma* (Figure 2d); significantly decreased the SS in the aboveground and belowground tissues of *M. dodecandrum* and the aboveground tissue of *D. dichotoma* (Figure 2c); and significantly decreased the NSC concentrations of *M. dodecandrum* while significantly increasing the NSC concentration in the aboveground tissue of *L. gracile* (Figure 2e).

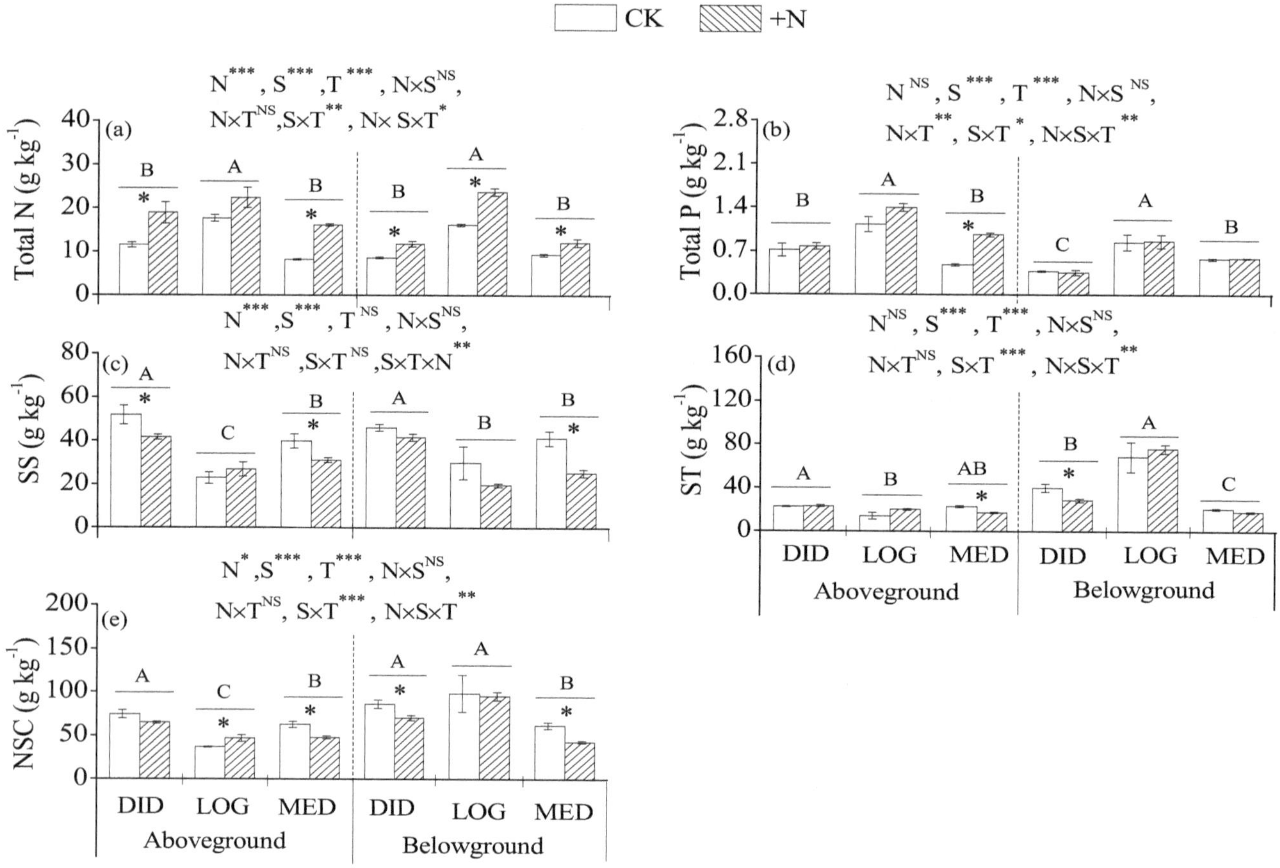

Figure 2. Total nitrogen (**a**), total phosphorus (**b**), soluble sugar (**c**), starch (**d**) and nonstructural ccarbohydrates (**e**) concentrations in aboveground and belowground tissues of the three understory plant species in the Chinese fir plantation treated by nitrogen addition (+N) or in the control (CK). Note: Values shown are the means ± SE (n = 4). The asterisks (*) indicate significant differences at the $p < 0.05$ level between the control and N-treatment plots within a single species. Different capital letters indicate significant differences ($p < 0.05$) among the three species within the same tissue based on Duncan's multiple range test. Total N = total nitrogen, Total P = total phosphorus, SS = soluble sugar, ST = starch, NSC = nonstructural carbohydrates; N = nitrogen addition, S = species, T = tissues. ^{NS}, not significant; * $p < 0.05$; ** $p < 0.01$; *** $p < 0.001$. DID: *D. dichotoma*, LOG: *L. gracile*, MED: *M. dodecandrum*.

3.3. The Ratios of N, P, and NSCs in Plant Tissues

Similarly, N addition, plant species, and plant tissue had significant interactive effects on the ratios of N, P, and the NSCs of the major understory plants (Figure 3). Nitrogen addition did not affect the N/P ratio in the aboveground tissue of any species, but it significantly increased the belowground N/P ratio in all plant tissues (Figure 3a). For *M. dodecandrum*, both the NSC/N and NSC/P ratios significantly decreased in the aboveground and belowground tissues due to N addition; for *D. dichotoma*, only the NSC/N ratio in the aboveground tissue significantly deceased with N addition; and for *L. gracile*, both the NSC/N and NSC/P ratios were unaffected by N addition (Figure 3b,c).

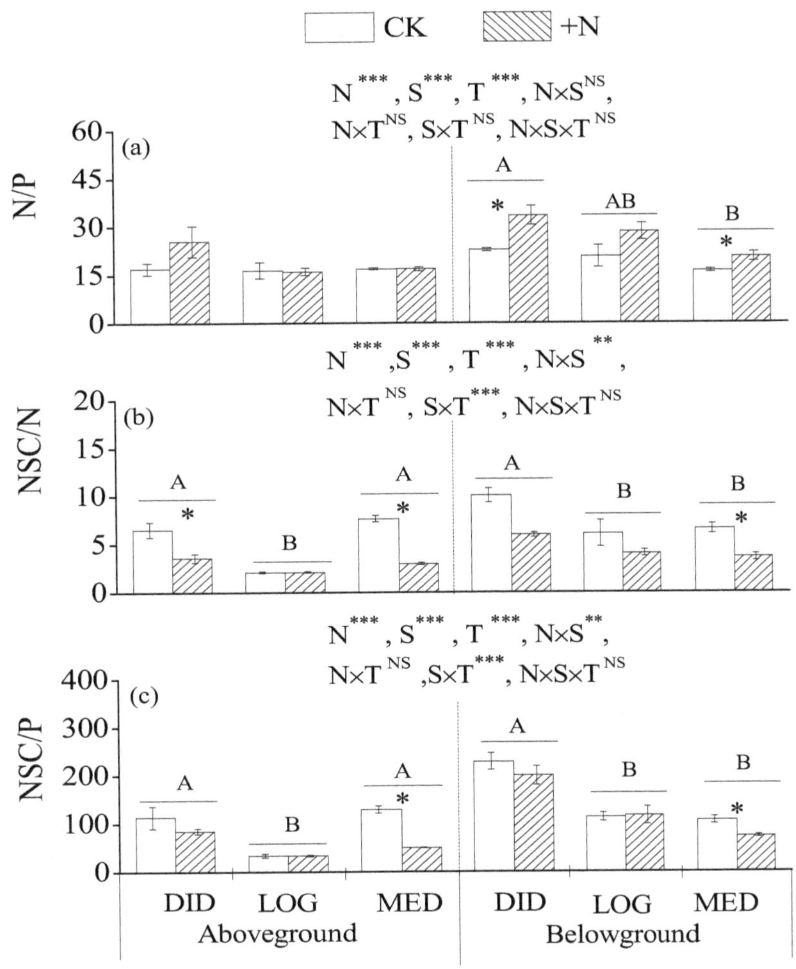

Figure 3. N/P (**a**), NSC/N (**b**), NSC/P (**c**) in aboveground and belowground tissues of the three understory plants in the Chinese fir forest plantation treated by nitrogen addition (+N) or in the control (CK). Note: Values shown are the means ± SE ($n = 4$). The asterisks (*) indicate significant differences at the $p < 0.05$ level between the control and N-treatment plots within a single species. Different capital letters indicate significant differences ($p < 0.05$) among the three species within the same tissue based on Duncan's multiple range test. N = nitrogen, P = phosphorus, NSC = nonstructural carbohydrates; N = nitrogen addition, S = species, T = tissues. ^{NS}, not significant; * $p < 0.05$; ** $p < 0.01$; *** $p < 0.001$. DID: *D. dichotoma*, LOG: *L. gracile*, MED: *M. dodecandrum*.

3.4. Linkages between Plant Tissue Acquirement and Rhizosphere Soil Supply

The tissue N concentration was generally and positively correlated with rhizosphere soil NH_4^+-N and/or NO_3^--N for each of the three understory plants (Table 3). The belowground tissue P concentration was significantly and positively correlated with rhizosphere soil available P for *D. dichotoma* and *L. gracile* but not for *M. dodecandrum*, while the aboveground tissue P concentration was significantly and positively correlated with rhizosphere soil NO_3^--N for *L. gracile* and *M. dodecandrum* but not for *D. dichotoma* (Table 3). The belowground tissue NSC concentration was negatively correlated with rhizosphere soil NH_4^+-N for *M. dodecandrum* and *D. dichotoma* and positively correlated with rhizosphere soil available P for *L. gracile*, while the aboveground tissue NSC concentration was positively correlated with rhizosphere soil NH_4^+-N and NO_3^--N for *L. gracile* and negatively correlated with rhizosphere soil NH_4^+-N for *M. dodecandrum* (Table 3). Likewise, the correlations among the ratios of N/P, NSC/N, NSC/P and soil available nutrients varied with plant species and tissues (Table 3).

Table 3. The correlation coefficients ($n = 8$) between plant tissue and rhizosphere soil nutrients for each of the three understory plant species in the Chinese fir plantation.

Plant Variables	N	P	NSCs	N/P	NSCs/N	NSCs/P
			Belowground Tissue			
Dicranopteris dichotoma						
NH_4^+-N	0.88 **	−0.15 ns	−0.74 *	0.81 *	−0.89 **	−0.41 ns
NO_3^--N	0.93 **	0.18 ns	−0.53 ns	0.61 ns	−0.81 *	−0.53 ns
Available P	0.29 ns	0.93 **	0.22 ns	−0.53 ns	−0.04 ns	−0.66 ns
Lophatherum gracile						
NH_4^+-N	0.95 **	0.04 ns	−0.04 ns	0.60 ns	−0.50 ns	0.10 ns
NO_3^--N	0.64 ns	0.77 *	0.16 ns	−0.30 ns	−0.12 ns	−0.54 ns
Available P	−0.33 ns	0.76 *	0.73 *	0.81 *	0.01 ns	−0.84 **
Melastoma dodecandrum						
NH_4^+-N	0.68 ns	0.66 ns	−0.79 *	0.60 ns	−0.78 *	−0.83 *
NO_3^--N	0.25 ns	0.11 ns	−0.62 ns	0.25 ns	−0.55 ns	−0.59 ns
Available P	0.06 ns	0.15 ns	−0.16 ns	−0.12 ns	−0.17 ns	0.03 ns
			Aboveground Tissue			
Dicranopteris dichotoma						
NH_4^+-N	0.75 *	0.16 ns	−0.57 ns	0.57 ns	−0.76 *	−0.39 ns
NO_3^--N	0.90 **	0.16 ns	−0.60 ns	0.71 *	−0.83 *	−0.42 ns
Available P	0.51 ns	−0.16 ns	−0.15 ns	−0.32 ns	−0.02 ns	0.56 ns
Lophatherum gracile						
NH_4^+-N	0.59 ns	0.61 ns	0.71 *	−0.10 ns	0.11 ns	−0.06 ns
NO_3^--N	0.84 **	0.75 *	0.86 **	0.03 ns	−0.31 ns	−0.06 ns
Available P	−0.20 ns	0.37 ns	−0.14 ns	0.33 ns	−0.55 ns	−0.53 ns
Melastoma dodecandrum						
NH_4^+-N	0.85 **	0.89 **	−0.88 **	−0.30 ns	−0.92 **	−0.94 **
NO_3^--N	0.77 *	0.83 *	−0.45 ns	−0.38 ns	−0.70 ns	−0.72 *
Available P	0.52 ns	0.51 ns	−0.64 ns	−0.58 ns	−0.56 ns	0.10 ns

NSCs = nonstructural carbohydrates; ns, not significant, $p > 0.05$; * $p < 0.05$; ** $p < 0.01$.

4. Discussion

4.1. Links between Plant Tissue Acquirement and Rhizosphere Soil Supply

Similar to the results in many previous studies [3,30], soil available N, including NH_4^+-N and NO_3^--N, increased with N addition in the form of NH_4NO_3. Consequently, in our study, the N concentrations in aboveground and belowground tissues were found to be higher in response to N addition for the three studied species. This result was consistent with those from previous studies in forest ecosystems [31–33]. Gurmesa found that plants could take up excessive N even in a N-saturated forest [34], which supported our results, as our plantation is located in a N-rich ecosystem within a region with extreme N deposition [35–37]. Thus, the higher soil N supply increased the N concentrations in aboveground and belowground tissues in response to N addition.

Surprisingly, the available P concentration and activities of both enzymes (NAG and AP) in rhizosphere and bulk soils of the three species generally did not change with N addition. Dong et al. (2015) also found the soil total P did not alter by N addition, while organic matter and total N increased due to three years' N addition of 100 kg hm^{-2} $year^{-1}$ in the same site of this study [26]. These results indicated that the intensity and duration of the simulated N addition might still be within the buffer range of the soil P supply in the studied plantation [3]. Thus, unsurprisingly, the P concentration in plant tissues was generally unaffected by N addition in our study. However, the P concentration in aboveground tissue of *M. dodecandrum* did increase with N addition. In general, the effect of N addition on plant P may be altered via P supply, P uptake and P resorption [38–40]. In our study, soil available P and belowground tissue P showed minimal responses to N addition. Some studies suggested that N addition decreased P uptake by reducing root biomass and inhibiting mycorrhiza growth [10,41]. Other studies found that N addition decreased the extraradical hyphae of Chinese fir mycorrhiza, but did not alter fine root biomass in a subtropical plantation forest [42]. Thus, we speculate that P resorption might be a potential mechanism increasing aboveground P concentration in *M. dodecandrum*. Compared with the other two species (a sun fern and a neutral grass), *M. dodecandrum* is a creeping shrub with traits characteristic of shade plants and is thus better suited to growing in an understory environment (Figure A1). To meet the requirements for growth, *M. dodecandrum* (the only studied species that did not exhibit decreased biomass with N addition) might improve P resorption to overcome the P insufficiency in soils and increase its aboveground P concentration to maintain N/P stoichiometric homeostasis.

As expected, the ratio of mineral N to available P in soils and the N/P ratio in belowground tissues generally increased with N addition, which partially supported our first hypothesis that N addition alters the balance between N and P in soils and understory plants. However, the N/P ratio in the aboveground tissue was not affected by N addition in any species, which seems counter to our first hypothesis of an imbalance between N and P in understory plants driven by N addition. The divergent patterns of the N/P ratio between aboveground and belowground tissues indicated that the understory plants might maintain N/P homeostasis in aboveground tissue to meet the requirement for leaf photosynthesis and plant growth [13,43], regardless of the alteration of the N/P ratio in plant belowground tissues due to N addition.

To achieve N/P homeostasis in aboveground tissue, the three understory species might use different mechanisms, such as the resorption in *M. dodecandrum* mentioned above, which should be further studied. The N/P ratio in plant tissue has been widely used as a diagnostic tool for evaluating nutrient limitation in terrestrial ecosystems [44,45]. Previous studies observed that the N/P ratio in tree leaves decreased with N deposition, and this indicated that N deposition might aggravate P limitation of plant growth in forest ecosystems [46]. In our study, the N/P ratio of aboveground tissue was generally less than 16, except for in *D. dichotoma* treated by N addition, and the N/P ratio in aboveground tissue was unchanged by N addition. These results suggested that N addition might not aggravate the P limitation of understory plant growth in this subtropical plantation.

4.2. Effects of N Addition on Understory Plant Growth and NSC Allocation

Understory plant growth might be altered by N addition through a direct effect of nutrient supply and an indirect effect of the overstory canopy via competition for resources such as light and water [3,14,31]. In our study, the biomass of two dominant understory species (*D. dichotoma* and *L. gracile*) significantly decreased with N addition. A recent study found that a moderate supply of N stimulated the understory vegetation growth in a tropical forest because N inputs satisfied plant demands for N [14]. However, some research suggested that understory vegetation productivity was mainly limited by light in a boreal forest, and N addition increased shading by the tree canopy (thus, less light was available to understory plants) [12]. Some other studies also found that plant growth was inhibited by N addition due to the aggravation of P limitation, with a mismatch in N and P stoichiometry [47,48].

Based on the responses of soil and plant nutrients to N addition discussed above, N addition likely did not aggravate the P limitation of understory plant growth in this subtropical plantation. Therefore, the observed decline in understory plant biomass may be most likely caused by increased shading by stimulated overstory tree growth due to N addition. Compared with overstory trees, understory plants are more easily limited by light [10]. Our data also indicated that N addition promoted the growth and litter production of Chinese fir (the canopy tree) (Figure 1) and thus could lead to a decrease in the light available to understory vegetation. The decrease in light availability might help explain the negative effect on understory vegetation growth in the N addition plots.

The NSCs in plant tissue are the products of photosynthesis and the main energy sources for plant growth [29]. Because NSCs can reflect plant response to light and nutrient availability [49,50], they may provide an effective way to reveal the underlying mechanism of N addition effects on understory plants driven mainly by light or nutrient resources. The storage of NSCs is lower in shade than in sun environments because carbohydrate synthesis is often limited by lower light availability [29]. In our study, NSCs, especially SS, concentrations in understory plants significantly decreased with N addition. However, previous studies found that carbohydrate reserves increased with elevated N supply when N inputs satisfied plant demand for N and increased the photosynthetic capacity [16,17]. There are several potential reasons for the reduction in NSC concentration of understory plants in response to N addition in our study. First, N addition may have increased the plant growth and decreased the accumulation of NSCs because the exogenous N supply stimulated the synthesis of amino acids and amide compounds to suppress the accumulation of carbohydrates for protein synthesis when N was in excess [49]. Second, an excessive concentration of N in aboveground tissue may have resulted in inorganic N toxicity, which may have downregulated the photosynthetic capacity [14]. Third, the light available to the understory plants may have significantly decreased because the tree canopy cover increased under the elevated N supply [51,52]. Our study found that the soil available N was strongly correlated with the N concentration in the tissues of the three understory species, which indicated that the deposited N might not have been excessive or toxic. We also found negative correlations between soil NH_4^+-N and tissue NSCs in *D. dichotoma* and *M. dodecandrum*, but both soil NH_4^+-N and NO_3^--N were positively correlated with aboveground tissue NSCs in *L. gracile*. Therefore, our results further indicated that the growth of understory species might be limited by light availability but not dominated by N (excess or toxicity).

In addition, the ratio of NSCs to nutrients reflects the relationship between the nutrients and the production of NSCs and their use efficiencies [53]. In our study, the ratios of NSCs to nutrients (NSC/N and NSC/P) in plant tissues generally decreased with N addition, implying that each unit of N and each unit of P produced fewer NSCs. The decrease in the NSC/N and NSC/P ratios in tissues also suggested that light availability may limit the photosynthetic rate and growth of understory plants in response to N addition [14,31].

Our results revealed that the three understory species were not consistent in their response to N addition, although they grow well in acid soils. *D. dichotoma* was a high light-demanding species, and *M. dodecandrum* was a shade tolerance species, but *L. gracile* was a duality species that can survive in low light and high light environment. Nitrogen addition has a direct influence on overstory vegetation that changes the light available reaching the understory. It is the reason that explains the different responses to N addition among the three understory species. First, decreased in the biomass of high light-demanding *D dichotoma* due to N addition is a consequence of lower light availability induced by the promoted growth of the Chinese fir tree canopy. This result is consistent with that of other research in that N fertilization has a significant influence on forest tree canopies, which can significantly reduce the light available to the understory plants [51,52]. Second, N addition promoted the growth of the Chinese fir tree canopy, leading to *L. gracile* being light limited as well as increasing the belowground nutrient accumulation and microbial activity of *L. gracile*. Compared with the sun and neutral plants (*D. dichotoma* and *L. gracile*), the shade plant (*M. dodecandrum*) showed a stronger capacity to synchronously increase N and P levels and decrease SS and ST pools in its tissues to maintain productivity and adapt to the shadier environment in the N addition treatment. In contrast, P and NSCs in the sun and neutral plants showed a weaker response to the N addition treatment, which led to a decrease in plant biomass with N addition. Furthermore, the correlations among soil-available nutrients, tissue nutrients, and NSCs differed among species, which further indicated species-specific mechanisms in response to N addition due to the differences in nutrient- and light-related traits among the three understory species.

5. Conclusions

Nitrogen addition did not lead to a mismatch in N and P stoichiometry in the aboveground tissues of understory plants, although N addition increased N availability in soils and plants. However, N addition decreased NSCs, especially SS, in two of the three studied species and the NSC/N and NSC/P ratios of all three species. Meanwhile, the aboveground biomass of *D. dichotoma* and *L. gracile* significantly decreased after four years of simulated N addition, and the biomass of *M. dodecandrum* did not respond to N addition. These results suggested that N addition might inhibit the growth of understory plants through decreasing the nonstructural carbohydrates and light availability indirectly rather than by changing nutrients and N/P stoichiometry directly, although species-specific responses to N addition occurred in the Chinese fir plantation. The limitation of available light to understory species through the facilitation of N addition on overstory canopy growth may be the underlying mechanism, and thinning should therefore be used to improve understory vegetation biomass and other potential functions to mitigate the adverse effects caused by N deposition on understory plant species.

Author Contributions: F.C. was responsible for funding acquisition and resources. F.C. and G.G.W. conceptualized the study. F.W. performed the data curation and investigation. F.C. and H.W. participated in the design of the study. X.F. and W.B. supervised the experimental process. F.W. wrote the original draft. R.M. reviewed and edited the manuscript. All authors read and approved the final manuscript.

Acknowledgments: We thank Xiu-Lan Zhang, Zhang-Min Li and Gao-Yang Wu for their field work and sample analysis.

Appendix A

Table A1. General properties of the Chinese fir forest plantation before experimental treatment in 2011.

Variables	Control	Nitrogen Addition	*T*-test
Soil			
Bulk density (g cm^{-3})	1.22 ± 0.03	1.25 ± 0.03	ns
pH	4.43 ± 0.04	4.40 ± 0.08	ns
Organic carbon (g kg^{-1})	21.20 ± 2.2	22.41 ± 1.62	ns
Total nitrogen (g kg^{-1})	1.29 ± 0.13	1.26 ± 0.09	ns
Total phosphorus (g kg^{-1})	0.29 ± 0.02	0.31 ± 0.03	ns
Stand			
Density (ha^{-1})	2250 ± 45	2150 ± 62	ns
Average DBH (cm)	12.2 ± 0.2	11.9 ± 0.3	ns
Average height (m)	8.5 ± 0.2	8.7 ± 0.2	ns

Data are cited from our previous study [3]; ns indicates not significant between the control and the nitrogen addition treatment at the $p < 0.05$ level. Values are the means ± SE ($n = 4$). DBH = diameter at breast height.

Table A2. Girth growth rates at breast height in the Chinese fir forest plantation in the nitrogen addition treatment and control from 2012 to 2015.

Variables	Control	Nitrogen Addition	*T*-test
Girth growth rate			
1-year (mm $year^{-1}$)	10.89 ± 0.69	13.69 ± 1.14	$p < 0.05$
2-year (mm $year^{-1}$)	15.13 ± 0.70	18.64 ± 1.18	$p < 0.05$
3-year (mm $year^{-1}$)	9.50 ± 0.57	10.66 ± 0.60	ns
4-year (mm $year^{-1}$)	13.72 ± 1.15	15.23 ± 1.14	ns
Average (mm $year^{-1}$)	12.31 ± 0.61	14.56 ± 0.88	$p < 0.05$

ns indicates not significant between the control and the nitrogen addition treatment at the $p < 0.05$ level. Values are the means ± SE ($n = 4$).

(**a**)

(**b**)

(**c**)

Figure A1. The morphological characteristics of the three species of understory plants in our study. (**a**) *Dicranopteris dichotoma* [54] (**b**) *Lophatherum gracile* [55] (**c**) *Melastoma dodecandrum* [56].

References

1. Galloway, J.N.; Townsend, A.R.; Erisman, J.W.; Bekunda, M.; Cai, Z.; Freney, J.R.; Sutton, M.A. Transformation of the nitrogen cycle, recent trends, questions, and potential solutions. *Science* **2008**, *320*, 889–892. [CrossRef] [PubMed]
2. Liu, X.; Zhang, Y.; Han, W.; Tang, A.; Shen, J.; Cui, Z.; Fangmeier, A. Enhanced nitrogen deposition over China. *Nature* **2013**, *494*, 459–462. [CrossRef] [PubMed]
3. Chen, F.S.; Niklas, K.J.; Liu, Y.; Fang, X.M.; Wan, S.Z.; Wang, H. Nitrogen and phosphorus additions alter nutrient dynamics but not resorption efficiencies of Chinese fir leaves and twigs differing in age. *Tree Physiol.* **2015**, *35*, 1106–1117. [CrossRef] [PubMed]
4. Lu, X.; Mao, Q.; Gilliam, F.S.; Luo, Y.; Mo, J. Nitrogen deposition contributes to soil acidification in tropical ecosystems. *Glob. Chang. Biol.* **2014**, *20*, 3790–3801. [CrossRef] [PubMed]

5. Wang, M.; Zhang, W.W.; Li, N.; Liu, Y.Y.; Zheng, X.B.; Hao, G.Y. Photosynthesis and growth responses of *Fraxinus mandshurica* Rupr. seedlings to a gradient of simulated nitrogen deposition. *Ann. For. Sci.* **2018**, *75*, 1. [CrossRef]
6. Kumar, P.; Chen, H.Y.; Thomas, S.C.; Shahi, C. Linking resource availability and heterogeneity to understory species diversity through succession in boreal forest of Canada. *J. Ecol.* **2018**, *106*, 1266–1276. [CrossRef]
7. Bobbink, R.; Hicks, K.; Galloway, J.; Spranger, T.; Alkemade, R.; Ashmore, M.; Emmett, B. Global assessment of nitrogen deposition effects on terrestrial plant diversity: A synthesis. *Ecol. Appl.* **2010**, *20*, 30–59. [CrossRef]
8. Gilliam, F.S.; Hockenberry, A.W.; Adams, M.B. Effects of atmospheric nitrogen deposition on the herbaceous layer of a central Appalachian hardwood forest. *J. Torry. Bot. Soc.* **2006**, *133*, 240–254. [CrossRef]
9. Lu, X.; Mo, J.; Gilliam, F.S.; Zhou, G.; Fang, Y. Effects of experimental nitrogen additions on plant diversity in an old-growth tropical forest. *Glob. Chang. Biol.* **2010**, *16*, 2688–2700. [CrossRef]
10. Gilliam, F.S. Response of the herbaceous layer of forest ecosystems to excess nitrogen deposition. *J. Ecol.* **2006**, *94*, 1176–1191. [CrossRef]
11. Hurd, T.M.; Brach, A.R.; Raynal, D.J. Response of understory vegetation of Adirondack forests to nitrogen additions. *Can. J. For. Res.* **1998**, *28*, 799–807. [CrossRef]
12. Strengbom, J.; Näsholm, T.; Ericson, L. Light, not nitrogen, limits growth of the grass *Deschampsia flexuosa* in boreal forests. *Can. J. Bot.* **2004**, *82*, 430–435. [CrossRef]
13. Kou, L.; Chen, W.; Jiang, L.; Dai, X.; Fu, X.; Wang, H.; Li, S. Simulated nitrogen deposition affects stoichiometry of multiple elements in resource-acquiring plant organs in a seasonally dry subtropical forest. *Sci. Total. Environ.* **2018**, *624*, 611–620. [CrossRef] [PubMed]
14. Mao, Q.; Lu, X.; Mo, H.; Gundersen, P.; Mo, J. Effects of simulated N deposition on foliar nutrient status, N metabolism and photosynthetic capacity of three dominant understory plant species in a mature tropical forest. *Sci. Total. Environ.* **2018**, *610*, 555–562. [CrossRef] [PubMed]
15. Elser, J.J.; Bracken, M.E.; Cleland, E.E.; Gruner, D.S.; Harpole, W.S.; Hillebrand, H.; Smith, J.E. Global analysis of nitrogen and phosphorus limitation of primary producers in freshwater, marine and terrestrial ecosystems. *Ecol. Lett.* **2007**, *10*, 1135–1142. [CrossRef] [PubMed]
16. Ibrahim, M.H.; Jaafar, H.Z.; Rahmat, A.; Rahman, Z.A. The relationship between phenolics and flavonoids production with total non-structural carbohydrate and photosynthetic rate in *Labisia pumila* Benth. under high CO_2 and nitrogen fertilization. *Molecules* **2011**, *16*, 162–174. [CrossRef] [PubMed]
17. Xiao, L.; Liu, G.; Li, P.; Xue, S. Nitrogen addition has a stronger effect on stoichiometries of non-structural carbohydrates, nitrogen and phosphorus in *Bothriochloa ischaemum* than elevated CO_2. *Plant Growth Regul.* **2017**, *83*, 325–334. [CrossRef]
18. Lü, X.T.; Reed, S.; Yu, Q.; He, N.P.; Wang, Z.W.; Han, X.G. Convergent responses of nitrogen and phosphorus resorption to nitrogen inputs in a semiarid grassland. *Glob. Chang. Biol.* **2013**, *19*, 2775–2784. [CrossRef]
19. Knox, K.J.E.; Clarke, P.J. Nutrient availability induces contrasting allocation and starch formation in resprouting and obligate seeding shrubs. *Funct. Ecol.* **2005**, *19*, 690–698. [CrossRef]
20. Xiong, Y.; Xu, X.; Zeng, H.; Wang, H.; Chen, F.; Guo, D. Low Nitrogen Retention in Soil and Litter under Conditions without Plants in a Subtropical Pine Plantation. *Forests* **2015**, *6*, 2387–2404. [CrossRef]
21. Kou, L.; Zhang, X.; Wang, H.; Yang, H.; Zhao, W.; Li, S. Nitrogen additions inhibit nitrification in acidic soils in a subtropical pine plantation: Effects of soil pH and compositional shifts in microbial groups. *J. For. Res.* **2017**, *29*, 1–10. [CrossRef]
22. Lü, C.; Tian, H. Spatial and temporal patterns of nitrogen deposition in China: Synthesis of observational data. *J. Geophys. Res. Atmos.* **2007**, *112*, 10–15. [CrossRef]
23. Hu, X.F.; Chen, F.S.; Wine, M.L.; Fang, X.M. Increasing acidity of rain in subtropical tea plantation alters aluminum and nutrient distributions at the root-soil interface and in plant tissues. *Plant Soil* **2017**, *417*, 261–274. [CrossRef]
24. Woodruff, D.R.; Meinzer, F.C. Water stress, shoot growth and storage of non-structural carbohydrates along a tree height gradient in a tall conifer. *Plant Cell Environ.* **2011**, *34*, 1920–1930. [CrossRef] [PubMed]
25. Allen, S.E. *Chemical Analysis of Ecological Materials*; Blackwell Scientific Publications: Oxford, UK, 1989.
26. Dong, W.Y.; Zhang, X.Y.; Liu, X.Y.; Fu, X.L.; Chen, F.S.; Wang, H.M.; Wen, X.F. Responses of soil microbial communities and enzyme activities to nitrogen and phosphorus additions in Chinese fir plantations of subtropical China. *Biogeosciences* **2015**, *12*, 5537–5546. [CrossRef]

27. Dubois, M.; Gilles, K.A.; Hamilton, J.K.; Rebers, P.T.; Smith, F. Colorimetric method for determination of sugars and related substances. *Anal. Chem.* **1956**, *28*, 350–356. [CrossRef]
28. Liu, J.F.; Arend, M.; Yang, W.J.; Schaub, M.; Ni, Y.Y.; Gessler, A.; Li, M.H. Effects of drought on leaf carbon source and growth of European beech are modulated by soil type. *Sci. Rep.* **2017**, *7*, 42462. [CrossRef]
29. Song, X.; Peng, C.; Zhou, G.; Gu, H.; Li, Q.; Zhang, C. Dynamic allocation and transfer of non-structural carbohydrates, a possible mechanism for the explosive growth of Moso bamboo (*Phyllostachys heterocycla*). *Sci. Rep.* **2016**, *6*, 25908. [CrossRef] [PubMed]
30. Zhu, B.; Panke-Buisse, K.; Kao-Kniffin, J. Nitrogen fertilization has minimal influence on rhizosphere effects of smooth crabgrass (Digitaria ischaemum) and bermudagrass (Cynodon dactylon). *J. Plant Ecol.* **2014**, *8*, 390–400. [CrossRef]
31. Mao, Q.; Lu, X.; Wang, C.; Zhou, K.; Mo, J. Responses of understory plant physiological traits to a decade of nitrogen addition in a tropical reforested ecosystem. *For. Ecol. Manag.* **2017**, *401*, 65–74. [CrossRef]
32. Nordin, A.; Näsholm, T.; Ericson, L. Effects of simulated N deposition on understorey vegetation of a boreal coniferous forest. *Funct. Ecol.* **1998**, *12*, 691–699. [CrossRef]
33. Talhelm, A.F.; Pregitzer, K.S.; Burton, A.J. No evidence that chronic nitrogen additions increase photosynthesis in mature sugar maple forests. *Ecol. Appl.* **2011**, *21*, 2413–2424. [CrossRef] [PubMed]
34. Gurmesa, G.A.; Lu, X.; Gundersen, P.; Mao, Q.; Zhou, K.; Fang, Y.; Mo, J. High retention of ^{15}N-labeled nitrogen deposition in a nitrogen saturated old-growth tropical forest. *Glob. Chang. Bio.* **2016**, *22*, 3608–3620. [CrossRef] [PubMed]
35. Jia, Y.; Yu, G.; He, N.; Zhan, X.; Fang, H.; Sheng, W.; Wang, Q. Spatial and decadal variations in inorganic nitrogen wet deposition in China induced by human activity. *Sci. Rep.* **2014**, *4*, 3763. [CrossRef] [PubMed]
36. Ma, Z.; Zhang, X.; Zhang, C.; Wang, H.; Chen, F.; Fu, X.; Lei, Q. Accumulation of residual soil microbial carbon in Chinese fir plantation soils after nitrogen and phosphorus additions. *J. For. Res.* **2018**, *29*, 953–962. [CrossRef]
37. Tang, Y.; Zhang, X.; Li, D.; Wang, H.; Chen, F.; Fu, X.; Yu, G. Impacts of nitrogen and phosphorus additions on the abundance and community structure of ammonia oxidizers and denitrifying bacteria in Chinese fir plantations. *Soil. Biol. Biochem.* **2016**, *103*, 284–293. [CrossRef]
38. Jones, A.G.; Power, S.A. Field-scale evaluation of effects of nitrogen deposition on the functioning of heathland ecosystems. *J. Ecol.* **2012**, *100*, 331–342. [CrossRef]
39. Kou, L.; Jiang, L.; Fu, X.; Dai, X.; Wang, H.; Li, S. Nitrogen deposition increases root production and turnover but slows root decomposition in *Pinus elliottii* plantations. *New. Phytol.* **2018**, *218*, 1450–1461. [CrossRef] [PubMed]
40. Yang, H. Effects of nitrogen and phosphorus addition on leaf nutrient characteristics in a subtropical forest. *Trees* **2018**, *32*, 383–391. [CrossRef]
41. Nakaji, T.; Fukami, M.; Dokiya, Y.; Izuta, T. Effects of high nitrogen load on growth, photosynthesis and nutrient status of *Cryptomeria japonica* and *Pinus densiflora* seedlings. *Trees* **2001**, *15*, 453–461.
42. Schulze, E.D. Air pollution and forest decline in a spruce (*Picea abies*) forest. *Science* **1989**, *244*, 776–783. [CrossRef] [PubMed]
43. Li, L.; McCormack, M.L.; Chen, F.; Wang, H.; Ma, Z.; Guo, D. Different responses of absorptive roots and arbuscular mycorrhizal fungi to fertilization provide diverse nutrient acquisition strategies in Chinese fir. *For. Ecol. Manag* **2019**, *433*, 64–72. [CrossRef]
44. Güsewell, S. N: P ratios in terrestrial plants: Variation and functional significance. *New Phytol.* **2004**, *164*, 243–266. [CrossRef]
45. Kang, H.; Xin, Z.; Berg, B.; Burgess, P.J.; Liu, Q.; Liu, Z.; Liu, C. Global pattern of leaf litter nitrogen and phosphorus in woody plants. *Ann. For. Sci.* **2010**, *67*, 811. [CrossRef]
46. Li, Y.; Niu, S.; Yu, G. Aggravated phosphorus limitation on biomass production under increasing nitrogen loading: A meta-analysis. *Glob. Chang. Biol.* **2016**, *22*, 934–943. [CrossRef] [PubMed]
47. Yuan, Z.Y.; Chen, H.Y. Decoupling of nitrogen and phosphorus in terrestrial plants associated with global changes. *Nat. Clim. Chang.* **2015**, *5*, 465. [CrossRef]
48. Yue, K.; Fornara, D.A.; Yang, W.; Peng, Y.; Li, Z.; Wu, F.; Peng, C. Effects of three global change drivers on terrestrial C: N: P stoichiometry: A global synthesis. *Glob. Chang. Biol.* **2017**, *23*, 2450–2463. [CrossRef]

49. Ai, Z.M.; Xue, S.; Wang, G.L.; Liu, G.B. Responses of non-structural carbohydrates and C: N: P Stoichiometry of *Bothriochloa ischaemum* to nitrogen addition on the Loess Plateau, China. *J. Plant Growth Regul.* **2017**, *36*, 714–722. [CrossRef]
50. Ai, Z.; He, L.; Xin, Q.; Yang, T.; Liu, G.; Xue, S. Slope aspect affects the non-structural carbohydrates and C: N: P stoichiometry of *Artemisia sacrorum* on the Loess Plateau in China. *Catena* **2017**, *152*, 9–17. [CrossRef]
51. Hedwall, P.O.; Nordin, A.; Brunet, J.; Bergh, J. Compositional changes of forest-floor vegetation in young stands of Norway spruce as an effect of repeated fertilisation. *For. Ecol. Manag.* **2010**, *259*, 2418–2425. [CrossRef]
52. Hedwall, P.O.; Strengbom, J.; Nordin, A. Can thinning alleviate negative effects of fertilization on boreal forest floor vegetation? *For. Ecol. Manag.* **2013**, *310*, 382–392. [CrossRef]
53. Li, M.H.; Xiao, W.F.; Wang, S.G.; Cheng, G.W.; Cherubini, P.; Cai, X.H.; Zhu, W.Z. Mobile carbohydrates in Himalayan treeline trees I. Evidence for carbon gain limitation but not for growth limitation. *Tree Physiol.* **2008**, *28*, 1287–1296. [CrossRef] [PubMed]
54. Dicranopteris dichotoma. Available online: http://frps.eflora.cn/frps/Dicranopteris%20dichotoma (accessed on 12 September 2018).
55. Lophatherum gracile. Available online: http://frps.eflora.cn/frps/Lophatherum%20gracile (accessed on 12 September 2018).
56. Melastoma dodecandrum. Available online: http://frps.eflora.cn/frps/Melastoma%20dodecandrum (accessed on 12 September 2018).

8

Sampling Method and Tree-Age Affect Soil Organic C and N Contents in Larch Plantations

Huimei Wang [1], Wenjie Wang [1,2,*] and Scott X. Chang [3]

1 Key Laboratory of Forest Plant Ecology, Northeast Forestry University, Harbin 150040, China; whm0709@hotmail.com
2 Northeast Institute of Geography and Agricultural Ecology, Chinese Academy of Sciences, Changchun 130102, China
3 Department of Renewable Resources, University of Alberta, 442 Earth Science Building, Edmonton, AB T6G 2E3, Canada; scott.chang@ualberta.ca
* Correspondence: wjwang225@hotmail.com

Academic Editor: Timothy A. Martin

Abstract: We currently have a poor understanding of how different soil sampling methods (pedogenetic horizon versus fixed-depth) influence the evaluation of soil properties. Here, 159 soil profiles were sampled from larch (*Larix gmelinii*) plantations in northeast China using both the pedogenetic horizon and fixed-depth sampling methods. Analysis of variance was used to determine how sampling method influences the assessment of the spatial variation in the concentration and storage of soil organic C (SOC) and N (SON), as well as how these properties are affected by tree age-group (<20, 20–40, and >40 years). In both the 20 cm (surface) and 80 cm (whole profile) sampling depths, pedogenetic sampling resulted in 1.2- to 1.4-fold higher SOC and SON concentrations than fixed-depth sampling. Surface soil nutrient storage between the two sampling methods was not significantly different, but was it was 1.2-fold higher ($p < 0.05$) with pedogenetic sampling than with fixed-depth sampling in the whole soil profile. For a given error limit in SOC and SON assessments, fixed-depth sampling had a 60%~90% minimum sampling intensity requirement compared with pedogenetic horizon sampling. Additionally, SOC was 1.1- to 1.3-fold greater in the >40 years age-group than in the <20 years age-group ($p < 0.05$), while SON was the highest in the 20–40 years age-group ($p < 0.05$). The total amount of SOC and nutrients in soil is fixed regardless how you sample, it is the different assumptions and different ways of extrapolation from samples to the population that cause sampling by horizon versus fixed depth to lead to different conclusions. Our findings highlight that soil sampling method and tree age-group affect the determination of the spatial variation of SOC and SON and future soil assessments should control for methodological differences.

Keywords: *Larix gmelinii*; tree-age effect; degraded farmland; Northeast China; soil nutrient variation; soil heterogeneity; pedogenetic horizon; fixed-depth sampling

1. Introduction

Accurate evaluation of soil properties is important for species-to-location matching during afforestation, precision agroforestry [1–3], and the investigation of forest ecological functions such as carbon (C) sequestration and nutrient dynamics [4]. Soil heterogeneity occurs horizontally and vertically [5]; thus, the lack of a standardized sampling procedure to assess soil fertility and C inventory could affect conclusions drawn from different studies. Therefore, research is needed to understand how sampling method influences the evaluation of the inter-site variations (i.e., interaction between sampling method and site variation) of soil nutrients. Accounting for the effect of sampling method could improve comparability among studies [6,7], particularly in terms of between-site soil fertility assessments [8].

Forest development with age can increase C storage and recover nutrient levels in soils degraded by long-term agricultural cultivation [9–12]. Soils can range from being a C sink (11 to 238 g·C·m^{-2}·year^{-1}) [4,13–15] to a source of atmospheric CO_2 (0.1 to 14.1 g·C·m^{-2}·year^{-1}) [11,16]. Additionally, studies examining post-afforestation conditions have reported both the depletion of soil nutrients [4,10,17–19] and the accumulation of soil organic N (SON) with increasing soil organic C (SOC) [9,12,20] over time (e.g., tree age effect). Previous land use, the tree species planted, and soil nutrient supply are thought to explain the contradictions in tree age effects on soil properties after afforestation [9–12,17–20]. Furthermore, forest age itself [21], soil depth, and other site-related variations [22–25] have been used to describe the spatial and temporal changes in nutrient accumulation. However, differences in sampling methods (fixed-depth [4,11,13] versus pedogenetic horizon [26–28] sampling) are a potentially powerful explanation for between-study differences in soil properties [29,30] that has not been fully studied. Indeed, comprehensive research including soil sampling method, sampling site, sampling depth, and tree age would provide a more holistic view of how different sampling methods alter evaluations of SOC and SON.

The objective of this study was to determine how soil sampling method, sampling site, sampling depth, and tree age independently and interactively influence SOC and SON, particularly variation across the soil profile in different soil depths and the effect of tree age. The investigation took advantage of widespread larch plantations (4.5 million hectares, or 70% of the total plantation area) available in northeastern China [11,31]. We hypothesized that soil sampling methods could greatly affect the evaluations of spatial changes of and tree-age effects on SOC and N, independent of soil depth inclusion and unit expression of storage or concentration. Our results may contribute to improving inter-study comparisons of SOC and SON measurements, leading to a broader study applicability and improved interpretation of conclusions drawn across disparate reports.

2. Materials and Methods

2.1. Study Site and Soil Sampling

All sampling was conducted in *Larix gmelinii* plantations located in northeast China. This region (45°20′–47°14′ N, 127°30′–128°55′ E) has a continental monsoon climate with mean annual temperatures ranging from −0.3 to 2.6 °C and mean annual precipitation ranging from 676 to 724 mm. The soil is generally a typical Dark-Brown Forest Soil, with occasional eluviation features. The soil in Chinese Soil Taxonomy is Mollic Bori-udic Cambosols or Albic Bori-udic Cambosols [32]; it is a Eutroboralf in the US Soil Taxonomy and a Halic Luvisol in the FAO (Food and Agriculture Organization) system [33]. Soil pH ranges from 4.9 to 6.5. The details of the study sites have been previously described [11].

A total of 159 plots were sampled from four sites: the Laoshan experimental station (LS; 57 plots), the Maoershan experimental farm (MES; 37 plots), the Dongshan forest farm (DS; 30 plots), and the Daqingchuan forest farm (DQC; 35 plots).

Fixed-depth and pedogenetic horizon sampling were employed for soil sampling. Soil cutting rings with a 100 cm^3 volume were used for sampling soil from a soil pit. A soil pit was excavated in each of four subplots established in each of the 159 20 m× 20 m plots (for a total of 636 soil pits), and the samples from the four pits from the same horizon/layer of the same sampling method were mixed to form a composite sample for that layer and plot. Soil samples were collected from one or two sides of the pit for sampling the same layer or depth. Those composite samples from a plot make up the samples for a soil profile for each plot. Such composite samples better represent the soil from a plot. Under fixed-depth sampling, the soil was collected from four depths: 0–20, 20–40, 40–60, and 60–80 cm. Under pedogenetic horizon sampling, soil samples were collected from the A, B, and C horizons. The first horizon contains partially humified organic matter, giving the soil a darker color than soils from lower horizons. The second horizon has undergone sufficient changes during soil genesis, with maximum accumulation of materials such as silicate clay, metal (e.g., Fe and Al) oxides, and organic material. The third horizon comprises unconsolidated material below the solum (A and B horizons) that is little affected by soil-forming processes and lacks pedogenic development.

A total of 1067 composite soil samples were collected (pedogenetic: 113 plots × 3 samples with one each from horizons A, B, and C, plus 46 plots with only A and B horizons, for a total of 431 samples; fixed-depth: 159 plots × 4 samples from the four depths = 636 samples). At the time of sampling, the soil samples were placed in cloth soil-sampling bags for transportation back to the laboratory. To compare the effect of sampling method, the surface soil layer (0–20 cm) and the whole profile soil (0–80 cm) as two sampling depths were examined in this study. In other words, within the 0–20 cm and 0–80 cm sampling depths, we assessed the impact of sampling method on the assessment of SOC and SON.

Tree age was determined using an increment borer (Zhonglinweiye, Beijing, China) on at least five trees in each plantation plot. Fifty-six plots contained trees younger than 20 years (<20 years age-group), with an average tree age of 8.1 ± 7.1 years (SD, standard deviation). The second age-group (20–40 years) had 73 plots, with an average tree age of 25.5 ± 5.3 years. The third age-group (>40 years) had 30 plots and an average tree age of 44.8 ± 2.3 years.

2.2. Bulk Density, SOC, and SON Measurements

Once the soil samples were brought back to the laboratory, they were air-dried in a ventilated room until they reached a constant weight. Cutting rings with a 100 cm^3 volume were used for intact bulk density measurement, and bulk density was determined based on the oven-dried mass and volume (400 cm^3 of soil from four cutting rings as a composite sample) of each site. In the fixed-depth sampling, we measured one soil bulk density at each soil depth (0–20, 20–40, 40–60, and 60–80 cm) of each site. In the pedogenetic sampling, we measured one soil bulk density in each horizon (A, B, and C) of each site, and in some places, the soil bulk density at some transitional horizons (e.g., AB horizon) was also measured. Soil was ground to <0.25 mm for measuring SOC and SON concentrations with a heated dichromate/titration and the semimicro-Kjeldahl method, respectively. Details about those methods are available in a previous publication of our group [11].

2.3. Storage Calculation

The storage of SOC and SON was calculated with the following equation:

$$\text{Storage } (g{\cdot}m^{-2}) = a \times \rho_b \times depth \times (1 - V_{\text{gravel}}), \tag{1}$$

where a is SOC or SON concentration ($g{\cdot}kg^{-1}$), ρ_b is soil bulk density ($g{\cdot}cm^{-3}$ = $Mg{\cdot}m^{-3}$, Mg: million gram), and V_{gravel} is the fraction of gravel volume (determined via water displacement) in the soil.

2.4. Assessing the Effect of Sampling Method on Tree Age Effect and Spatial Variation in SOC and SON

Analysis of variance (ANOVA) was used to determine whether soil sampling method affected the spatial variation (among study sites) of SOC and SON and differences among tree age-groups. Specifically, we examined whether study site and age-group interacted with sampling method (all were included in the model as fixed factors). The concentration and storage of SOC and SON at 0–20 and 0–80 cm soil were analyzed. Duncan post-hoc tests were then conducted to identify the specific parameter (SOC concentration, SOC storage, SON concentration, and SON storage) that was affected by site, age-group, and sampling method. Marginal means were estimated and pairwise comparisons (LSD test) were used to separate the means across predictor variables, and their interactions with sampling method (site × method and age-group × method). All of these analyses were performed in SPSS 17.0 (SPSS Inc., Armonk, NY, USA).

2.5. Number of Samples Required for Estimating SOC and SON at Specific Error Limits

As a way of quantifying differences between the two soil sampling methods, statistical analyses [34,35] were conducted to compare the number of sampling points (n) required for estimating mean SOC and SON at a site within 5%, 10%, 20%, and 30% of its actual value, at the 95% probability level for both datasets (pedogenetic and fixed-depth). The following equation was used:

$$n = \frac{t_\alpha^2 s^2}{D^2}, \quad (2)$$

where t_α is the Student's t statistic with degrees of freedom at the α probability level, s is standard deviation, and D is the specified error limit. Only the SOC and SON concentration data were used as patterns in the storage data were similar.

3. Results

3.1. Differences between Sites, Sampling Methods, and Age-Groups

The SOC and SON concentrations in both the surface and whole profile soil differed significantly among sites ($p < 0.001$) and age-groups ($p < 0.05$), and between sampling methods ($p < 0.001$). However, SOC and SON storage did not follow the same pattern. For example, no difference between sampling methods was observed in surface soil SOC storage ($p > 0.05$; Table 1).

Table 1. ANOVA result on the influence of sampling method, age-group, and study site on soil organic C (SOC) and N (SON) concentrations, storage, in surface (0–20 cm), and the whole profile soil (0–80 cm). Bold fonts indicate statistical significance ($p < 0.05$).

	Concentration (g·kg^{-1})				Storage (kg·m^{-2})			
	SOC		SON		SOC		SON	
	F	p-Value	F	p-Value	F	p-Value	F	p-Value
Surface soil (0–20 cm)								
Site	104.31	**<0.001**	119.5	**<0.001**	131.26	**<0.001**	118.77	**<0.001**
Age-group	8.50	**<0.001**	5.51	**0.004**	11.17	**<0.001**	4.13	**0.017**
Sampling method	15.29	**<0.001**	15.50	**<0.001**	1.65	0.200	1.83	0.177
Site × age-group	0.51	0.771	1.76	0.122	0.62	0.682	2.38	**0.039**
Site × method	8.08	**<0.001**	6.18	**<0.001**	4.71	**0.003**	2.12	0.098
Age-group × method	2.68	0.070	0.35	0.708	2.46	0.087	0.06	0.946
Site × age-group × method	0.75	0.586	0.66	0.656	0.67	0.647	0.41	0.839
Whole profile soil (0–80 cm)								
Site	87.56	**<0.001**	95.16	**<0.001**	135.95	**<0.001**	122.43	**<0.001**
Age-group	9.74	**<0.001**	11.32	**<0.001**	9.39	**<0.001**	9.25	**<0.001**
Sampling method	37.45	**<0.001**	25.57	**<0.001**	28.60	**<0.001**	10.42	**0.001**
Site × age-group	1.30	0.262	1.64	0.149	1.46	0.203	1.94	0.088
Site × method	6.37	**<0.001**	3.25	**0.022**	13.76	**<0.001**	4.89	**0.002**
Age-group × method	0.18	0.839	0.11	0.892	0.18	0.832	0.47	0.625
Site × age-group × method	0.33	0.897	0.49	0.782	0.21	0.958	0.48	0.790

The highest SOC and SON concentrations were usually observed at DQC or DS, while the lowest values were found in LS (Table 2). Inter-site variation (the highest/lowest ratio) was 2.8 and 3.5 for surface soil SOC and SON, respectively, whereas inter-site variation was 2.9 and 3.2 for SOC and SON, respectively, in the whole profile soil. The highest SOC and SON storage in surface and whole profile soils were found at DS and DQC, respectively, with the lowest values at LS. The ratios of the highest to the lowest values for SOC and SON storage in the surface soil were 2.3 and 2.6, respectively, while those for the whole profile soil were 2.5 and 2.7, respectively (Table 2).

Pedogenetic sampling generally yielded higher SOC and SON concentrations (Table 2). Surface and whole profile soil SOC and SON concentrations were 1.2- to 1.4-fold higher in pedogenetic than in fixed-depth sampling (Table 2). Different patterns were found in their storage. Whole profile soil storage of both SOC and SON was 1.2-fold higher under pedogenetic than under fixed-depth sampling ($p < 0.05$), but the difference was only 1.05- to 1.07-fold ($p > 0.05$) for the surface soil (Table 2).

Table 2. Differences in soil organic C (SOC) and N (SON) concentration and storage across four sites, three age-groups, and two soil sampling methods.

Item	Site	Concentration ($g \cdot kg^{-1}$)	Storage ($kg \cdot m^{-2}$)	Age-Group (Year)	Concentration ($g \cdot kg^{-1}$)	Storage ($kg \cdot m^{-2}$)	Method	Concentration ($g \cdot kg^{-1}$)	Storage ($kg \cdot m^{-2}$)
	DQC	**50.3 c**	**8.96 c**	<20	**35.5 a**	**6.83 a**	FD	**36.3 a**	7.35 a
SOC	DS	**54.5 c**	**9.95 c**	20–40	**40.7 b**	**7.65 b**	PE	**43.8 b**	7.71 a
0–20	LS	**19.4 a**	**4.29 a**	>40	**45.2 b**	**8.30 b**			
cm	MES	**39.9 b**	**6.61 b**						
	Highest/lowest	2.8	2.3	>40/<20	1.3	1.2	PE/FD	1.2	1.05
	DQC	**23.8 c**	**21.1 c**	<20	**18.5 a**	**16.6 a**	FD	**17.5 a**	**16.2 a**
SOC	DS	**28.6 d**	**24.1 d**	20–40	**22.02 b**	**18.7 b**	PE	**23.6 b**	**19.7 b**
0–80	LS	**10.0 a**	**9.71 a**	>40	**21.4 ab**	**18.8 b**			
cm	MES	**19.4 b**	**16.4 b**						
	Highest/lowest	2.9	2.5	>40/<20	1.2	1.1	PE/FD	1.4	1.2
	DQC	**3.80 c**	**0.68 c**	<20	**2.58 a**	**0.51 a**	FD	**2.6 a**	0.52 a
SON	DS	**3.70 c**	**0.67 c**	20–40	**3.02 b**	**0.57 b**	PE	**3.2 b**	0.56 a
0–20	LS	**1.10 a**	**0.26 a**	>40	**3.05 b**	**0.55 ab**			
cm	MES	**2.80 b**	**0.55 b**						
	Highest/lowest	3.5	2.6	20–40/<20	1.2	1.1	PE/FD	1.2	1.08
	DQC	**1.80 c**	**1.66 c**	<20	**1.35 a**	**1.21 a**	FD	**1.30 a**	**1.19 a**
SON	DS	**1.90 c**	**1.51 bc**	20-40	**1.69 b**	**1.42 b**	PE	**1.70 b**	**1.37 b**
0–80	LS	**0.60 a**	**0.61 a**	>40	**1.37 a**	**1.19 a**			
cm	MES	**1.60 b**	**1.39 b**						
	Highest/lowest	3.2	2.7	20–40/<20	1.3	1.2	PE/FD	1.3	1.2

Note: DQC, Daqingchuan site; DS, Dongshan site; LS, Laoshan site; MES, Maoershan site; FD, fixed-depth sampling; PE, pedogenetic horizon sampling. Different letters in the same column and sampling depth indicate significant differences. Means that are significantly different among the treatment under comparison are highlighted.

Age-group differences increased linearly from <20, to 20–40, and >40 years groups in SOC concentration and storage in the surface soil, SON concentration in the surface soil, and SOC storage in the whole profile soil (Table 2). Forests in the 20–40 years age-group exhibited peak whole profile soil.

SOC and SON concentrations, as well as whole profile soil and surface soil SON storage (Table 2). In general, the >40 years group had 1.1- to 1.3-fold higher SOC than the <20 years age-group, while the 20–40 years group tended to have higher SON (Table 2).

3.2. SOC and SON Concentrations Across Study Sites: Interaction with Sampling Method

Marked site by sampling method interaction was found for SOC and SON concentrations in the surface and whole profile soil (Table 1; Figure 1). Both surface and whole profile soil SOC concentrations were higher with pedogenetic than with fixed-depth sampling. Inter-site differences were also found (Figure 1). The highest increases from one sampling method to another in surface soil SOC concentrations generally occurred in DQC (52%), while moderate increases (7%–10%) were observed in DS, LS, and MES (Figure 1). The whole profile soil (0–80 cm) typically accompanied higher SOC in pedogenetic sampling than in fixed-depth sampling. For example, SOC concentration at DS was 8% higher with pedogenetic sampling than with fixed-depth sampling in the surface soil, whereas it was 38% higher when the whole profile soil was evaluated. Furthermore, at LS, SOC in the surface soil was 10% higher, but that in the whole profile soil was 20% higher in pedogenetic than in fixed-depth sampling (Figure 1).

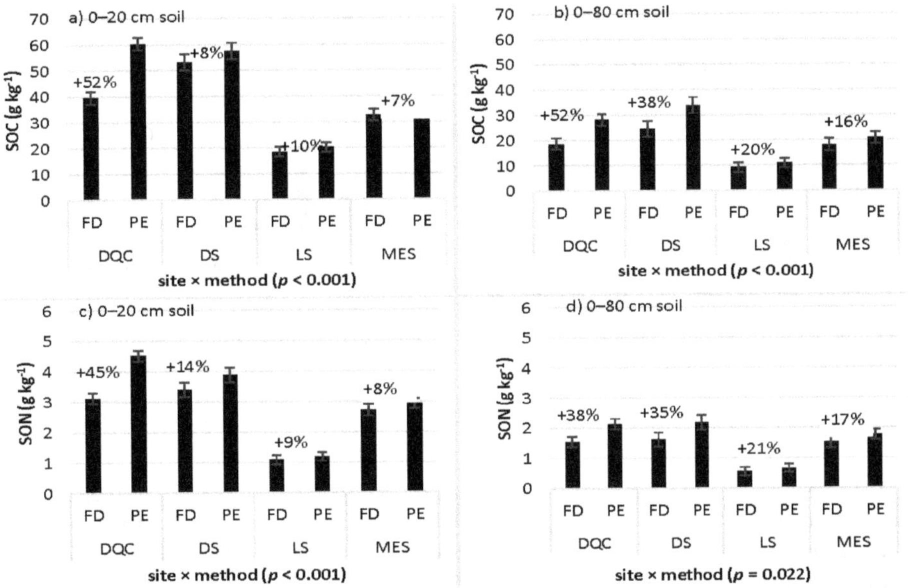

Figure 1. Sampling method influences soil organic C (SOC) and N (SON) concentrations depending on the sampling site (site × sampling method interaction). Data are presented for the two different sampling depths (surface soil: 0–20 cm and whole profile soil: 0–80 cm). A plus sign (+) indicates that the pedogenetic method resulted in higher values than the fixed-depth method, while a minus sign (−) indicates the opposite result. DQC, Daqingchuan site; DS, Dongshan site; LS, Laoshan site; MES, Maoershan site; FD, fixed-depth sampling; PE, pedogenetic horizon sampling. The error bars represent standard error of the mean. 0–20 cm soil for SOC (**a**); 0–80 cm soil for SOC (**b**); 0–20 cm soil for SON (**c**); 0–80 cm soil for SON (**d**).

Sampling influence on SON concentration was similar to that on SOC. Pedogenetic sampling led to higher SON in both surface and whole profile soil as compared with fixed-depth sampling, again with differences among sites (Figure 1). The largest increases (38%–45%) of pedogenetic over fixed-depth sampling occurred at DQC, whereas the other sites had relatively smaller sampling-induced changes

(14%–35% at DS, 9%–21% at LS, and 8%–17% at MES; Figure 1). In the surface soil, pedogenetic sampling resulted in 14% (DS), 9% (LS), and 8% (MES) increases in SON concentration as compared with fixed-depth sampling, whereas the increases were 35% (DS), 21% (LS), and 17% (MES) when the whole profile soil was evaluated (Figure 1).

3.3. SOC and SON Storage across Study Sites: Interaction with Sampling Method

The interaction between sampling method and study site significantly affected SOC and SON storage in the two sampling depths, except SON storage in the surface soil (Table 1; Figure 2). At DQC, pedogenetic sampling resulted in 16%–22% higher SOC and SON storage compared with fixed-depth sampling in the surface soil; and 30%–44% higher SOC and SON storage when the whole profile soil was considered. At DS, there was a <6% difference in SOC and SON storage in the surface soil, and a 14%–27% increase in the whole profile soil from pedogenetic to fixed-depth sampling. Additionally, SOC and SON storage values were similar under both methods at LS and MES (Figure 2).

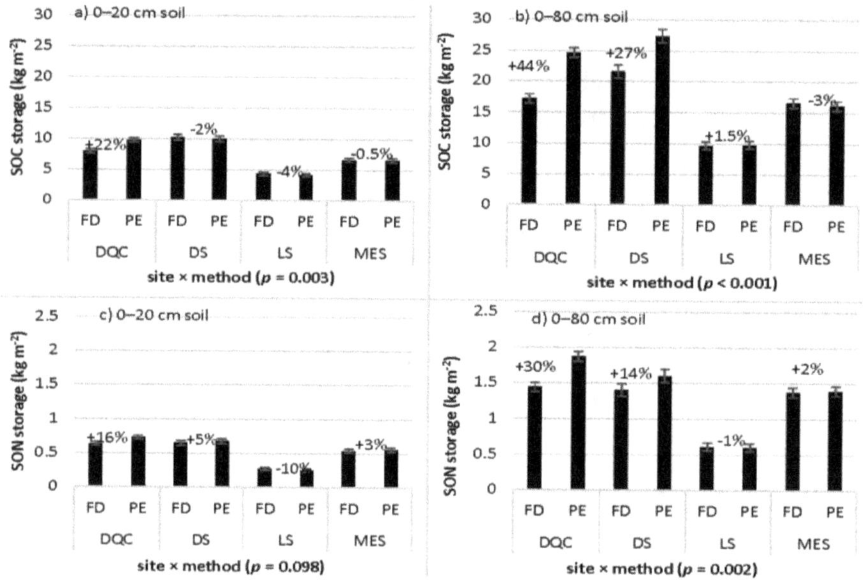

Figure 2. The influence of sampling method on soil organic C (SOC) and N (SON) storage across study sites (site×method interaction) and sampling depth (surface soil: 0–20 cm, whole profile soil: 0–80 cm). A plus sign (+) indicates that the pedogenetic method resulted in higher values than the fixed-depth method, while a minus sign (−) indicates the opposite result. DQC, Daqingchuan site; DS, Dongshan site; LS, Laoshan site; MES, Maoershan site; FD, fixed-depth sampling; PE, pedogenetic horizon sampling. The error bars represent standard error of the data. 0–20 cm soil for SOC (**a**); 0–80 cm soil for SOC (**b**); 0–20 cm soil for SON (**c**); 0–80 cm soil for SON (**d**).

3.4. Age-Group Differences in SOC and SON Concentrations: Interaction with Sampling Method

Age-group effects were similar between the two sampling methods for both SOC and SON concentrations (age-group* method interaction: $p > 0.05$; Table 1 and Figure 3). Moreover, age-group related differences were apparent in surface versus whole profile soil (Figure 3).

Both pedogenetic and fixed-depth sampling found sharp increases (6%–34%) in surface soil SOC concentrations from the <20 years to the 20–40 years age-group, which then moderately increased (9%–10%) from the latter to the >40 years age-group. Similarly, both fixed-depth and pedogenetic sampling found sharp increases (11%–22%) in SON concentration in the surface soil from the <20 years to the 20–40 years age-group, also followed by a small increase (2%) to the >40 years age-group (Figure 3).

Unlike in the surface soil, SOC and SON concentrations in the whole profile soil were lower in the >40 years age-group compared with the 20–40 years age-group (Figure 3). Both sampling methods found sharp increases (19%–28%) in SOC concentration in the 20–40 years age-group from the <20 years age-group, but 4%–5% decreases from the 20–40 years age-group to the >40 years group. Both sampling methods also found sharp increases (25%–27%) in SON concentration in the 20–40 years age-group from the <20 years age-group, followed by 17%–21% decreases in the >40 years age-group from the 20–40 years age-group (Figure 3).

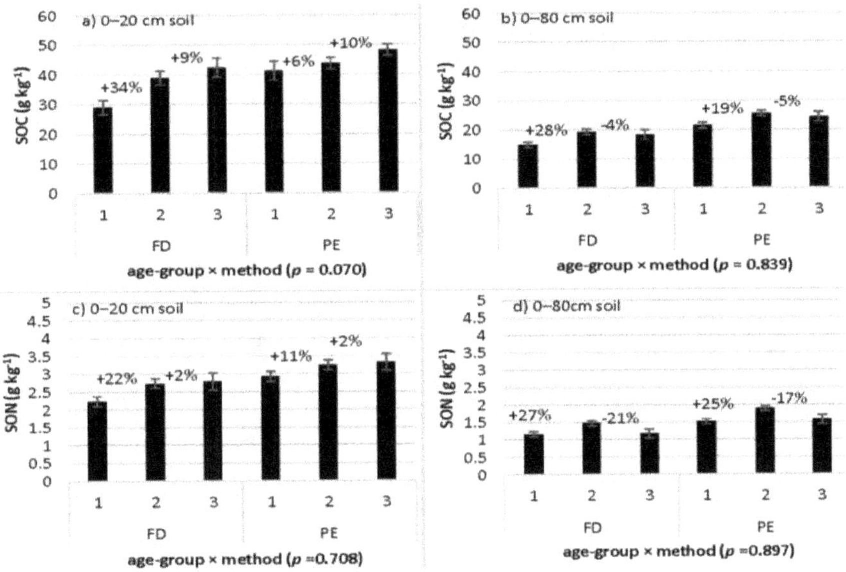

Figure 3. The interaction of age-group and sampling method affected soil organic C (SOC) and N (SON) concentrations, as well as their variation between surface (0–20 cm) and whole profile soil (0–80 cm). A plus sign (+) indicates that the pedogenetic method resulted in higher values than the fixed-depth method, while a minus sign (−) indicates the opposite result. DQC, Daqingchuan site; DS, Dongshan site; LS, Laoshan site; MES, Maoershan site; FD, fixed-depth sampling; PE, pedogenetic horizon sampling. The error bars represent standard error of the data. 0–20 cm soil for SOC (**a**);. 0–80 cm soil for SOC (**b**); 0–20 cm soil for SON (**c**); 0–80 cm soil for SON (**d**).

Owing to non-significant interactions between the age-group and sampling method (Figure 3), the age-group effect could be expressed as group means of the two sampling methods (Table 2). For the surface soil, marked increases in SOC and SON concentrations were found in the 20–40 years age-group from the <20 years age-group, while no marked changes were found between the 20–40 years and >40 years age-groups. However, different patterns were found in the 0–80 cm soil, with marked decreases of SON concentrtion from the 20–40 years age-group to the >40 years age-group (Table 2).

3.5. Age-Group Differences in SOC and SON Storage: Interaction with Sampling Method

Similar to SOC and SON concentrations, the sampling method did not alter age-group effects on SOC and SON storage (non-significant age-group × method interaction; Table 1 and Figure 4). Both fixed-depth and pedogenetic sampling revealed increases (11%–13%) in SON storage from the <20 years to the 20–40 years age-group, then small decreases (2%–3%) from the 20–40 years to the >40 years age-group. In addition, both sampling methods found sharp increases (7%–24%) in SOC storage from the <20 years to the 20–40 years age-group, then small increases (7%) from the 20–40 years to the >40 years age-group. Age-group differences in the whole profile soil SOC and SON storage clearly contrasted those in the surface soil (Figure 4). For example, there were sharp increases (15%–17%) in SOC storage from the <20 years to the 20–40 years age-group, but then no obvious changes (0%–2% decreases) between the 20–40 years and the >40 years age-groups regardless of the sampling method used. Furthermore, there were sharp increases (15%–20%) in SON storage from the <20 years to the 20–40 years age-group, and 13%–20% decreases from the 20–40 years to the >40 years age-group regardless of the sampling method used (Figure 4).

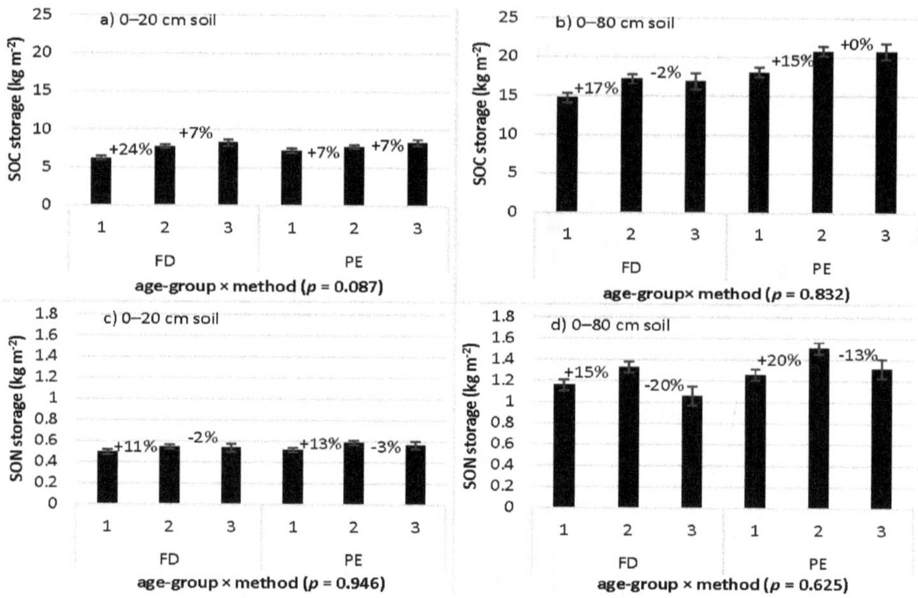

Figure 4. The interaction of age-group and sampling method on variation in soil organic C (SOC; upper two panels) and N (SON; lower two panels) storage, and differences between surface (0–20 cm; **a** and **c**) and whole profile soil (0–80 cm; **b** and **d**). A plus sign (+) indicates higher values in an older group than in the immediately younger one (e.g., >40 years versus 20–40 years), while a minus sign (−) indicates the opposite. 1 = <20 years age-group; 2 = 20–40 years age-group; 3 = >40 years age-group. FD, fixed-depth sampling; PE, pedogenetic horizon sampling. The error bars represent standard error of the data. 0–20 cm soil for SOC (**a**); 0–80 cm soil for SOC (**b**); 0–20 cm soil for SON (**c**); 0–80 cm soil for SON (**d**).

There were moderate increases in SOC storage and sharper decreases in SON storage from the 20–40 years to the >40 years age-group when the whole profile soil was considered (Figure 4). For example, surface soil SOC storage in the >40 years age-group was 7% higher than that in the 20–40 years age-group, whereas the percentage difference was a decrease of 0%–2% in the whole profile soil. Moreover, surface soil SON storage in the >40 years age-group was 2%–3% lower than that in the 20–40 years age-group, but the percentage difference increases to 13%–20% in the whole profile soil (Figure 4).

Owing to non-significant interactions (Figure 4), the age-group effect could be expressed as pooled means of two sampling methods (Table 2). For the surface soil, marked increases in SOC and SON storage were usualy found from the <20 years to the 20–40 years age-group, while no marked changes were found between the 20–40 years and >40 years age-groups. However, different patterns were found in the 0–80 cm soil, with marked decreases of SON storage in the >40 years age-group (Table 2).

3.6. *Minimum Sample Number Required for Estimating SOC and SON at Specific Error Limits: Effects of Sampling Method*

Fewer sampling points are required with fixed-depth sampling than with pedogenetic sampling, across multiple error limits (Figure 5). At least 538 samples should be collected for analyzing whole profile soil SOC concentration under pedogenetic sampling at the 5% error limit, whereas 37% fewer points are required with fixed-depth sampling to secure the same precision. This pattern of requiring more data points for pedogenetic sampling holds as the error limit increases (Figure 5). Similarly, under the same error limit, at least 543 pedogenetic samples are needed to analyze whole profile soil SON concentrations, compared with 480 samples for in fixed-depth sampling (12% fewer sampling points required) (Figure 5).

Similar results were observed for whole profile soil SOC and SON storage. At the 5% limit, pedogenetic sampling and fixed-depth required 387 and 233 samples, respectively, to estimate SOC storage, as well as 395 and 364 samples for SON storage estimation. At other error limits, these general patterns again held (Figure 5). Outcomes for surface-soil SOC and SON storage and concentration data were similar to the whole profile soil results (data not shown).

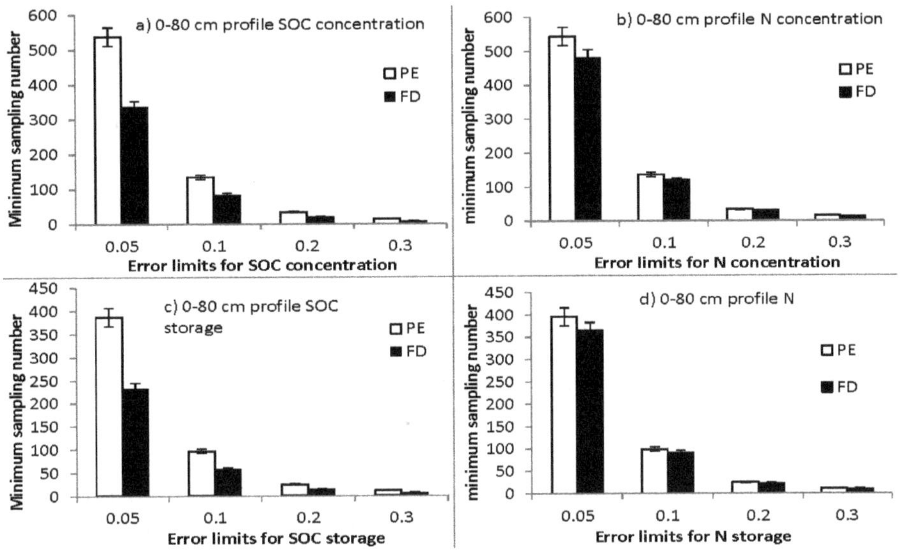

Figure 5. Effects of sampling method on minimum sampling number at different error limits to estimate differences in soil organic C (SOC) and soil organic N (SON) concentration and storage. PE, pedogenetic sampling; FD, fixed-depth sampling. Vertical bars represent a 5% error. 0–80 cm soil profile SOC concentration (**a**); 0–80 cm soil profile SON concentration(**b**); 0–80 cm soil profile SOC storage (**c**); 0–80 cm soil profile SON storage (**d**).

4. Discussion

4.1. Age-Group Effect on SOC and SON Is Dependent on Sampling Depth but Not Sampling Method

Compared with younger forests, older forests can have both higher and lower SOC and SON [11,14–16], but few have examined how sampling methods may affect such evaluations. Here, we demonstrated that sampling methods did not alter the estimate of the age-group effect on SOC and SON in larch plantations (Table 1, Figures 3 and 4). Pedogenetic horizon sampling in previous studies has revealed large increases (31.3 to 48.2 g$\cdot$kg^{-1}) in soil organic matter (SOM) concentration due to thinning [36], as well as large SOM variations (184 to 510 g$\cdot$kg^{-1}) within different larch forests [37]. Using fixed-depth sampling, SOM was found to increase from 47.7 to 51.6 g$\cdot$kg^{-1} after a second rotation of a 36-years old larch forest at the same site [38]. Therefore, differences in sampling methods do not explain contradictory patterns of SOC and SON variation over time in the studied plantations. As seen in previous reports [9–12,17,18,20], prior land use, tree species planted, and soil nutrient supply may all affect SOC and SON dynamics after afforestation.

The patterns of tree age effect on SOC and SON was strongly dependent on soil sampling depth under both sampling methods. Under a deeper soil sampling depth, the SOC and SON estimates (storage and concentration) were accompanied by moderate SOC increases or sharper SON decreases from the 20–40 years to the >40 years age-group. Previous soil studies in larch forests were focused on the more dynamic surface soil (0–20 cm), which also contains the highest root density [39,40]. However, this focus may have overlooked the importance of deep soil, which is significant given that SOM and nutrient storage in deep soils are important for forest vegetation with deep root systems [41,42], such as the larch trees studied in this paper. A recent, 12 years chronosequence study in larch plantations showed that SOC initially decreased before increasing with stand age [43]. Larch plantations in northeast China could accumulate SOC at a rate that ranges from 57.9 to 139.4 $g \cdot m^{-2} \cdot year^{-1}$ in the top 20 cm of soil [11]. Moreover, researchers also noted a slight increase in surface soil N storage (~0.33 $kg \cdot m^{-2}$), which decreased from 0.11–0.16 to 0.06–0.11 $kg \cdot m^{-2}$ in the 20–60 cm soil during larch forest development [11]. Larch tree growth could cause divergent changes in SIC (soil inorganic carbon) and SOC levels, particularly in terms of their vertical distribution, and therefore, these effects should be fully considered in SIC-rich calcareous soils [44]. Larch reforestation could markedly affect the temporal dynamics in the concentration, storage, and vertical distribution of most soil nutrients, as indicated by findings showing eight SOC, SON, P, and K related parameters with differences between surface and subsurface soils [21]. These studies suggest that deeper soil C and nutrient depletion was possible. In contrast, this study suggests that larch plantation establishment in northeast China could sequester C in the mineral soil without depleting SON, at least in the surface soil of stands in the >40 years age-group. However, the inclusion of deeper soil layers could strongly modify these patterns to indicate possible depletion of SOC and SON with stand age.

4.2. Sampling Method Effects on SOC and SON Differ between Their Concentration and Storage

Understanding SOC and SON spatial variation is important in order to improve intensive forest management [2], but the influence of pedogenetic versus fixed-depth sampling on the assessment of SOC and SON spatial variation has not been statistically assessed [29,30,45,46]. This study demonstrated that sampling method significantly affected ($p < 0.05$) the identified size of inter-site variations of SOC and SON. Moreover, such sampling effects differed between concentration and storage of SOC and SON, as well as across the two soil sampling depths. These effects suggest that the effect of sampling method and its interaction with soil sampling depth on nutrient storage or concentration should be considered in future research, particularly when comparing results across different studies.

The A and B horizons in this study were generally thicker than 20 cm, with the A horizon at LS, MES, DS, and DQC averaging 17.8, 27.0, 40.5, and 25.0 cm, respectively, whereas the corresponding B horizon averaged 55.2, 66.1, 69.2, and 49.6 cm. Thus, fixed-depth sampling with a 20-cm increment may collect mixed A and B horizon samples. This corroborates with the literature that if soil horizons with different SOC concentrations are mixed during fixed-depth sampling, SOC and SON concentrations could be much higher in fixed-depth sampling than in pedogenetic sampling [47]. Others have also reported that SOC stock variability decreases under fixed-depth sampling as compared with pedogenetic horizon sampling [29]. Therefore, fixed-depth sampling could result in higher SOC and SON estimates than pedogenetic sampling depending on the depth of the soil sampled and also sampling intensities at different horizons [29,47].

Effects of sampling method on SOC and SON have been reported previously [29,30,45], but quantitative assessment on the effect of sampling method on inter-site variation and differences between sampling depths is lacking. In the upper 30 cm of ploughed Gleysols, VandenBygaart et al. found that pedogenetic horizon sampling reduced the variability in SOC stock [45]. Sampling by pedogenetic horizon is also recommended over fixed-depth when monitoring SOC stock variation in hydromorphic soils of the agricultural landscape [46], as well as when studying pedogenetic processes controlling SOC stocks. In contrast, fixed-depth sampling is preferred for determining regional SOC stock [29]. Yet another report [30] indicated that fixed-depth sampling tends to overestimate cultivation-induced SOC depletion, and thus they recommended the use of pedogenetic soil information (e.g., soil classes) in SOC stock calculation and C sequestration when assessing the impact of land-use change. Our findings differed from these reports; we showed that pedogenetic sampling typically resulted in much higher SOC and SON than fixed-depth sampling (Table 2), which was 1.05- to 1.2-fold higher for SOC and SON storage and 1.2- to 1.4-fold higher for SOC and SON concentration. Pedogenetic sampling was also used in several major datasets in China (e.g., the 1979–1982 national survey) and has since become the main sampling method for nationwide evaluation of SOC and SON budgets [48–50]. Currently, however, more studies use the fixed-depth method for estimating SOC and SON storage in China [11,21,51]. This study provides a statistical assessment of differences in SOC and SON between two major sampling methods, which should be useful for meta-analysis and soil fertility diagnoses across multiple locations.

4.3. *Lower SOC and SON Spatial Variations Requiring Lower Sampling Intensity in Fixed-Depth Sampling*

Many studies have addressed the effect of forest development on SOC and SON dynamics [11,21], inter-site and vertical soil variations [29,52], as well as the minimum sample number required for accurate assessment [20–23]. The minimum number of samples required should be determined for detecting a specified change in soil properties [34,53]. For designing efficient soil sampling methods, Conant et al. found that differences in the order of 2.0 $Mg{\cdot}C{\cdot}ha^{-1}$ could be detected through collecting and analyzing samples from at least five (tilled) or two (forest) microplots in Tennessee [5]. Compared with the large number required as was identified in this paper (300–600 samples), a proper sampling procedure, such as the microplot averaging method, could strongly decrease the number of samples required for accurate evaluation [5]. Moreover, the smaller variation with the fixed-depth sampling indicates that this method requires a lower sampling intensity (10%–40% lower) for a given precision as compared with pedogenetic horizon sampling in assessing changes in SOC and SON over time. Average SOC and SON changing rates are smaller relative to their total amount in the soil, resulting in a small signal to noise ratio [5]. The methodological effect on the minimum sample number required, together with previous findings on proper soil sampling design [5], may increase the precision of SON and SOC evaluations for a given error limit.

5. Conclusions

Larch plantation establishment on degraded farmland increased SOC and SON storage in the 20–40 years age-group, and sampling method (fixed-depth versus pedogenetic horizon sampling) did not alter this pattern ($p > 0.05$). Furthermore, pedogenetic horizon sampling resulted in higher SOC and SON storage and inter-site variations than fixed-depth sampling. At a given error limit for SOC and SON estimation, the minimum number of sampling points required for fixed-depth sampling was 60%~90% of that required for pedogenetic horizon sampling. These sampling method effects can help interpret differences in data collected in disparate studies (e.g., for meta-analysis), and improve our understanding of the assessment of spatial variations of SOC and SON.

Acknowledgments: This study was supported financially by the National Key Research and Development Program (2016YFA0600802), Fundamental Research Funds for the Central Universities (2572016EAJ1&DL13EA03-03), China's National Foundation of Natural Sciences (41373075, 31100457), and the "111" project (B16010). Thanks are due to De'an Xia from Forestry College, Northeast Forestry University for his kind support with statistical analysis and data explanation.

Author Contributions: W.W. conceived and designed the experiment; H.W. performed the experiment and analyzed the data; and H.W., W.W. and SC wrote the manuscript.

References

1. Rushton, B.T. Matching tree species to site conditions in reclamation. In *Evaluation of Alternatives for Restoration of Soil and Vegetation on Phophatic Clay Settling Ponds*; Odum, H.T., Rushton, B.T., Paulic, M., Everett, S., McClanahan, T.R., Munroe, M., Wolfe, R.W., Eds.; Florida Institute of Phosphate Research: Bartow, FL, USA, 1991.
2. Kravchenko, A.N. Influence of spatial structure on accuracy of interpolation methods. *Soil Sci. Soc. Am. J.* **2003**, *67*, 1564–1571. [CrossRef]

3. Franzen, D.W.; Hopkins, D.H.; Sweeney, M.D.; Ulmer, M.K.; Halvorson, A.D. Evaluation of soil survey scale for zone development of site-specific nitrogen management. *Agron. J.* **2002**, *94*, 381–389. [CrossRef]
4. Wang, H.-M.; Wang, W.-J.; Chen, H.; Zhang, Z.; Mao, Z.; Zu, Y.-G. Temporal changes of soil physic-chemical properties at different soil depths during larch afforestation by multivariate analysis of covariance. *Ecol. Evol.* **2014**, *4*, 1039–1048. [CrossRef] [PubMed]
5. Conant, R.T.; Smith, G.R.; Paustian, K. Spatial variability of soil carbon in forested and cultivated sites: Implications for change detection. *J. Environ. Qual.* **2003**, *32*, 278–286. [CrossRef] [PubMed]
6. Guo, L.B.; Gifford, R.M. Soil carbon stocks and land use change: A meta analysis. *Glob. Chang. Biol.* **2002**, *8*, 345–360. [CrossRef]
7. Laganiere, J.; Angers, D.A.; Pare, D. Carbon accumulation in agricultural soils after afforestation: A meta-analysis. *Glob. Chang. Biol.* **2010**, *16*, 439–453. [CrossRef]
8. Zinkevičius, R. Influence of soil sampling for precision fertilizing. *Agron. Res.* **2008**, *6*, 423–429.
9. Morris, S.J.; Bohm, S.; Haile-Mariam, S.; Paul, E.A. Evaluation of carbon accrual in afforested agricultural soils. *Glob. Chang. Biol.* **2007**, *13*, 1145–1156. [CrossRef]
10. Berthrong, S.T.; Jobbagy, E.G.; Jackson, R.B. A global meta-analysis of soil exchangeable cations, pH, carbon, and nitrogen with afforestation. *Ecol. Appl.* **2009**, *19*, 2228–2241. [CrossRef] [PubMed]
11. Wang, W.J.; Qiu, L.; Zu, Y.G.; Su, D.X.; An, J.; Wang, H.Y.; Zheng, G.Y.; Sun, W.; Chen, X.Q. Changes in soil organic carbon, nitrogen, pH and bulk density with the development of larch (*Larix gmelinii*) plantations in china. *Glob. Chang. Biol.* **2011**, *17*, 2657–2676.
12. Li, D.; Niu, S.; Luo, Y. Global patterns of the dynamics of soil carbon and nitrogen stocks following afforestation: A meta-analysis. *New Phytol.* **2012**, *195*, 172–181. [CrossRef] [PubMed]
13. Springsteen, A.; Loya, W.; Liebig, M.; Hendrickson, J. Soil carbon and nitrogen across a chronosequence of woody plant expansion in North Dakota. *Plant Soil* **2010**, *328*, 369–379. [CrossRef]
14. Covington, W. Changes in forest floor organic matter and nutrient content following clear cutting in northern hardwoods. *Ecology* **1981**, *62*, 41–48. [CrossRef]
15. Garten, J. Soil carbon storage beneath recently established tree plantations in Tennessee and South Carolina, USA. *Biomass Bioenergy* **2002**, *23*, 93–102. [CrossRef]
16. Wirth, C.; Czimczik, C.J.; Schulze, E.D. Beyond annual budgets: Carbon flux at different temporal scales in fire-prone siberian scots pine forests. *Tellus B* **2002**, *54*, 611–630. [CrossRef]
17. Kueffer, C.; Klingler, G.; Zirfass, K.; Schumacher, E.; Edwards, P.J.; Gusewell, S. Invasive trees show only weak potential to impact nutrient dynamics in phosphorus-poor tropical forests in the seychelles. *Funct. Ecol.* **2008**, *22*, 359–366. [CrossRef]
18. Cavard, X.; Bergeron, Y.; Chen, H.Y.H.; Pare, D. Effect of forest canopy composition on soil nutrients and dynamics of the understorey: Mixed canopies serve neither vascular nor bryophyte strata. *J. Veg. Sci.* **2011**, *22*, 1105–1119. [CrossRef]
19. Wei, X.; Shao, M.; Gale, W.; Li, L. Global pattern of soil carbon losses due to the conversion of forests to agricultural land. *Sci. Rep.* **2014**, *4*, 4062. [CrossRef] [PubMed]
20. Sauer, T.J.; James, D.E.; Cambardella, C.A.; Hernandez-Ramirez, G. Soil properties following reforestation or afforestation of marginal cropland. *Plant Soil* **2012**, *360*, 375–390. [CrossRef]
21. Wang, W.; Wang, H.; Zu, Y. Temporal changes in SOM, N, P, K, and their stoichiometric ratios during reforestation in China and interactions with soil depths: Importance of deep-layer soil and management implications. *For. Ecol. Manag.* **2014**, *325*, 8–17. [CrossRef]
22. Yim, M.H.; Joo, S.J.; Shutou, K.; Nakane, K. Spatial variability of soil respiration in a larch plantation: Estimation of the number of sampling points required. *For. Ecol. Manag.* **2003**, *175*, 585–588. [CrossRef]
23. Adachi, M.; Bekku, Y.S.; Konuma, A.; Kadir, W.R.; Okuda, T.; Koizumi, H. Required sample size for estimating soil respiration rates in large areas of two tropical forests and of two types of plantation in malaysia. *For. Ecol. Manag.* **2005**, *210*, 455–459. [CrossRef]
24. Laiho, R.; Penttilä, T.; Laine, J. Variation in soil nutrient concentrations and bulk density within Pearland forest sites. *Silva Fenn.* **2004**, *38*, 29–41. [CrossRef]
25. Kulmatiski, A.; Vogt, D.J.; Siccama, T.G.; Beard, K.H. Detecting nutrient pool changes in rocky forest soils. *Soil Sci. Soc. Am. J.* **2003**, *67*, 1282–1286. [CrossRef]
26. Verboom, W.H.; Pate, J.S. Evidence of active biotic influences in pedogenetic processes. Case studies from semiarid ecosystems of south-west Western Australia. *Plant Soil* **2006**, *289*, 103–121. [CrossRef]

27. Funakawa, S.; Hirooka, K.; Yonebayashi, K. Temporary storage of soil organic matter and acid neutralizing capacity during the process of pedogenetic acidification of forest soils in Kinki District, Japan. *Soil Sci. Plant Nutr.* **2008**, *54*, 434–448. [CrossRef]
28. Eger, A.; Almond, P.C.; Condron, L.M. Pedogenesis, soil mass balance, phosphorus dynamics and vegetation communities across a Holocene soil chronosequence in a super-humid climate, South Westland, New Zealand. *Geoderma* **2011**, *163*, 185–196. [CrossRef]
29. Grueneberg, E.; Schoening, I.; Kalko, E.K.V.; Weisser, W.W. Regional organic carbon stock variability: A comparison between depth increments and soil horizons. *Geoderma* **2010**, *155*, 426–433. [CrossRef]
30. Wiesmeier, M.; Sporlein, P.; Geuss, U.; Hangen, E.; Haug, S.; Reischl, A.; Schilling, B.; von Lutzow, M.; Kogel-Knabner, I. Soil organic carbon stocks in Southeast Germany (Bavaria) as affected by land use, soil type and sampling depth. *Glob. Chang. Biol.* **2012**, *18*, 2233–2245. [CrossRef]
31. Sun, Z.H.; Jin, G.Z.; Mu, C.C. *Study on the Keeping Long-Term Productivity of Larix Olgensis Plantation*; Science Press: Beijing, China, 2009.
32. Chen, Z.; Gong, Z.; Zhang, G.; Zhao, W. Correlation of soil taxa between Chinese Soil Genetic Classification and Chinese Soil Taxonomy on various scales. *Soils* **2004**, *36*, 584–595.
33. Gong, Z.; Chen, Z.; Luo, G.; Zhang, G.; Zhao, W. China Soil Classification Reference System. *Soils* **1999**, 57–63.
34. Petersen, R.; Calvin, L. Sampling. In *Methods of Soil Analysis. Part 1. Physical and Mineralogical Methods. Agronomy Monograph 9*, 2nd ed.; Klute, A., Ed.; American Society of Agronomy, Soil Science Society of America: Madison, WI, USA, 1986; pp. 33–51.
35. Robertson, G.P.; Coleman, D.C.; Bledsoe, C.S.; Sollins, P. *Standard Soil Methods for Long-Term Ecological Research*; Oxford University Press: Oxford, UK, 1999; p. 480.
36. Chen, X.Q. The effect of community structure of artificial larch wood on physic-chemical properties of Baijing soil. *J. Northeast For. Univ.* **1986**, *14*, 113–116.
37. Niu, X.; Wei, J.S.; Zhou, M.; Liu, B. Study on the soil organic matter under *Larix gmelinii* forest. In *Soil Fertilizer Society of Inner Mongolia 2007 Conference on Soil Fertilizer and Sustainable Development*; Inner Mogolica Publisher: Hohhot, China, 2007; Volume 2007, pp. 1145–1149.
38. Yan, D.; Wang, J.; Yang, M. Tendency of Soil Degradation in the Pure Larch Plantations. *Chin. J. Ecol.* **1997**, *16*, 62–66.
39. Kajimoto, T.; Osawa, A.; Matsuura, Y.; Abaimov, A.P.; Zyryanova, O.A.; Kondo, K.; Tokuchi, N.; Hirobe, M. Individual-based measurement and analysis of root system development: Case studies for *Larix gmelinii* trees growing on the permafrost region in Siberia. *J. For. Res.* **2007**, *12*, 103–112. [CrossRef]
40. Liu, S.R.; Wang, W.Z.; Wang, M.Q. The characteristics of energy in the formative process of net primary productivity of larch artificial forest ecosystem. *Acta Phytoecol. Geobot. Sin.* **1992**, *16*, 209–219.
41. Jobbágy, E.G.; Jackson, R.B. The vertical distribution of soil organic carbon and its relation to climate and vegetation. *Ecol. Appl.* **2000**, *10*, 423–436. [CrossRef]
42. Rumpel, C.; Kögel-Knabner, I. Deep soil organic matter—A key but poorly understood component of terrestrial c cycle. *Plant Soil* **2011**, *338*, 143–158. [CrossRef]
43. Wang, C.M.; Shao, B.; Wang, R. Carbon sequestration potential of ecosystem of two main tree species in Northeast China. *Acta Ecol. Sin.* **2010**, *30*, 1764–1772.
44. Wang, W.; Su, D.; Qiu, L.; Wang, H.; An, J.; Zheng, G.; Zu, Y. Concurrent changes in soil inorganic and organic carbon during the development of larch, *Larix gmelinii*, plantations and their effects on soil physicochemical properties. *Environ. Earth Sci.* **2012**, *69*, 1559–1570. [CrossRef]
45. VandenBygaart, A.J.; Gregorich, E.G.; Angers, D.A.; McConkey, B.G. Assessment of the lateral and vertical variability of soil organic carbon. *Can. J. Soil Sci.* **2007**, *87*, 433–444. [CrossRef]
46. VandenBygaart, A.J. Monitoring soil organic carbon stock changes in agricultural landscapes: Issues and a proposed approach. *Can. J. Soil Sci.* **2006**, *86*, 451–463. [CrossRef]
47. Palmer, C.J.; Smith, W.D.; Conkling, B.L. Development of a protocol for monitoring status and trends in forest soil carbon at a national level. *Environ. Pollut.* **2002**, *116* (Suppl. 1), S209–S219. [CrossRef]
48. Wu, H.B.; Guo, Z.T.; Peng, C.H. Land use induced changes of organic carbon storage in soils of China. *Glob. Chang. Biol.* **2003**, *9*, 305–315. [CrossRef]
49. Yan, X.Y.; Cai, Z.C.; Wang, S.W.; Smith, P. Direct measurement of soil organic carbon content change in the croplands of China. *Glob. Chang. Biol.* **2011**, *17*, 1487–1496. [CrossRef]

50. Yang, Y.H.; Fang, J.Y.; Ji, C.J.; Ma, W.H.; Mohammat, A.; Wang, S.F.; Wang, S.P.; Datta, A.; Robinson, D.; Smith, P. Widespread decreases in topsoil inorganic carbon stocks across China's grasslands during 1980s-2000s. *Glob. Chang. Biol.* **2012**, *18*, 3672–3680. [CrossRef]
51. Lv, H.; Wang, W.; He, X.; Xiao, L.; Zhou, W.; Zhang, B. Quantifying tree and soil carbon stocks in a temperate urban forest in Northeast China. *Forests* **2016**, *7*, 200. [CrossRef]
52. Wang, Q.; Wang, W. Grsp amount and compositions: Importance for soil functional regulation. In *Fulvic and Humic Acids: Chemical Composition, Soil Applications and Ecological Effects*; Barrett, K.D., Ed.; Nova Science Publishers, Inc.: Hauppauge, NY, USA, 2015.
53. Johnson, C.E.; Johnson, A.H.; Huntington, T.G. Sample size requirements for the determination of changes in soil nutrient pools. *Soil Sci.* **1990**, *150*, 637–644. [CrossRef]

Understanding the Fate of Applied Nitrogen in Pine Plantations of the Southeastern United States Using ^{15}N Enriched Fertilizers

Jay E. Raymond *, Thomas R. Fox and Brian D. Strahm

Department of Forest Resources and Environmental Conservation, Virginia Polytechnic Institution and State University, Blacksburg, VA 24061, USA; trfox@vt.edu (T.R.F.); brian.strahm@vt.edu (B.D.S.)
* Correspondence: jayer11@vt.edu

Academic Editors: Scott X. Chang and Xiangyang Sun

Abstract: This study was conducted to determine the efficacy of using enhanced efficiency fertilizer (EEFs) products compared to urea to improve fertilizer nitrogen use efficiency (FNUE) in forest plantations. All fertilizer treatments were labeled with ^{15}N (0.5 atom percent) and applied to 100 m^2 circular plots at 12 loblolly pine stands (*Pinus taeda* L.) across the southeastern United States. Total fertilizer N recovery for fertilizer treatments was determined by sampling all primary ecosystem components and using a mass balance calculation. Significantly more fertilizer N was recovered for all EEFs compared to urea, but there were generally no differences among EEFs. The total fertilizer N ecosystem recovery ranged from 81.9% to 84.2% for EEFs compared to 65.2% for urea. The largest amount of fertilizer N recovered for all treatments was in the loblolly pine trees (EEFs: 38.5%–49.9%, urea: 34.8%) and soil (EEFs: 30.6%–38.8%, urea: 28.4%). This research indicates that a greater ecosystem fertilizer N recovery for EEFs compared to urea in southeastern pine plantations can potentially lead to increased FNUE in these systems.

Keywords: ^{15}N; forest fertilization; nitrogen cycle; plantation forestry; enhanced efficiency fertilizers

---

1. Introduction

Loblolly pine (*Pinus taeda* L.) is the most widely planted and commercially valuable tree species in the United States [1,2], with large areas in the southeastern United States managed intensively in plantations. Although loblolly pine stemwood can exceed 10 $m^3 \cdot ha^{-1} \cdot year^{-1}$ in intensively managed plantations [3], growth of many stands is less due to low levels of plant available nitrogen (N) and phosphorous (P) in the soil [4,5]. Nitrogen deficiencies that occur when plant available N in the soil is inadequate to meet tree N demand [6–8] translate to low leaf areas, decreased photosynthetic capacity, and hence reduced growth [9,10]. Temporal patterns in N availability often lead to N deficiencies developing during later parts of the rotation [3,11,12]. Following disturbance, such as harvesting, plant N availability in the soil is high due to N mineralization of organic matter [13,14]. Yet, as the stand develops, plant N availability decreases because of increasing N immobilization in the ecosystem [15], and N fertilization is often required to maintain forest productivity in mid-rotation stands [6]. The average growth response in mid-rotation southeastern pine plantations averages 3 $m^3 \cdot ha^{-1} \cdot year^{-1}$ over the 8 years following fertilization [3].

However, less than 30% of N applied in fertilizer is taken up by trees [16–20]. The low fertilizer N uptake by trees is likely due to several factors including: (1) N loss from the system; (2) N immobilization; and (3) N application that is asynchronous to seasonal plant N demand [21–26]. Forest soils supporting pine plantations in the southeastern United States contain large natural quantities of N, typically ranging from 2 to 7 $Mg \cdot ha^{-1}$ [27–29]. Because of the large amount of N in forest soils, it is difficult to follow the fate, cycling and uptake of fertilizer N which typically adds only 150–250 kg·ha^{-1} of N to the system [28–30].

Fertilizers labeled with ^{15}N can be used to trace the fate of fertilizer N through the ecosystem over time [31–34]. Studies using ^{15}N have improved the understanding of applied N cycling in ecosystems since the 1950s in both agriculture [35] and forested [36] systems. Recent research in forested ecosystems using ^{15}N tracer techniques have focused on understanding N cycling in natural systems [37–39] or the effects of chronic N deposition from industrialization [40,41]. Fertilizers labeled with ^{15}N have also been used to determine fertilizer N uptake and nitrogen use efficiency (FNUE) [21,22,36,42–44].

Urea (46-0-0) is the most commonly used N fertilizer in southern forestry due to its high N content and low cost per unit of applied N [4,5]. However, large N losses following urea fertilization can occur due to ammonia (NH_3) volatilization depending on the interactions of weather and edaphic factors [45–50]. Volatilization, combined with leaching and denitrification, reduce the amount of fertilizer N remaining in the system, and may decrease fertilizer N availability for plant uptake and hence FNUE [51–54].

Enhanced efficiency N fertilizers (EEFs) were developed to reduce N loss and increase N availability [55–65]. The EEFs can be divided into slow release (SRN), controlled release (CRN) and stabilized (SNF) N fertilizers [55–63]. The SRN products slowly release fertilizer N due to microbial decomposition [56]. The CRN products have coatings around the fertilizer N to alter rate, pattern and duration of fertilizer N release [56,57]. The SNF products have compounds to inhibit rapid fertilizer N transformations to less stable forms [61]. The different attributes of the various N containing EEF products increases the flexibility of N fertilization under diverse conditions to optimize plant N uptake and increase FNUE when compared to urea.

Our overall objective was to determine ecosystem uptake of fertilizer N and determine if N uptake was greater for EEFs compared to urea. In this study, we compared fertilizer N uptake in southeastern pine ecosystems following fertilization with urea and three enhanced efficiency fertilizers after a spring application to determine if there were differences among treatments for: (1) total ecosystem fertilizer N recovery; and (2) ecosystem partitioning of fertilizer N.

2. Materials and Methods

2.1. Experimental Design

This study was established as a complete-block design with five fertilizer treatments. Twelve sites were chosen from a network of existing fertilizer and thinning trials, and each site served as a block. At each site, five 100 m^2 circular plots were installed prior to fertilization, and each site was fertilized with a single fertilizer treatment on the same day between 26 March and 8 April 2012.

2.2. Site Description

All sites were considered mid-rotation, with stand ages ranging from 8 to 15 years (Figure 1). The understory of the sites ranged from no understory to encompassing 25% of the plot. Selected climate, physical and stand characteristics are detailed in Table 1.

Table 1. Selected climate, physical and stand characteristics of pine stands in the southeastern United States selected to evaluate ecosystem partitioning of fertilizer N following application of urea or enhanced efficiency N fertilizers enriched with ^{15}N. MAP = Mean annual precipitation. MAT = Mean annual temperature.

State	Latitude	Longitude	Alt (m)	MAP (cm)	MAT (°C)	Physiographic Region	Soil Taxonomic Class	Drainage Class	Trees plot^{-1}	Trees ha^{-1}	Ht (m)	DBH (cm)
VA	37.445331	78.662917	60	109	13	Piedmont	fine, mixed, subactive, mesic Typic Hapludults	Well	8	880	9.1	15.1
SC	34.450000	80.505383	29	107	16	Sandhills	thermic coated Typic Quartzipsamments	Excessively	15	1560	12.5	16.1
GA1	33.625317	82.801183	35	118	16	Piedmont	fine kaolinitic thermic Rhodic Kandiudults	Well	19	1800	7.2	8.6
GA2	31.339978	81.857283	1	114	18	Atlantic Coastal Plain	sandy siliceous thermic Aeric Alaquods	Poorly	15	1460	14.7	16.5
GA3	31.299333	81.847217	1	114	18	Atlantic Coastal Plain	loamy, siliceous, subactive, thermic Arenic Paleaquults	Somewhat Poorly	13	1340	13.6	15.1
FL	30.205267	83.866817	0.6	142	20	Eastern Gulf Coastal Plain	loamy siliceous superactive thermic Aquic Arenic Hapludalfs	Somewhat Poorly	16	1580	10.2	13.4
MS	31.066717	89.602467	26	152	19	Western Gulf Coastal Plain	coarse loamy siliceous subactive thermic Typic Paleudults	Well	21	2160	12.5	13.8
LA1	31.337017	93.182783	28	147	19	Western Gulf Coastal Plain	fine smectitic thermic Albaquic Hapludalfs	Moderately Well	7	720	14.9	19.3
LA2	31.013333	93.422600	28	147	19	Western Gulf Coastal Plain	fine loamy siliceous subactive thermic Plinthic Paleudults	Well	6	780	14.8	17.5
LA3	30.560533	90.727650	0.9	160	19	Western Gulf Coastal Plain	fine silty mixed active thermic Typic Glossaqualfs	Poorly	23	2380	12.0	13.3
OK	34.029333	94.825017	42	136	16	Western Gulf Coastal Plain	fine silty mixed active thermic Aquic Paleudalfs	Moderately well	15	1580	3.0	4.0
TX	31.13255	94.462533	32	127	20	Western Gulf Coastal Plain	fine loamy siliceous semiacitve thermic Oxyaquic Glossudalfs	Moderately well	13	1360	13.9	11.2

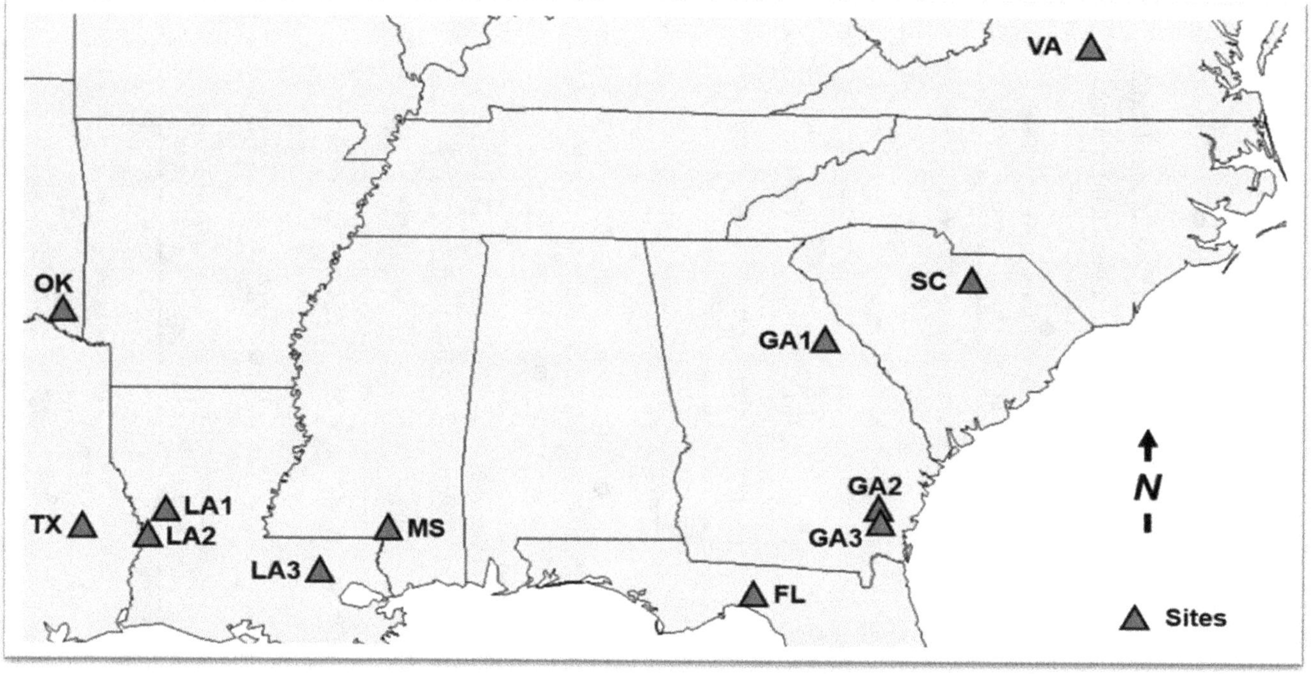

Figure 1. Location map of 12 mid-rotation pine stands across the southeastern United States selected to evaluate ecosystem partitioning of fertilizer N following the application of urea or enhanced efficiency N fertilizers enriched with ^{15}N.

2.3. Fertilizer Treatments

The five fertilizer treatments used in this study were: (1) urea; (2) urea impregnated with N-(n-Butyl) thiophosphoric triamide (NBPT); (3) urea impregnated with NBPT and coated with monoammonium phosphate (CUF); (4) polymer coated urea (PCU); and (5) a control treatment with no fertilizer added. Urea (46-0-0) was used because it is the most common N fertilizer applied in the southeastern United States. The enhanced efficiency fertilizers (EEFs) tested in this study were developed to reduce NH_3 volatilization and release fertilizer N slowly to the environment. The NBPT treatment (46-0-0) added N-(n-butyl) thiophosphoric triamide at a rate of 26.7% by weight to urea granules to inhibit urease activity. The CUF treatment (39-9-0) also added NBPT to urea granules, which was then coated with an aqueous binder solution of boron and copper sulfate to slow N release. A final coating of monoammonium phosphate was added to provide P. The PCU (44-0-0) treatment encapsulated urea granules with a polymer coating containing pores designed to slowly release N (~80%) over 120 days. All N treatments were applied at an equivalent rate of 224 $kg{\cdot}N{\cdot}ha^{-1}$. Because the CUF treatment had P in a coating, P was applied in the other fertilizer treatments at the equivalent rate of 28 $kg{\cdot}P{\cdot}ha^{-1}$ as triple superphosphate (TSP). The urea in all treatments was enriched with the stable isotope ^{15}N (0.5 atom percent). Each fertilizer treatment was broadcast applied by hand in individual 100 m^2 circular plots at each site. Due to high rates of volatilization and the impact this process has on isotopic fractionation, a fractionation factor of 1.029 was used for each fertilizer treatment as detailed in Högberg [66].

2.4. Field Sampling

The center of each of the 100 m^2 circular plots was located between two co-dominant loblolly pine trees in areas with similar stand, soil and landscape characteristics. Immediately prior to each treatment application, the height and diameter breast height (DBH) of all trees greater than 2.54 cm DBH were measured. The sapling, shrub, vine and herbaceous strata were estimated and individual species in each respective strata was composite sampled. The forest floor (O horizon = Oi + Oe + Oa) was collected with a circular sampler from 4 random locations in the plot and composited. Two mineral soil depth increments (0–15 cm, 15–30 cm) were randomly sampled from 8 locations in the plot with a push tube sampler and composited. Roots were sampled to a depth of 20 cm at 4 random locations in the plot with a bulk density corer and composited. Soil bulk density cores were taken from the 0–15 cm and 15–30 cm depth increments from the center of each plot. After this sampling was completed, two 1 m^2 circular mesh litterfall traps were placed randomly in the plot to sample litterfall from the year after N fertilization.

Fertilizer treatments were randomly applied to the 100 m^2 circular plots (March 26 to April 8). All primary ecosystem components in each plot were resampled at the end of the growing season following fertilization between November 1st and March 31st using similar sampling procedures previously detailed. Litterfall was collected once from the two litterfall traps at the end of the growing season and composited for each individual plot. One of the two central crop trees was selected and felled for sampling. All components of the crop tree sampled were weighed in the field on the same day the tree was felled to obtain field green weights, and subsamples were brought to the laboratory to obtain dry weights. In the field, a 2.54 cm cookie was taken from the tree stem at DBH, height to live crown (HLC), and height to mid crown (HMC). The tree stem was cut into 1.2 m lengths for weighing. The canopy (foliage, fine branches, coarse branches) was randomly placed in 3 piles in the field and each pile was weighed. The foliage, fine branches (branches with foliage attached) and coarse branches (branches with no foliage attached) were separated and also weighed in the field. One of the three canopy piles was randomly selected and returned to the laboratory for analysis. The sapling stratum, if present, was sampled by 2.54 cm diameter classes for individual species categories. Shrubs, vines and herbaceous species were sampled from a randomly selected 3.13 m^2 area of the plot. Individual shrub species were sampled in their entirety, with vines and herbaceous species composite sampled in their respective strata.

2.5. Laboratory Procedures

All samples were dried in a forced air oven at 60 °C. In the laboratory, subsamples of bark, wood from the current year of growth (CGR), and the wood of growth rings prior to fertilizer treatment (PGR) were taken from the stem cookies collected at DBH. Litterfall was separated into pine needles, deciduous leaves, fine branches, coarse branches, bark and unidentifiable litterfall. The forest floor was sieved through a 6 mm sieve and the mineral soil was sieved through a 2 mm sieve. Root samples were elutriated and divided into fine (<2 mm) and coarse (>2 mm) size fractions, dried and weighed.

After drying, all organic material samples were coarse ground in a Wiley Mill to pass a 2 mm sieve. The organic samples were then homogenized to a fine powder with a ball mill (Retsch® Mixer Mill MM 200, Haan, Germany) for 1 min at 25 revolutions per second (rps), whereas all mineral soil samples were ball milled for 2 min at 25 rps. After ball milling, individual homogenized samples were put in separate tin capsules and weighed on a Mettler-Toledo© MX5 microbalance (Mettler-Toledo, Inc., Columbus, OH, USA). These individually weighed samples were analyzed to determine the $^{15}N/^{14}N$ isotope ratio and total N on a coupled elemental analysis-isotope ratio mass spectrometer (IsoPrime 100 EA-IRMS, Isoprime© Ltd., Manchester, UK) at the Forest Soils and Plant Nutrition Laboratory at Virginia Polytechnic Institute and State University (Virginia Tech). All grinding, ball milling and weighing equipment were cleaned after each sample with ethanol to reduce contamination.

2.6. Calculation of Fertilizer N Recovery

The amount of total fertilizer N recovered in each ecosystem component from the labeled ^{15}N fertilizer was calculated using a mass balance tracer technique that compared individual ecosystem component ^{15}N prior to and 1 year after N fertilization [32,67,68]. Once the fertilizer N recovery for each individual ecosystem component was determined, the fertilizer N recovery for each individual component was summed on a per plot basis to calculate total fertilizer N recovery for the entire plot. The fertilizer N recovery value for the individual loblolly pine sampled in each plot was multiplied by the number of loblolly pine trees in each individual plot to obtain the total fertilizer N recovery for loblolly pine trees on an individual plot basis. The difference between the amount of fertilizer N applied to the plot and the amount of fertilizer N recovered after a single growing season was considered lost from the system.

2.7. Statistical Analysis

Total fertilizer N recovery, expressed as a percentage of fertilizer N applied, was analyzed using a general linear model (GLM) analysis of variance with SAS® 9.4 (SAS Institute Inc., Cary, NC, USA). Percent data was arcsin transformed prior to analysis. Percent fertilizer N recovery (%) was the response variable for the model, fertilizer treatment (CUF, NBPT, PCU, urea, control) was the fixed effects, and site was a random effect. Total fertilizer N recovery, expressed as a percentage of fertilizer N applied, for individual ecosystem components were also analyzed using a GLM analysis of variance, except for analysis of individual mineral soil depth increments (0–15 cm, 15–30 cm) which were analyzed as a repeated measures analysis of variance. Significance levels were set at $\alpha = 0.05$ and the $p > |t|$ values for the treatment means were tested. All post-hoc analysis was conducted with Tukey's HSD.

3. Results

Fertilization increased nitrogen concentrations ($g \cdot kg^{-1}$) in several of the ecosystem components. Foliar N mean ($\pm$ SEM) concentrations increased from 12.5 ± 0.4 $g \cdot kg^{-1}$ in the control to between 13.2 ± 0.5 $g \cdot kg^{-1}$ and 14.2 ± 0.5 $g \cdot kg^{-1}$ for fertilizer treatments, with a significant difference between the control and urea (14.2 ± 0.5 $g \cdot kg^{-1}$) (Table 2). The fine branch N concentrations increased from the control (5.1 ± 0.3 $g \cdot kg^{-1}$) to CUF (7.3 ± 0.6 $g \cdot kg^{-1}$) and NBPT (6.7 ± 0.3 $g \cdot kg^{-1}$), while the coarse branch N concentration increased between the control (2.8 ± 0.2 $g \cdot kg^{-1}$) to urea (3.9 ± 0.4 $g \cdot kg^{-1}$). For the stem, the N concentration of the bark increased between the control (2.1 ± 0.2 $g \cdot kg^{-1}$) and NBPT (2.9 ± 0.1 $g \cdot kg^{-1}$), while the N concentration for the growth ring for the year after fertilization (CGR) increased from 1.8 ± 0.1 $g \cdot kg^{-1}$ for the control to between 2.3 ± 0.1 $g \cdot kg^{-1}$ and 2.7 ± 0.1 $g \cdot kg^{-1}$ for all fertilizer treatments. There were also minor effects of N fertilization on N concentration for fine or coarse roots, litterfall and the mineral soil. The forest floor N concentration was greater for both CUF (8.1 ± 0.6 $g \cdot kg^{-1}$) and PCU (8.7 ± 0.6 $g \cdot kg^{-1}$) compared to the control (6.6 ± 0.6 $g \cdot kg^{-1}$).

Table 2. The mean fertilizer N recovery (% of applied fertilizer N), $\delta^{15}N$ (‰), and N concentration ($g \cdot kg^{-1}$) for individual ecosystem components of pine stands in the southeastern United States selected to evaluate ecosystem partitioning of fertilizer N for urea or enhanced efficiency N containing fertilizers enriched with ^{15}N.

Ecosystem Component	Treatment	Fertilizer N Recovery (% of Applied N)	$\delta^{15}N$ values (‰)	N Concentration ($g \cdot kg^{-1}$)
Foliage	Control	0.0 a	−2.4 (0.5) a	12.5 (0.4) a
	CUF	10.7 (1.1) bc	118.3 (11.2) b	14.1 (0.8) ab
	NBPT	14.8 (1.8) c	126.1 (8.5) b	13.9 (0.6) ab
	PCU	8.1 (1.1) b	101.6 (9.8) b	13.2 (0.5) ab
	Urea	11.0 (1.7) bc	124.5 (11.2) b	14.2 (0.5) b
Fine Branches	Control	0.0 a	−2.0 (0.6) a	5.1 (0.3) a
	CUF	3.2 (0.4) b	110.2 (10.2) b	7.3 (0.6) b
	NBPT	4.1 (0.5) c	111.8 (7.4) b	6.7 (0.3) b
	PCU	2.6 (0.5) b	98.7 (10.5) b	6.1 (0.5) ab
	Urea	2.9 (0.5) b	108.4 (9.67) b	6.2 (0.4) ab

Table 2. *Cont.*

Ecosystem Component	Treatment	Fertilizer N Recovery (% of Applied N)	$\delta^{15}N$ values (‰)	N Concentration ($g \cdot kg^{-1}$)
Coarse Branches	Control	0.0 a	−1.8 (0.4) a	2.8 (0.2) a
	CUF	2.9 (0.4) bc	72.6 (9.3) b	3.7 (0.3) ab
	NBPT	3.0 (0.6) c	78.9 (8.2) b	3.6 (0.4) ab
	PCU	2.6 (0.4) b	73.7 (8.6) b	3.3 (0.3) ab
	Urea	3.2 (1.0) bc	69.3 (6.1) b	3.9 (0.4) b
Bark	Control	0.0 a	−2.1 (0.4) a	2.1 (0.2) a
	CUF	0.6 (0.2) b	22.1 (2.1) b	2.5 (0.2) ab
	NBPT	0.8 (0.2) b	22.5 (2.1) b	2.9 (0.1) b
	PCU	0.5 (0.1) b	20.5 (2.4) b	2.4 (0.1) a
	Urea	0.4 (0.1) ab	20.8 (2.1) b	2.5 (0.2) ab
Current year Growth Ring (CGR)- year of fertilization	Control	0.0 a	−1.9 (0.4) a	1.8 (0.1) a
	CUF	2.2 (0.4) bc	80.0 (8.0) b	2.4 (0.5) bc
	NBPT	3.1 (0.6) c	88.1 (5.8) b	2.7 (0.1) c
	PCU	1.7 (0.3) b	71.9 (7.6) b	2.3 (0.1) b
	Urea	1.9 (0.3) b	78.9 (7.5) b	2.3 (0.1) b
Previous year growth rings (PGR)—growth prior to fertilization	Control	0.0 a	−1.9 (0.4) a	1.4 (0.1) a
	CUF	4.4 (0.8) bc	40.2 (3.9) b	1.6 (0.1) a
	NBPT	5.1 (0.7) c	40.3 (1.3) b	1.6 (0.1) a
	PCU	3.7 (0.4) bc	37.1 (3.9) b	1.6 (0.1) a
	Urea	2.9 (0.5) b	34.8 (5.0) b	1.4 (0.1) a
Fine roots (<2 mm)	Control	0.0 a	−0.3 (0.9) a	9.4 (0.9) a
	CUF	19.2 (1.8) c	36.1 (3.1) b	10.5 (0.6) a
	NBPT	16.2 (1.5) c	33.2 (3.8) b	10.4 (0.4) a
	PCU	16.4 (1.8) c	36.0 (3.5) b	10.3 (0.5) a
	Urea	10.8 (1.4) b	31.2 (5.1) b	9.7 (0.5) a
Coarse roots (>2 mm)	Control	0.0 a	−0.1 (0.37) a	6.0 (0.6) a
	CUF	3.8 (1.4) b	18.3 (1.6) b	6.2 (0.5) a
	NBPT	2.7 (0.6) ab	19.0 (3.2) b	6.3 (0.5) a
	PCU	2.9 (0.9) b	15.4 (2.2) b	6.0 (0.5) a
	Urea	1.7 (0.5) ab	13.9 (3.1) b	6.0 (0.5) a
Litterfall	Control	0.0 a	−3.0 (1.2) a	7.4 (0.3) a
	CUF	1.6 (0.3) b	48.5 (1.4) b	8.0 (0.3) a
	NBPT	1.4 (0.2) b	55.8 (2.1) b	8.3 (0.4) a
	PCU	1.8 (0.3) b	55.6 (1.6) b	8.1 (0.3) a
	Urea	1.7 (0.2) b	55.1 (1.4) b	8.2 (0.2) a
Forest Floor (Organic horizon: Oi + Oe + Oa)	Control	0.0 a	−1.9 (0.4) a	6.6 (0.6) a
	CUF	3.0 (1.4) a	62.9 (6.8) b	8.1 (0.6) b
	NBPT	2.8 (1.3) a	59.3 (5.8) b	8.0 (0.6) ab
	PCU	3.8 (0.9) b	91.1 (9.9) c	8.7 (0.6) b
	Urea	1.2 (0.4) a	55.2 (6.3) b	8.0 (0.7) ab
0–15 cm Mineral Soil	Control	0.0 a	3.3 (0.7) a	0.8 (0.1) a
	CUF	23.1 (3.6) b	21.8 (3.6) b	0.6 (0.1) b
	NBPT	22.4 (2.7) b	22.3 (3.5) b	0.6 (0.1) ab
	PCU	24.2 (5.1) b	21.0 (4.1) b	0.7 (0.1) ab
	Urea	15.9 (2.6) b	15.9 (2.1) b	0.6 (0.1) ab
15–30 cm Mineral Soil	Control	0.0 a	5.4 (0.4) a	0.4 (0.1) a
	CUF	6.7 (1.1) a	13.2 (1.5) b	0.4 (0.0) a
	NBPT	5.4 (0.9) a	11.7 (1.0) b	0.4 (0.1) a
	PCU	10.7 (3.4) b	16.3 (3.0) b	0.4 (0.0) a
	Urea	11.3 (3.1) b	15.2 (2.4) b	0.4 (0.1) a

Different letters represent significant differences at $\alpha = 0.05$. Numbers in parentheses represent the standard error of the mean ($n = 12$).

The mean (±SEM) $\delta^{15}N$ (‰) values at the end of the first growing season after treatment application for all tree and soil ecosystem components were greater for fertilizer treatments compared to the control (Table 2). The mean $\delta^{15}N$ values of loblolly pine trees for fertilizer treatments was greatest in the foliage (101.6‰ ± 9.8‰ to 126.1‰ ± 8.5‰), fine branches (98.7‰ ± 10.5‰ to 111.8‰ ± 7.4‰), and CGR (71.9 ± 7.6‰ to 88.1 ± 5.8‰). The $\delta^{15}N$ values in litterfall (48.5‰ ± 1.4‰ to 55.8‰ ± 2.1‰) for the year immediately following fertilization of all fertilized treatments was lower than the foliage mean $\delta^{15}N$ values. The lowest $\delta^{15}N$ values for the loblolly pine components for fertilized treatments were in the coarse roots (13.9‰ ± 3.1‰ to 19.0‰ ± 3.2‰), bark (20.5‰ ± 2.4‰ to 22.5‰ ± 2.1‰), fine roots (31.2‰ ± 5.1‰ to 36.1‰ ± 3.1‰) and stemwood produced in the years prior to fertilization (PGR) (34.8‰ ± 5.0‰ to 40.3 ± 1.3‰). The $\delta^{15}N$ value of the soil for the fertilized treatments was greatest in the forest floor (55.2‰ ± 6.3‰ to 91.1‰ ± 9.9‰), with PCU significantly greater than other fertilizer treatments. The surface 0–15 cm mineral soil ranged from 15.9‰ ± 2.1‰ to 22.3‰ ± 3.5‰ for fertilizer treatments, while the 15–30 cm mineral soil depth increment ranged from 11.7‰ ± 1.0‰ to 16.3‰ ± 3.0‰.

There were several significant differences in the mean (±SEM) fertilizer N recovery for individual ecosystem components (Figure 2, Table 2). The fertilizer N recovered in the loblolly pine trees was greatest in the foliage ranging from 8.1% ± 1.1% to 14.8% ± 1.8% of the N applied. More fertilizer N was recovered in the foliage for NBPT (14.8% ± 1.8%) compared to PCU (8.1% ± 1.1%). More fertilizer N was also recovered for NBPT (4.1% ± 0.5%) than the other fertilizer treatments (2.6% ± 0.5% to 3.2% ± 0.4%) for fine branches, and coarse branches for NBPT (3.0% ± 0.6%) compared to PCU (2.6% ± 0.5%). Fertilizer N recovery was greater for NBPT (3.1% ± 0.6%) for CGR than PCU (1.7% ± 0.3%) and urea (1.9 ± 0.3%), and the PGR for NBPT (5.1% ± 0.7%) compared to urea (2.9% ± 0.5%). For belowground loblolly pine biomass, more fertilizer N was recovered for EEFs (16.2% ± 1.5% to 19.2% ± 1.8%) compared to urea (10.8% ± 1.4%) in fine roots. In the soil, there was greater fertilizer N recovery for PCU (3.8% ± 0.9%) compared to the other fertilizer treatments (1.2% ± 0.4% to 3.0% ± 1.4%) in the forest floor, and for both PCU (10.7% ± 3.4%) and urea (11.3% ± 3.1%) compared to CUF (6.7% ± 1.1%) and NBPT (5.4% ± 0.9%) in the 15–30 cm mineral soil.

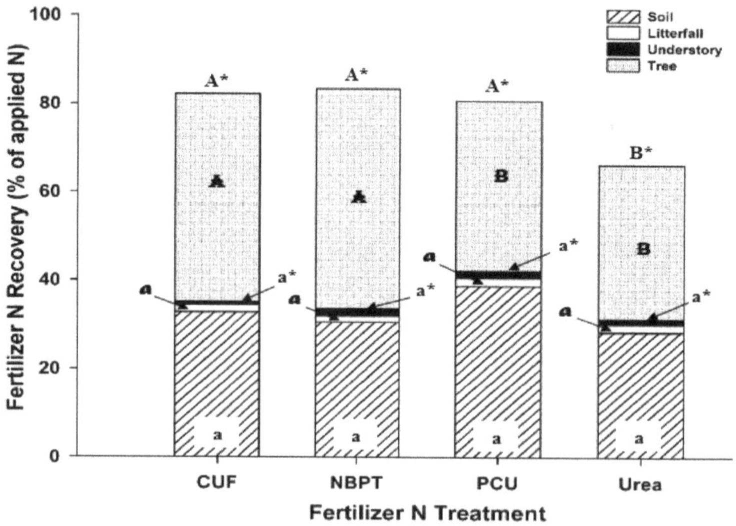

Figure 2. The total fertilizer N recovery (% of fertilizer N applied) of the major ecosystem components (Tree—loblolly pine aboveground + belowground biomass, understory, litterfall, and soil (O horizon + mineral soil- 0 to 30 cm) for pine stands in the southeastern United States selected to evaluate ecosystem partitioning of fertilizer N of urea or enhanced efficiency N fertilizers enriched with ^{15}N. Data represents fertilizer application dates for spring (March–April 2012). Different letters represent significant differences at α = 0.05. Different letter fonts represent comparisons among treatments between same ecosystem components (soil, litterfall, understory, tree). n = 12.

Differences between fertilizer treatments also occurred when individual ecosystem components were combined into primary components (tree, soil) (Figure 2). For the loblolly pine canopy (foliage, fine branches, coarse branches), fertilizer N recovery for NBPT (22.0%) was greater than PCU (13.3%). For the stem (bark, CGR, PGR), NBPT (9.1%) was greater than both PCU (5.9%) and urea (5.2%). When all aboveground biomass components are combined (canopy, stem), NBPT (31.0%) was greater than both PCU (19.2%) and urea (22.3%). For belowground biomass (total roots), all EEFs (CUF = 23.0%, NBPT = 18.9%, PCU = 19.3%) were greater than urea (12.5%). When all pine components were combined (canopy, stem, roots), both CUF (46.9%) and NBPT (49.9%) had a greater fertilizer N recovery compared to PCU (38.5%) and urea (34.8%). For the entire soil (forest floor, 0–30 cm mineral soil), the percentage of recovery was similar for EEFs (30.6% to 38.8%) and urea (28.4%). When all ecosystem components were combined to determine the total ecosystem recovery of fertilizer N, all the EEFs (CUF = 81.9%, NBPT = 84.2%, PCU = 79.8%) had a greater recovery total fertilizer N recovery compared to urea (65.2%).

4. Discussion

This study evaluated differences in the ecosystem retention and crop tree uptake of applied fertilizer N between urea and three enhanced efficiency fertilizers in mid-rotation pine plantations of the southeastern United States. The sites in this study covered the entire southeastern United States region where loblolly pine is planted to improve the understanding of the ultimate fate of fertilizer N in these systems to augment the results of numerous site specific studies. The primary hypotheses tested in this study were: (1) whether there were differences in the total fertilizer N recovery among conventional and enhanced efficiency fertilizers; and (2) if there were differences in the ecosystem partitioning of fertilizer N among conventional and enhanced efficiency N fertilizers. The overall primary objective was to improve fertilizer N use efficiency in southeastern pine plantations to increase the productivity and efficiency of these systems in a sustainable, environmentally responsible approach.

There were significant differences in the total ecosystem fertilizer N recovery among individual fertilizer treatments (CUF, NBPT, PCU, urea). The total ecosystem fertilizer N recovery was greater for all enhanced efficiency fertilizers (CUF, NBPT, PCU) compared to urea, with no differences between individual EEFs. The primary reason for lower fertilizer N recovery for urea compared to EEFs was likely due to initially large ammonia (NH_3) volatilization losses from the urea treatment compared to the EEFs immediately following fertilization. Raymond et al. [49] compared NH_3 volatilization losses of the same fertilizer treatments used in this study and found higher losses with urea (26%–49%) compared to all EEFs (4%–26%). Other studies in pine plantations in the southeastern United States have also shown fertilizer N losses through the NH_3 volatilization pathway after urea fertilization exceeding 25%, with lower losses using various EEF products [47,48,64–66,69,70]. Interestingly, when the results for NH_3 volatilization reported by Raymond et al. [49] 15 days following fertilization for the same fertilizer treatments used in this study are included in the treatments in respect to mass balance accounting of ecosystem fertilizer N recovery on a per plot basis, total ecosystem recovery of fertilizer N for each treatment exceeds 90%. This evidence indicates that greater fertilizer N loss via NH_3 volatilization immediately following fertilization, combined with other minor initial fertilizer N loss pathways (denitrification, leaching) for urea compared to EEFs, translates to less fertilizer N remaining in the ecosystem for urea. Although the results from this study are similar to those of other tracer studies in a variety of ecosystems [39,40,71,72], results from this study indicated a generally higher fertilizer N uptake by trees for the EEFs.

Results from numerous studies specific to loblolly pine plantations and fertilizer N show generally lower fertilizer N uptake (<30%) by loblolly pine trees [21–25]. Several potential explanations for this low fertilizer N uptake by loblolly pine plantations exist. First, many studies in loblolly plantations have used urea as the fertilizer N source. Because high fertilizer N losses can occur immediately following fertilization with urea [46–49], less fertilizer N remains in the ecosystem and hence less fertilizer N is available for plant uptake. As previously stated, this explanation partially explains the results for the fertilizer N recovery in the loblolly pine trees in this study (35%) which is near the upper end reported in the literature. The fertilizer N recovery for the EEF treatments used in this study had higher fertilizer N recovery for the entire tree compared to urea, ranging from 39% to 50%. Clearly, the lower amount of fertilizer N remaining in the system when urea is used as a fertilizer source affects the quantity, and hence intensity, of fertilizer N in the soil over the growing season that becomes available for uptake by the desired crop trees. Second, the application of fertilizer N for this study in the spring (March-April) may have been more synchronous to the seasonal growth patterns of desired crop trees and could explain higher recovery rates of fertilizer N for all treatments found in this study. For example, results from Blazier et al. [25] had higher fertilizer N recovery after a spring and summer fertilization compared to a winter application, when operational N fertilization is traditionally conducted in southeastern pine plantations in an effort to minimize high NH_3 volatilization losses. Although losses were high for urea compared to the EEFs, recovery for the entire tree for fertilizer N was greater compared to most fertilizer studies. This result may indicate that although losses can be high in the spring, conditions for N demand and uptake by the crop tree for a readily available N source may still be high. Low fertilizer N uptake by trees after fertilization during the dormant season (winter) in the south may cause other loss pathways, such as denitrification [51] and leaching [53] to become more important than NH_3 volatilization and also contribute to lower fertilizer N recovery in studies. Third, many studies only account for N uptake in the foliage. Although foliage in loblolly pine trees has a high amount of N in the tree, other portions of the tree also contain N. If only N in the foliage is measured, N uptake may be underestimated.

There were also significant differences in ecosystem partitioning of fertilizer N among treatments. Analysis of the percent of fertilizer N recovered in individual ecosystem components (foliage, fine branches, soil, etc.) showed numerous differences among fertilizer treatments. Despite differences among fertilizer treatments in the percentage of fertilizer N recovered in different ecosystem components, most fertilizer N for all treatments was recovered in either the loblolly pine trees or soil. There was a difference for the fertilizer N recovered in the entire tree between NBPT and CUF and urea and PCU, yet no significant differences occurred in the soil. Although there were no significant differences in the percentage of fertilizer N in the soil, there was 10% more fertilizer N recovered (NS) for PCU compared to urea. This difference may have important implications for the long term fate and cycling of fertilizer N for PCU, which provides a more gradual constant release compared to the other fertilizers. Additionally, significantly higher amounts of fertilizer N were found in the fine roots of all EEFs compared to urea. This additional amount of fertilizer N was likely the result of a higher amount of fertilizer N in the soil and potentially greater root uptake for EEFs compared to urea.

Although results from this research showed a greater ecosystem recovery for each EEF compared to urea, several primary questions remain concerning the long-term fate of each of these fertilizer treatments. For example, significantly more fertilizer N was recovered in the forest floor for PCU than other treatments. Additionally, although not significant, there was a high proportion of fertilizer N recovered for all EEFs in the 0–15 cm mineral soil compared to urea. A shift in the ecosystem recovery of fertilizer N compared to urea in the ecosystem raises the question of whether additional gains in fertilizer nitrogen use efficiency for the system for EEFs will increase if it becomes bioavailable in the future of the stand or becomes immobilized in the system for the stand rotation.

Results from agroecosystems have shown that the majority of fertilizer N that is not incorporated into plant biomass during the initial growing season after fertilization becomes immobilized in the soil [73]. Similar results specific to fertilizer N immobilization in the soil have also been shown in forested systems [74–78]. Despite these results, laboratory experiments using a bioassay approach with soil collected from long term ^{15}N tracer experiments show a measureable uptake of ^{15}N labeled fertilizers by seedlings from both forest floor and mineral soil sources a decade after fertilization [79]. The mechanisms governing this disconnect between field and laboratory studies will require additional examination. Yet, if it is found that the additional fertilizer N becomes bioavailable to seedlings after the harvesting of the remaining stand, this finding could translate to additional gains in fertilizer nitrogen use efficiency in these managed pine ecosystems. At certain managed pine ecosystem sites, especially those which have been thinned, fertilizer management may include an initial fertilizer N application at stand establishment or shortly after. Yet, if the increase in existing fertilizer N remaining in the soil is found to be bioavailable, a management system may be altered to forgo the initial N fertilization while still maintaining high levels of productivity.

Extending research for this study to monitor the long term fate of the ^{15}N enriched fertilizer treatments will be needed to answer the questions specific to the bioavailability of the fertilizer N remaining in the soil. Answering whether the increase in fertilizer N in the soil with EEFs will translate to a higher N availability for trees in these intensively managed pine ecosystems in the future or become immobilized and/or leached from the system is an important question to continue improving the economic viability and environmental stewardship of fertilization in these systems. If future research for the ^{15}N plots used in this study find a majority of the fertilizer N immobilized in the soil for EEFs, as has been found with other studies using traditional N fertilizers in forest ecosystems, the primary question becomes whether fertilizer N application rates can be reduced by a percentage when compared to the traditional rates used in forestry while maintaining productivity. Results indicating either a higher fertilizer N availability for extended periods from the soil using EEFs and/or a reduction in fertilizer N application rates while maintaining productivity would assist in improving the FNUE efficiency of these pine plantation systems.

This research has continued to refine our understanding of the differences in the ecosystem fate and cycling of fertilizer N between enhanced efficiency and urea fertilizers through the use of stable isotopes in southeastern pine plantations. Results from this study are from a broad geographic region, and indicate similar cycling patterns and differences between enhanced efficiency N containing fertilizers compared to urea after a spring application. The results from this study provide forest managers a level of confidence that the enhanced efficiency products used in this research after a spring fertilization will have a generally higher fertilizer N ecosystem retention when compared with urea across the southeastern United States where pine plantations are intensively managed. The results from this study will continue to assist forest managers in improving fertilizer nitrogen use efficiency in pine plantations across the southeastern United States.

5. Conclusions

The principal results of this study were: (1) there was a greater amount of fertilizer N recovered from the ecosystem for all EEFs compared to urea; and (2) there were differences in the ecosystem partitioning of fertilizer N among treatments. The reduced ecosystem loss of fertilizer N for all EEFs compared to urea may translate to an increased quantity of fertilizer N remaining in the system that could be available for crop tree N uptake and increase the FNUE for the EEFs used in this study compared to urea. The increased fertilizer N retention in the soil for all EEFs, although not significant, may also contribute to increasing FNUE over the course of the stand rotation if the increased quantity of soil N translates to an increase in intensity of N supply. Additional research will be required to determine the long term fate of EEFs in forest plantations, and whether this added quantity of soil N from fertilization using EEFs translates to increased productivity. The results from this study will continue to improve site specific management and increase the efficiency of southeastern pine plantations. The use of enhanced efficiency fertilizers, if found to be economically viable, could provide managers the flexibility to apply fertilizer N under a variety of conditions to improve the synchronicity of fertilizer application and plant N demand.

Acknowledgments: This research was primarily supported by the Forest Productivity Cooperative (FPC), the National Science Foundation Center for Advanced Forest Systems (CAFS) and the United States Department of Agriculture National Institute of Food and Agriculture McIntire-Stennis Capacity Grant. We thank David Mitchem for assistance in the laboratory, Andy Laviner and Eric Carbaugh for field assistance and the numerous work study and seasonal student employees for both field and laboratory assistance.

Author Contributions: "Thomas Fox, Jay Raymond and Brian Strahm conceived and designed the experiments. Jay Raymond performed the experiments. Jay Raymond analyzed the data; Thomas Fox and Brian Strahm contributed reagents/materials/analysis tools; Jay Raymond wrote the paper.

References

1. Schultz, R.P. *The Ecology and Culture of Loblolly Pine (Pinus taeda L.)*; Agriculture Handbook #713; USDA Forest Service: Washington DC, USA, 1997; p. 493.
2. US Department of Agriculture (USDA), Forest Service. *Future of America's Forest and Rangelands: Forest Service 2010 Resources Planning Act Assessment*; US Department of Agriculture (USDA), Forest Service: Washington DC, USA, 2012; p. 198. Available online: http://www.fs.fed.us/research/publications/gtr/gtr_wo87.pdf (accessed on 5 February 2016).
3. Fox, T.R.; Jokela, E.J.; Allen, H.L. The development of pine plantation silviculture in the southern United States. *J. For.* **2007**, *105*, 337–347.
4. Allen, H.L. Forest fertilizers: Nutrient amendment, and productivity, and environmental impact. *J. For.* **1987**, *85*, 37–46.
5. Fox, T.R.; Allen, H.L.; Albaugh, T.J.; Rubilar, R.; Carlson, C.A. Tree nutrition and forest fertilization of pine plantations in the southern United States. *South. J. Appl. For.* **2007**, *31*, 5–11.
6. Miller, H.G. Forest fertilization: Some guiding concepts. *Forestry* **1981**, *54*, 157–167. [CrossRef]
7. Chapin, F.S.; Vitousek, P.M.; Van Cleve, K. The nature of nutrient limitation in plant-communities. *Am. Nat.* **1986**, *127*, 48–58. [CrossRef]
8. Vitousek, P.M.; Howarth, R.W. Nitrogen limitation on land and in the sea—How can it occur? *Biogeochemistry* **1991**, *13*, 87–115. [CrossRef]
9. Linder, S. Responses to water and nutrients in coniferous ecosystems. In *Potentials and Limitations of Ecosystems Analysis*; Schulze, E.D., Wolfer, H.Z., Eds.; Springer-Verlag: New York, NY, USA, 1987; pp. 180–202.
10. LeBauer, D.S.; Treseder, K.K. Nitrogen limitation of net primary productivity in terrestrial ecosystems is globally distributed. *Ecology* **2008**, *89*, 371–379. [CrossRef] [PubMed]
11. Switzer, C.L.; Nelson, L.E. Nutrient accumulation and cycling in loblolly pine (*Pinus taeda* L.) plantation ecosystems: The first 20 years. *Soil Sci. Soc. Am. Proc.* **1972**, *36*, 143–147. [CrossRef]
12. Wells, C.G.; Jorgensen, J.R. Effect of intensive harvesting on nutrient supply and sustained productivity. In Proceedings of the Impacts of Intensive Harvesting on Forest Nutrient Cycling, Syracuse, NY, USA, 1979; pp. 212–230.
13. Vitousek, P.M.; Matson, P.A. Intensive harvesting and site preparation decrease soil nitrogen availability in young plantations. *South. J. Appl. For.* **1985**, *9*, 120–125.
14. Fox, T.R.; Burger, J.A.; Kreh, R.E. Effects of site preparation on nitrogen dynamics in the southern Piedmont. *For. Ecol. Manag.* **1986**, *15*, 241–256. [CrossRef]
15. Birk, E.M.; Vitousek, P.M. Nitrogen availability and nitrogen use efficiency in loblolly pine stands. *Ecology* **1986**, *67*, 69–79. [CrossRef]
16. Pritchett, W.L.; Smith, W.H. Fertilizer response in young slash pine stands. *Soil Sci. Soc. Am. J.* **1972**, *36*, 660–663. [CrossRef]
17. Martin, S.W.; Bailey, R.L.; Jokela, E.J. Growth and yield predictions for lower Coastal Plain slash pine plantations fertilized at mid-rotation. *South. J. Appl. For.* **1999**, *23*, 39–45.
18. Amateis, R.L.; Liu, J.; Ducey, M.J.; Allen, H.L. Modeling response to midrotation nitrogen and phosphorus fertilization in loblolly pine plantations. *South. J. Appl. For.* **2000**, *24*, 207–212.
19. Carlson, C.A.; Fox, T.R.; Allen, H.L.; Albaugh, T.J.; Rubilar, R.A.; Stape, J.L. Growth responses of loblolly pine in the southeast united states to midrotation applications of nitrogen, phosphorus, potassium, and micronutrients. *For. Sci.* **2014**, *60*, 157–169. [CrossRef]
20. Rojas, J.C. Factors influencing responses of loblolly pine stands to fertilization. Ph.D. Dissertation, North Carolina State University, Raleigh, NC, USA, 2005.
21. Mead, D.J.; Pritchett, W.L. Fertilizer movement in a slash pine ecosystem. *Plant Soil* **1975**, *43*, 451–465. [CrossRef]

22. Melin, J.; Nommik, H.; Lohm, U.; Flower-Ellis, J. Fertilizer nitrogen budget in a Scots pine ecosystem attained by using root-isolated plots and ^{15}N technique. *Plant Soil* **1983**, *74*, 249–263. [CrossRef]
23. Johnson, D.W.; Todd, D.E. Nitrogen fertilization of young yellow poplar and loblolly pine plantations at differing frequencies. *Soil Sci. Soc. Am. J.* **1988**, *52*, 1468–1477. [CrossRef]
24. Albaugh, T.J.; Allen, H.L.; Dougherty, P.M.; Johnsen, K.H. Long term growth responses of loblolly pine to optimal nutrient and water resource availability. *For. Ecol. Manag.* **2004**, *192*, 3–19. [CrossRef]
25. Blazier, M.A.; Hennessey, T.C.; Dougherty, P.; Campbell, R. Nitrogen accumulation and use by a young loblolly pine plantation in southeast Oklahoma: Effects of fertilizer formulation and date of application. *South. J. Appl. For.* **2006**, *30*, 66–78.
26. Raison, R.J.; Khanna, P.K.; Connell, M.J.; Falkiner, R.A. Effects of water availability and fertilization on nitrogen cycling in a stand of Pinus radiate. *For. Ecol. Manag.* **1990**, *30*, 31–43. [CrossRef]
27. Kiser, L.C.; Fox, T.R. Soil accumulation of nitrogen and phosphorus following annual fertilization of loblolly pine and sweetgum on sandy sites. *Soil Sci. Soc. Am. J.* **2012**, *76*, 2278–2288. [CrossRef]
28. Villanueva, A.T. Impacts of Fertilization on Soil Properties in Loblolly Pine Plantations in the Southeastern United States. Master's Thesis, Virginia Polytechnic Institute and State University, Blacksburg, VA, USA, 2015.
29. Fisher, R.F.; Binkley, D. *Ecology and Management of Forest Soils*; John Wiley & Sons: New York, NY, USA, 2000.
30. Flint, C.M. Leaching of Nitrogen from the Rooting Zone of Douglas-Fir Forests Following Urea Fertilization and Potential Impacts on the Water Quality of Hood Canal. Master's Thesis, University of Washington, Seattle, WA, USA, 2007.
31. Knowles, R.; Blackburn, T.H. *Nitrogen Isotope Techniques*; Academic Press, Inc.: New York, NY, USA, 1993.
32. Nadelhoffer, K.J.; Fry, B. Nitrogen isotope studies in forest ecosystems. In *Stable Isotopes in Ecology and Environmental Science*; Lajtha, K., Michener, R.H., Eds.; Blackwell Scientific Publications: New York, NY, USA, 1994; pp. 22–44.
33. Robinson, D. δ15N as an integrator of the nitrogen cycle. *Trends Ecol. Evol.* **2001**, *16*, 153–162. [CrossRef]
34. Dawson, T.E.; Mambelli, S.; Plamboeck, A.H.; Templer, P.H.; TuSource, K.P. Stable isotopes in plant ecology. *Annu. Rev. Ecol. Syst.* **2002**, *33*, 507–559. [CrossRef]
35. Hauck, R.D.; Bystrom, M. *^{15}N-A Selected Bibliography for Agricultural Scientists*; The Iowa State University Press: Ames, IA, USA, 1971; p. 206.
36. Hauck, R.D. Nitrogen source requirements in different soil-plant systems. In *Forest Fertilization: Theory and Practice*; TVA National Fertilizer Development Center: Muscle Shoals, AL, USA, 1968; pp. 47–57.
37. Currie, W.S.; Nadelhoffer, K.J. Dynamic redistribution of isotopically labeled cohorts of nitrogen inputs in two temperate forests. *Ecosystems* **1999**, *2*, 4–18. [CrossRef]
38. Dinkelmeyer, H.; Lehmann, J.; Renck, A.; Trujillo, L.; Pereira da Silva, J., Jr.; Gebauer, G.; Kaiser, K. Nitrogen uptake from ^{15}N-enriched fertilizer by four tree crops in an Amazonian agroforest. *Agrofor. Syst.* **2003**, *57*, 213–224. [CrossRef]
39. Nadelhoffer, K.J.; Colman, B.P.; Currie, W.S.; Magill, A.; Aber, J.D. Decadal-scale fates of ^{15} N tracers added to oak and pine stands under ambient and elevated N inputs at the Harvard Forest (USA). *For. Ecol. Manag.* **2004**, *196*, 89–107. [CrossRef]
40. Tietema, A.; Emmett, B.A.; Gundersen, P.; Kjonaas, O.J.; Koopmans, C.J. The fate of ^{15}N-labelled nitrogen deposition in coniferous forest ecosystems. *For. Ecol. Manag.* **1998**, *101*, 19–27. [CrossRef]
41. Templer, P.H.; Mack, M.C.; Chapin, F.S., III; Christenson, L.M.; Compton, J.E.; Crook, H.D.; Currie, W.S.; Curtis, C.J.; Dail, D.B.; D'Antonio, C.M.; et al. Sinks for nitrogen inputs in terrestrial ecosystems: A meta-analysis of ^{15}N tracer field studies. *Ecology* **2012**, *93*, 1816–1829. [CrossRef] [PubMed]
42. Chang, S.X.; Preston, C.M.; McCullough, K.; Weetman, G.; Barker, J. Effect of understory competition on distribution and recovery of ^{15}N applied to a western red cedar—Western hemlock clear-cut site. *Can. J. For. Res.* **1996**, *26*, 313–321. [CrossRef]
43. Bubb, K.A.; Xu, Z.H.; Simpson, J.A.; Saffigna, P.G. Growth response to fertilization and recovery of ^{15}N-labelled fertilizer by young hoop pine plantations of subtropical Australia. *Nutr. Cycling Agroecosystems* **1999**, *54*, 81–92. [CrossRef]
44. Werner, A.T. Nitrogen Release, Tree Uptake, and Ecosystem Retention in a Mid-Rotation Loblolly Pine Plantation Following Fertilization with 15N-Enriched Enhanced Efficiency Fertilizers. Master's Thesis, Virginia Tech, Blacksburg, VA, USA, 2013.

45. Zhang, X.; Mauzerall, D.L.; Davidson, E.A.; Kanter, D.R.; Cai, R. The economic and environmental consequences of implementing nitrogen-efficient technologies and management practices in agriculture. *J. Environ. Qual.* **2015**, *44*, 312–324. [CrossRef] [PubMed]
46. Cabrera, M.L.; Kissel, D.E.; Vaio, N.; Craig, J.R.; Rema, J.A.; Morris, L.A. Loblolly pine needles retain urea fertilizer that can be lost as ammonia. *Soil Sci. Soc. Am. J.* **2005**, *69*, 1525–1531. [CrossRef]
47. Zerpa, J.L.; Fox, T.R. Controls on volatile NH_3 losses from loblolly pine plantations fertilized with urea in the southeast USA. *Soil Sci. Soc. Am. J.* **2011**, *75*, 257–266. [CrossRef]
48. Elliot, J.R.; Fox, T.R. Ammonia volatilization following fertilization with urea or ureaform in a thinned loblolly pine plantation. *Soil Sci. Soc. Am. J.* **2014**, *78*, 1469–1473. [CrossRef]
49. Raymond, J.E.; Fox, T.R.; Strahm, B.; Zerpa, J. Ammonia Volatilization following Nitrogen Fertilization with Enhanced Efficiency Fertilizers and Urea in Loblolly Pine Plantations of the Southern United States. *For. Ecol. Manag.* **2016**, *376*, 247–255. [CrossRef]
50. Engel, R.; Jones, C.; Wallander, R. Ammonia volatilization from urea and mitigation by by NBPT following surface application to cold soils. *Soil Sci. Soc. Am. J.* **2011**, *75*, 2348–2357. [CrossRef]
51. Shrestha, R.; Strahm, B.D.; Sucre, E.B. Nitrous oxide fluxes in fertilized *Pinus taeda* plantations across a gradient of soil drainage classes. *J. Environ. Qual.* **2014**, *43*, 1823–1832. [CrossRef] [PubMed]
52. Binkley, D.; Carter, R.; Allen, H.L. Nitrogen fertilization practices in forestry. In *Nitrogen Fertilization and the Environment*; Bacon, P., Ed.; Marcel Dekker: New York, NY, USA, 1995; pp. 421–441.
53. Aust, W.M.; Blinn, C.R. Forestry best management practices for timber harvesting and site preparation in the eastern United States: An overview of water quality and productivity research during past 20 years. *Water Air Soil Poll. Focus* **2004**, *4*, 5–36. [CrossRef]
54. Galloway, J.N.; Aber, J.D.; Erisman, J.W.; Seitzinger, S.P.; Howarth, R.W.; Cowling, E.B.; Cosby, B.J. The Nitrogen Cascade. *Bioscience* **2003**, *53*, 341–356. [CrossRef]
55. Chien, S.H.; Prochnow, L.I.; Cantarella, H. Recent developments of Fertilizer Production and Use to Improve Nutrient Efficiency and Minimize Environmental Impacts. In *Advances in Agronomy 102*; Elsevier Inc.: Newark, NJ, USA, 2009; pp. 267–322.
56. Trenkel, M.E. *Slow-and Controlled-Release and Stabilized Fertilizers: An option for Enhancing Nutrient Use Efficiency in Agriculture*; IFA, International Fertilizer Industry Association: Paris, France, 2010.
57. Azeem, B.; KuShaari, K.; Man, Z.B.; Basit, A.; Thanh, T.H. Review on materials & methods to produce controlled release coated urea fertilizer. *J. Control. Release* **2014**, *181*, 11–21. [PubMed]
58. Fertilizer Institute. Available online: https://www.tfi.org/introduction-fertilizer/nutrient-science/enhanced-efficiency-fertilizers (accessed on 11 October 2015).
59. AAPFCO. Association of American Plant Food Control Officials. 2011. Available online: http://www.aapfco.org/index.html (accessed on 1 October 2015).
60. Shaviv, A. *Controlled Release Fertilizers. IFA International Workshop on Enhanced-Efficiency Fertilizers, Frankfurt*; International Fertilizer Industry Association: Paris, France, 2005.
61. Hauck, R.D. Slow release and bio-inhibitor-amended nitrogen fertilizers. In *Fertilizer Technology and Use*, 3rd ed.; Engelstad, O.P., Ed.; SSSA: Madison, WI, USA, 1985; pp. 293–322.
62. Goertz, H.M. Controlled Release Technology. In *Kirk-Othmer Encyclopedia of Chemical Technology*; Controlled Release Technology (Agricultural): New York, NY, USA, 1993; pp. 251–274.
63. Gowariker, V.; Krishnamurthy, V.N.; Gowariker, S.; Dhanorkar, M.; Paranjape, K. *The Fertilizer Encyclopedia*; John Wiley & Sons, Inc.: Hoboken, NJ, USA, 2009.
64. Bremner, J.M.; Douglas, L.A. Inhibition of urease activity in soils. *Soil Biol. Biochem.* **1971**, *3*, 297–307. [CrossRef]
65. Bremner, J.M.; Chai, H.S. Evaluation of N-butyl phosphorothioic triamide for retardation of urea hydrolysis in soil. *Commun. Soil Sci. Plant Anal.* **1986**, *17*, 337–351. [CrossRef]
66. Högberg, P. ^{15}N natural abundance in soil-plant systems. *New Phytol.* **1997**, *137*, 179–203. [CrossRef]
67. Powlson, D.S.; Barraclough, D. Mineralization and assimilation in soil-plant systems. In *Nitrogen Isotope Techniques*; Academic Press, Inc.: New York, NY, USA, 1993; pp. 209–242.
68. Nadelhoffer, K.J.; Downs, M.R.; Fry, B.; Aber, J.D.; Magill, A.H.; Melillo, J.M. The Fate of ^{15}N-Labelled Nitrate Additions to a Northern Hardwood Forest in Eastern Maine, USA. *Oecologia* **1995**, *103*, 292–301. [CrossRef]
69. Kissel, D.; Cabrera, M.; Vaio, N.; Craig, J.; Rema, J.; Morris, L. Rainfall timing and ammonia loss from urea in a loblolly pine plantation. *Soil Sci. Soc. Am. J.* **2004**, *68*, 1744–1750. [CrossRef]

70. Kissel, D.; Cabrera, M.L.; Vaio, N.; Craig, J.R.; Rema, J.A.; Morris, L.A. Forest floor composition and ammonia loss from urea in a loblolly pine plantation. *Soil Sci. Soc. Am. J.* **2009**, *73*, 630–637. [CrossRef]
71. Burke, I.C.; Lauenroth, W.K. Ecosystem ecology at regional scales. *Ecology* **2002**, *83*, 305–306. [CrossRef]
72. Blanes, M.C.; Emmett, B.A.; Viñegla, B.; Carreira, J.A. Alleviation of P limitation makes tree roots competitive for N against microbes in a N-saturated conifer forest: A test through P fertilization and ^{15}N labelling. *Soil Biol. Biochem.* **2012**, *48*, 51–59. [CrossRef]
73. Preston, C.M. The availability of residual fertilizer nitrogen immobilized as clay-fixed ammonium and organic N. *Can. J. Soil Sci.* **1982**, *62*, 479–486. [CrossRef]
74. Foster, N.W.; Beauchamp, E.G.; Corke, C.T. Reactions of ^{15}N-labelled urea with Jack pine forest floor materials. *Soil Biol. Biochem.* **1985**, *17*, 699–703. [CrossRef]
75. Foster, N.W.; Beauchamp, E.G.; Corke, C.T. Immobilization of ^{15}N-labelled urea in a Jack pine forest floor. *Soil Sci. Soc. Am. J.* **1985**, *49*, 448–452. [CrossRef]
76. Hulm, S.C.; Killham, K. Response over two growing seasons of a Sitka spruce stand to ^{15}N-urea fertilizer. *Plant Soil.* **1990**, *124*, 65–72. [CrossRef]
77. NÔmmik, H.; Larsson, K. Effects of nitrogen source and placement on fertilizer ^{15}N enrichment in *Pinus sylvestris* foliage. *Scand. J. For. Res.* **1992**, *7*, 155–163. [CrossRef]
78. Preston, C.M.; Mead, D.J. Growth response and recovery of ^{15}N-fertilizer one and eight growing season after application to lodgepole pine in British Columbia. *For. Ecol. Manag.* **1994**, *65*, 219–229. [CrossRef]
79. Swanston, C.W.; Preston, C.M. Availability of residual fertilizer ^{15}N from forest floor and mineral soil to Douglas-fir seedlings ten years after fertilization. *Plant Soil* **2014**, *381*, 381–394. [CrossRef]

10

Fine Root Biomass Mediates Soil Fauna Community in Response to Nitrogen Addition in Poplar Plantations (*Populus deltoids*) on the East Coast of China

Haixue Bian [1,2], Qinghong Geng [1,2], Hanran Xiao [1,2], Caiqin Shen [3], Qian Li [4], Xiaoli Cheng [5], Yiqi Luo [6], Honghua Ruan [1,2] and Xia Xu [1,2,*]

1 College of Biology and the Environment, Nanjing Forestry University, Nanjing, Jiangsu 210037, China; haixue_bian@163.com (H.B.); gengqh0127@163.com (Q.G.); xiaobaisky123@163.com (H.X.); hhruan@njfu.edu.cn (H.R.)
2 Co-Innovation Centre for Sustainable Forestry in Southern China, Nanjing Forestry University, Nanjing, Jiangsu 210037, China
3 Dongtai Forest Farm, Yancheng, Jiangsu 224200, China; shencaiqin123@163.com
4 Advanced Analysis and Testing Center, Nanjing Forestry University, Nanjing, Jiangsu 210037, China; liqian.1987@outlook.com
5 Key Laboratory of Aquatic Botany and Watershed Ecology, Wuhan Botanical Garden, Chinese Academy of Sciences, Wuhan 430074, China; chengxiaoli@wbgcas.cn
6 Center for Ecosystem Science and Society, Northern Arizona University, Flagstaff, AZ 86011, USA; yiqi.luo@nau.edu
* Correspondence: xuxia.1982@njfu.edu.cn

Abstract: Soil fauna is critical for maintaining ecosystem functioning, and its community could be significantly impacted by nitrogen (N) deposition. However, our knowledge of how soil-faunal community composition responds to N addition is still limited. In this study, we simulated N deposition (0, 50, 100, 150, and 300 kg N ha^{-1} $year^{-1}$) to explore the effects of N addition on the total and the phytophagous soil fauna along the soil profile (0–10, 10–25, and 25–40 cm) in poplar plantations (*Populus deltoids*) on the east coast of China. Ammonium nitrate (NH_4NO_3) was dissolved in water and sprayed evenly under the canopy with a backpack sprayer to simulate N deposition. Our results showed that N addition either significantly increased or decreased the density (D) of both the total and the phytophagous soil fauna (D_{total} and D_p) at low or high N addition rates, respectively, indicating the existence of threshold effects over the range of N addition. However, N addition had no significant impacts on the number of groups (G) and diversity (H) of either the total or the phytophagous soil fauna (G_{total}, G_p and H_{total}, H_p). With increasing soil depth, D_{total}, D_p, G_{total}, and G_p largely decreased, showing that the soil fauna have a propensity to aggregate at the soil surface. H_{total} and H_p did not significantly vary along the soil profile. Importantly, the threshold effects of N addition on D_{total} and D_p increased from 50 and 100 to 150 kg N ha^{-1} $year^{-1}$ along the soil profile. Fine root biomass was the dominant factor mediating variations in D_{total} and D_p. Our results suggested that N addition may drive changes in soil-faunal community composition by altering belowground food resources in poplar plantations.

Keywords: soil fauna; N addition; soil profile; community structure; food resources; poplar plantations

1. Introduction

Nitrogen (N) deposition, which is predicted to continuously increase, has become a major global concern [1–3]. In terrestrial ecosystems, most of the deposited N eventually dissipates into soils [4].

N accumulation in soils could change the community structure of soil fauna and could thereby impact ecosystem functioning [5]. Collectively, the soil fauna is a major consumer and decomposer and is therefore an essential component of the forest ecosystem [6,7]. As the "engineer of the soil ecosystem", soil fauna plays an important role in dissolving residues and altering biogeochemical cycles [8,9]. However, our understanding of the responses of soil fauna community to increased levels of N input is still limited.

Increasing evidence has demonstrated that N addition that mimics natural N deposition can undoubtedly have profound impacts on soil fauna communities. Despite much research, however, no consensus exists regarding the impacts of N addition in soil fauna communities. For instance, N addition significantly increases the densities of both phytophagous soil fauna and total soil fauna, which could be attributed to enhanced ammonium (NH_4^+) production and nitrification processes from increased root exudation and organic matter input to the system [10]. In addition, Raub, et al. [11] confirmed the positive effect of N-rich food resources on the abundance of soil fauna. Conversely, a long-term N addition study reports that continuous N addition significantly reduces the density and taxa richness of the soil fauna due to the decreased allocation of carbon (C) to leaf and fine root litter, especially C from the roots associated with changes in rhizodeposition [12–14]. Additionally, some studies find no significant effects of N addition on the density of soil fauna due to reduced competitive exclusion of soil fauna and a short observation period [15]. A lack of impacts of N addition on the diversity of soil fauna has also been reported; results from the demands for N are similar among various soil fauna [16,17] or elevated intraguild predation [11]. The different responses of soil fauna communities to N addition—whether positive, negative, or non-significant—may primarily depend on the N addition rate [4]. The impact of N addition on soil fauna communities could change directionally and dramatically at a critical concentration [18,19], suggesting a threshold effect. The study of soil fauna community responses to N addition at a range of rates is therefore of great importance.

N addition could affect the vertical distribution of soil fauna, leading to major variations in the soil fauna community with increasing soil depth [20]. The density of soil fauna, for example, could appreciably increase in the topsoil layer due to N addition while exhibiting no significant changes in deep soil layers [21]. Soil depth is one of the most important factors influencing soil fauna communities due to variations in the soil's physical and chemical properties along the depth gradient [22,23]. Soil fauna may be more abundant in the subsoil, where they take refuge to avoid the high N concentrations in the upper soil layers under N addition [24]. Moreover, thresholds, if present, could shift from high to low along the soil profile [24]. Although N addition is reported to affect the soil fauna community, field-based studies of the responses of a soil fauna community to soil depth are still lacking, making predictions about soil food web changes with N deposition in plantations difficult.

The soil fauna community's responses to N addition are mediated by the quantity and quality of the food resources available, such as roots, litter, and fungi [25,26]. Typically, soil fauna density is correlated with root biomass [23], though previous studies in various ecosystems have shown that N addition can have a variety of impacts on fine root biomass, either stimulating the production of fine root biomass [27] or retarding it [28]. Moreover, root chemical traits are key factors in shaping soil fauna communities, and do so via mediating species-specific interactions and affecting the soil fauna at multiple trophic levels [29]. It is well established that plant root-derived nutrition, such as N, may strongly influence a soil fauna community [30]. Litter quantity and quality are two additional factors that have a major influence on the activity and composition of soil fauna [31]. High quantities of litter can stimulate an increase in soil fauna density, mainly due to concomitant increases in the food supply and the habitat space of the ecosystem [11,32]. In addition, the density and community parameters of soil fauna also largely depend on the chemical and physical qualities of the litter [31]. A previous study showed that the density of soil fauna could increase as the litter C:N ratio decreased [33]. Furthermore, some soil fauna prefer feeding on older, more decomposed litter, and this preference strongly depends on the structure of the litter, e.g., tensile strength and palatability [26]. Recent studies, however, have suggested that root-derived food resources can be much more effective in fueling belowground food webs than those that come from leaf litter (e.g., [30,34]). Roots as a food source for soil fauna are considered to be an essential component of the underground food web [35]. The quantity and quality of food resources are likely to change in response to N addition, mediating the effects of N addition on soil fauna communities.

China grows more than 7 million ha of poplar plantations. Biomass production and C fixation by poplars (*Populus*) thus play important roles in mediating global climate change [36]. Compared to the natural forest ecosystem, changes in the soil fauna community in response to N addition in poplar plantations are less studied, especially for the whole soil fauna community [19]. Over the decades, many studies have evaluated the impact of N addition on particular species of soil fauna [37,38], while few studies have reported whole soil fauna community responses to N addition. Studies focused on one or two soil fauna species are not sufficiently indicative of the changes in whole soil fauna community [39]. It is thus necessary to study the response of the whole soil fauna to N addition in poplar plantations. In this study, we aimed to: (1) explore the effects of N addition and soil depth on the whole soil fauna community, and (2) determine the primary factors controlling the responses of the whole soil fauna community across N treatments and the soil profile in a poplar plantation at the Dongtai Forest Farm, eastern China.

2. Materials and Methods

2.1. The Experimental Site and Design

Our experimental site is located at the Dongtai Forest Farm in Yancheng, Jiangsu Province, eastern China (120°49′ E, 32°52′ N). The farm is close to the Yellow Sea State Forest Park (on the coast of the Yellow Sea) and has a climate classified as Cfa (Humid subtropical climate) according to Köppen [40]. The mean annual temperature (MAT) is 13.7 °C and the mean annual precipitation (MAP) is 1051 mm. The soil of the forest farm is a desalting meadow and sandy soil with a pH value ~8 [41].

Our N addition experiment was established in May 2012. We chose 12-year-old pure poplar plantations (*Populus deltoids cv. 'I-35'*) with uniform site conditions and management measures as our plots. We used a randomized block design with a gradient of five levels of N addition (0, 50, 100, 150, and 300 kg N ha^{-1} $year^{-1}$) in three replicate blocks (25 × 190 m). Each N treatment subplot was 25 × 30 m with a 10 m buffer zone between any two adjacent subplots. The distance between any two adjacent blocks was at least 500 m. We chose a range of N addition rates because the ambient N deposition rate is about 50 kg N ha^{-1} $year^{-1}$ for this area [42] and Liu et al. [18] estimated that the critical loads for N deposition in Jiangsu province could be more than 200 kg N ha^{-1} $year^{-1}$. In each month of the growing season (May through October), one-sixth of the yearly amount of NH_4NO_3 was dissolved in 20 L water and sprayed evenly under the canopy by backpack sprayer to simulate natural N deposition. Each control subplot received 20 L of water.

2.2. Microclimate

Soil temperature (0–15 cm) and moisture (0–20 cm) were measured once or twice a month using a Delta-T WET-2 (UK) from January to December 2016. Litter was collected in each subplot from November to December 2016 with a 2 m × 2 m litter trap constructed of 2 mm mesh nylon cloth. We oven-dried the litter at 65 °C for 48 h and weighed it to calculate litter mass.

2.3. Measurement of C and N in Litter and Soil Samples

Litter samples for measuring C and N content were collected in October 2016, oven-dried, ground, filtered by sieving (0.5 mm), and analyzed with an elemental analyzer (PerkinElmer 2400 II, Waltham, MA, USA). Soil cores of 4 cm in diameter were taken in October 2016 from the 0–15 cm soil layer in each subplot. Soil samples were transferred to our lab, air-dried, ground, and dipped into 0.5 M hydrochloric acid (HCl) to remove carbonates [43,44]. We then measured soil organic C (SOC) and total N (TN) by combustion with an elemental analyzer (Elementar, Vario EL III, Elementar Analysen Systeme GmbH, Elementar Analysensysteme GmbH, Hanau, Germany).

2.4. Estimation of Fine Root Biomass

Soil cores were taken from the top 45 cm of soil in each subplot with polyvinyl chloride (PVC) tubes (with an inside diameter of 4.6 cm) in June 2016 to estimate the fine root biomass [45–47]. Soil cores were numbered and then transported to our laboratory at Nanjing Forestry University, where they were frozen at −20 °C before analyzing. The cores were separated into three sections by depth (0–10, 10–25, and 25–40 cm), then carefully washed by wet sieving (0.5 mm) under gently flowing water to remove attached soil and dark-brown/black debris. Collected root samples were separated into coarse (>2 mm in diameter) and fine (≤2 mm in diameter, live and dead) roots, oven-dried at 65 °C for 48 h and weighed to calculate the live and dead fine root biomass.

2.5. Soil Fauna Sampling and Identification

Soil samples were collected from the 0–10, 10–25, and 25–40 cm soil layers using a soil coring with a diameter of 4 cm. Different layers represent organic horizon (O horizon, 0–10 cm), eluvial horizon (A horizon, 10–25 cm), and deposition horizon (B horizon, 25–40 cm), respectively [48,49]. In June 2016, four soil cores were collected from each 25 × 30 m subplot and pooled together as a replicate sample. The soil samples were immediately shipped back to our laboratory. The soil fauna was then collected from each soil sample using Tullgren extractors (Tullgren Funnel Unit, BURKARD, BURKARD SCIENTIFIC Ltd., Uxbridge, UK) [50–53]. All collected fauna samples were preserved in 75% ethanol and then sorted under a dissecting microscope (Nikon Eclipse E200, Nikon Instech Co., Ltd., Tokyo, Japan). Soil fauna was identified according to Yin [54,55].

2.6. Statistical Analysis

To determine the structure of the soil fauna in our study site, we chose density, the number of groups, and the Shannon-Wiener diversity index to describe the characteristics and composition of the soil fauna of each sample. The soil fauna density (D) was calculated as the following:

$$D = \frac{N}{V}; \tag{1}$$

where D is the density of soil fauna, N is the total number of individuals, and V is the volume of the soil samples. The number of groups (G) was estimated as the number of groups of the fauna in the same order. The Shannon-Wiener diversity index (H) of each sample was estimated according to Whittaker [56]:

$$H = -\sum_{i=1}^{n} P_{\mathrm{i}} \ln P_{\mathrm{i}} \tag{2}$$

where H is the Shannon-Wiener diversity index of each sample, n is the number of groups of soil fauna, and P_{i} is the proportion of the number of individuals in the ith order to the total number of individuals, which could be calculated with the following equation:

$$P_{\mathrm{i}} = \frac{N_{\mathrm{i}}}{N} \tag{3}$$

where N_{i} is the number of individuals in the ith order.

Two-way ANOVA was used to examine the impact of N addition, soil layers, and their interactions on the fine root biomass (including live, dead, and total fine roots), density, number of groups, and biodiversity of phytophagous and total soil fauna. Repeated-measures ANOVA was performed to test the responses of litter mass, soil temperature, and moisture to N addition. One-way ANOVA was used to examine the responses of the litter C:N ratio, soil TN, and SOC to N addition. Linear regression analyses were performed to examine the relationships between soil fauna and fine root biomass (live, dead, and total), SOC, TN, mean annual soil temperature and between moisture and the soil temperature in June. All statistical analyses were performed using SPSS 22.0 for Windows (SPSS Inc., Chicago, IL, USA).

3. Results

3.1. Microclimate, Litter C and N, Soil Samples, and Fine Root Biomass

Soil temperature and moisture showed remarkable seasonal variations. Soil temperature increased from January to August and then decreased from August to December (Table 1, Figure S1a) and soil moisture was low in summer and high in spring and fall (Table 1, Figure S1b). N addition did not affect the C:N ratio of the litter, SOC, TN, soil temperature or moisture (all $p > 0.05$, Tables 1 and 2). While N addition had no significant effect on live, dead, or total fine root biomass (all $p > 0.05$), root biomass in general decreased with increasing soil depth (all $p < 0.01$, Table 3, Figure S2).

Table 1. Results of repeated-measures ANOVA for seasonal responses of litter mass, soil temperature, and moisture to nitrogen (N) addition (0, 50, 100, 150, and 300 kg N ha^{-1} $year^{-1}$). Statistical analysis was performed on log-transformed data. p levels are indicated beside the F values by the following: *: <0.05, **: <0.01, ***: <0.001, and no asterisks indicate $p > 0.05$. df represents degree of freedom.

Variables	N Addition (N)		Time (T)		N × T	
	df	*F*, *p*	*df*	*F*, *p*	*df*	*F*, *p*
Litter mass	4	0.41	1	311.77 ***	4	0.58
Soil temperature	4	0.49	15	1582.24 ***	60	0.29
Soil moisture	4	0.67	15	233.25 ***	60	0.38

Table 2. Results of one-way ANOVA for responses of litter carbon (C): N ratio, soil N (TN), and soil organic C (SOC) to N addition. Statistical analysis was performed on log-transformed data. p levels are indicated beside the F values by the following: *: <0.05, **: <0.01, ***: <0.001, and no asterisks indicate $p > 0.05$.

Variables	N Addition (N)	
	df	*F*, *p*
Litter C:N ratio	4	1.68
TN	4	0.96
SOC	4	1.67

Table 3. Results of two-way ANOVA for responses of the density (D), number of groups (G), Shannon-Wiener diversity index (H) of soil fauna and fine root biomass (live root biomass, dead root biomass, and total root biomass) to nitrogen (N) addition and soil depth (0–10, 10–25, and 25–40 cm). Statistical analysis was performed on log-transformed data. P levels are indicated beside the F values by the following: *: <0.05, **: <0.01, ***: <0.001, and no asterisks indicate $p > 0.05$.

Variables		N Addition (N)		Soil Depth (depth)		N × Depth	
		df	*F*, *p*	*df*	*F*, *p*	*df*	*F*, *p*
Total soil fauna	D_{total}	4	8.11 **	2	308.52 ***	8	2.77 *
	G_{total}	4	1.91	2	10.63 *	8	2.19
	H_{total}	4	1.33	2	6.35	8	1.78

Table 3. *Cont.*

Variables		N Addition (N)		Soil Depth (depth)		N × Depth	
		df	*F, p*	*df*	*F, p*	*df*	*F, p*
Phytophagous soil fauna	D_P	4	5.91 *	2	118.38 ***	8	1.75
	G_P	4	1.74	2	8.13 *	8	2.79*
	H_P	4	2.28	2	4.94	8	2.81
Fine root biomass	Live root biomass	4	1.84	2	13.42 ***	8	0.72
	Dead root biomass	4	2.54	2	8.9 **	8	1.39
	Total root biomass	4	1.64	2	17.32 ***	8	0.88

3.2. Soil Fauna Community as Affected by N Addition

In general, N addition and soil depth both played important roles in altering the density of the soil fauna. With increasing N addition rate, the density of the total soil fauna (D_{total}) significantly increased first, then declined to the levels of the controls ($p < 0.01$, Table 3, Figure 1a). In the 25–40 cm soil layer, for example, D_{total} increased from N_0 to N_3 and then declined from N_3 to N_4 (Figure 1a). In addition, D_{total} significantly decreased with increasing soil depth ($p < 0.001$, Table 3, Figure 1a). The interaction between N addition and soil depth had a significant impact on D_{total} ($p < 0.05$, Table 3). D_{total} increased from N_0 to N_1 and then declined with higher N addition (from N_1 to N_4) in the topsoil layer (0–10 cm). In the deep soil layers (10–25 and 25–40 cm), however, declines in D_{total} started at N_2 and N_3, respectively, compared to the topsoil layer. The density of phytophagous soil fauna (D_P) showed similar responses to N addition along the soil profile (Table 3, Figure 1b).

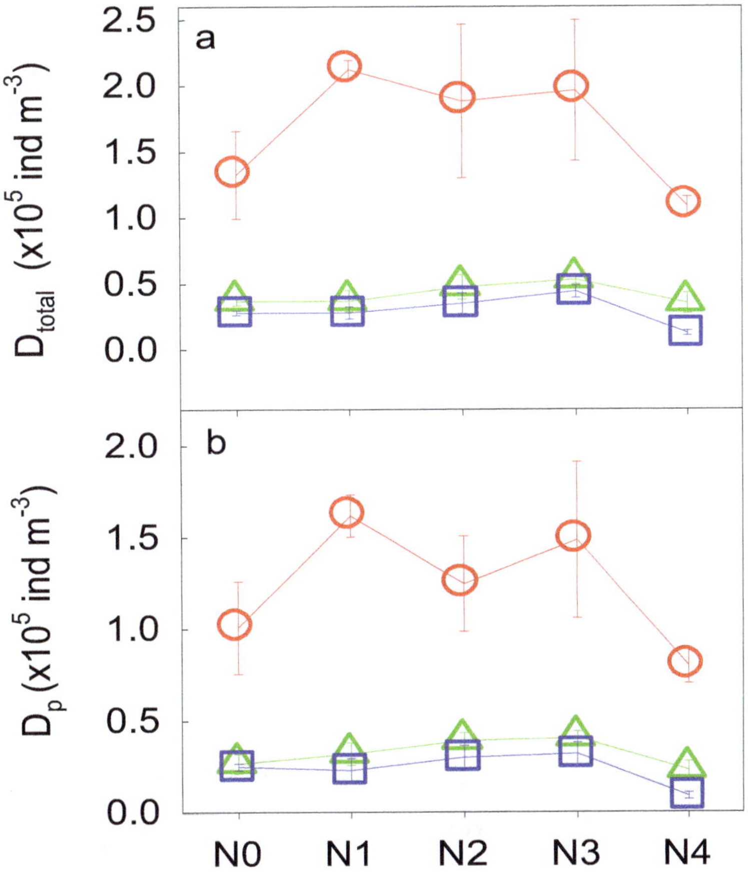

Figure 1. The density of total soil fauna (D_{total}, **a**) and phytophagous soil fauna (D_P, **b**) along the N addition gradient and the soil profile (red for 0–10 cm, green for 10–25 cm, and blue for 25–40 cm). N_0, N_1, N_2, N_3, N_4 indicate the gradient of N additions (N_0: 0 kg N ha^{-1} year^{-1}; N_1: 50 kg N ha^{-1} year^{-1}; N_2: 100 kg N ha^{-1} year^{-1}; N_3: 150 kg N ha^{-1} year^{-1}; N_4: 300 kg N ha^{-1} year^{-1}).

N addition had no significant impact on either the number of groups of total soil fauna (G_{total}) or the number of groups of phytophagous fauna (G_p) (all $p > 0.05$, Table 3, Figure 2). With increasing soil depth, in general, we found significant decreases in G_{total} and G_p (all $p < 0.05$, Table 3, Figure 2). Abrupt decreases in G_{total} and G_p at N_2 were observed in the topsoil layer (0–10 cm). We also found interactive effects between N addition and soil layer on G_p ($p < 0.05$, Table 3, Figure 2b). Along the soil profile, for instance, G_p reached its highest values at N_1, N_2, and N_3. For the Shannon-Wiener diversity index of the soil fauna, neither N addition nor soil depth had significant effects on the diversity of total soil fauna (H_{total}) or the diversity of phytophagous soil fauna (H_p) (all $p > 0.05$, Table 3, Figure 3).

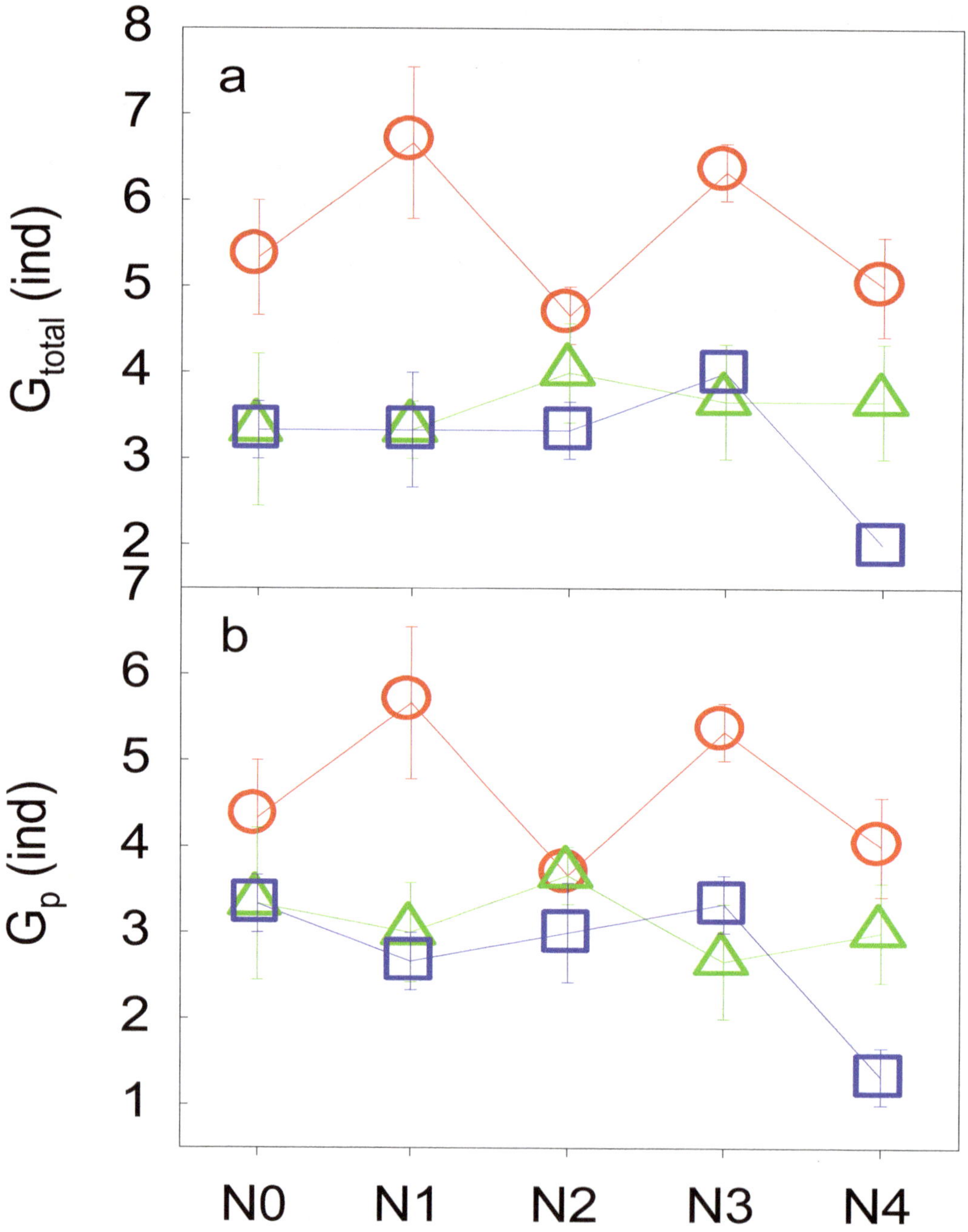

Figure 2. The number of groups of total soil fauna (G_{total}, **a**) and phytophagous soil fauna (G_p, **b**) along the N addition gradient and the soil profile. See Figure 1 for abbreviations.

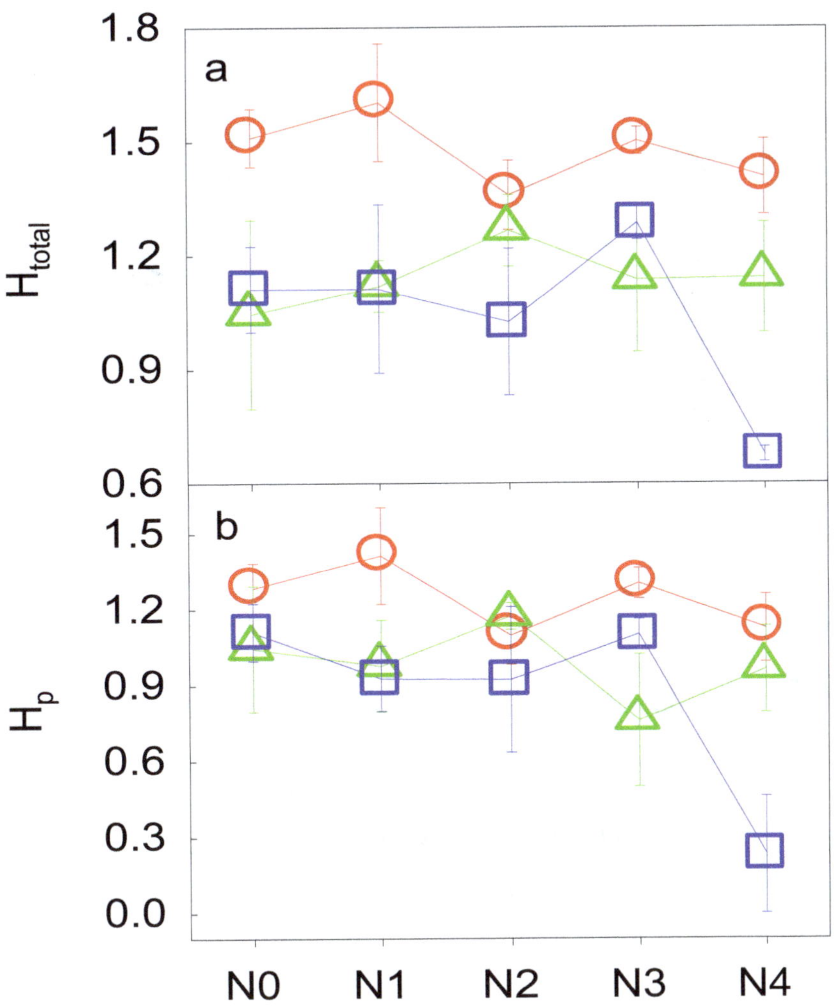

Figure 3. The Shannon-Wiener diversity index of total soil fauna (H_{total}, **a**) and phytophagous soil fauna (H_p, **b**) along the N addition gradient and the soil profile. See Figure 1 for abbreviations.

3.3. Fine Root Biomass Regulation of Soil Fauna Density

Variations in soil fauna densities (both D_p and D_{total}) were consistently regulated by fine root biomass (Table 4). We found that variations in D_p and D_{total} across N addition treatments and soil profile were positively correlated with the amount of live, dead, and total fine roots (all $p < 0.01$, Figures 4 and 5). The positive relationships remained even without the "outliers" — the very small values (both in fine root biomass on the X-axis as well as in the soil fauna density on the Y axis, all $p < 0.01$, Figures S3 and S4). Multifactor linear regression showed that the dominant factors controlling variations in D_p and D_{total} were the total and the dead fine root biomass across N addition treatments (all $p < 0.001$, Table 4).

Table 4. Results of multi-regression analysis of the density of phytophagous soil fauna (D_p) and total soil fauna (D_{total}) with fine root biomass (live, dead, and total), SOC, TN, mean annual soil temperature and moisture, and the soil temperature and moisture in June across N addition treatments. ***: $p < 0.001$.

Dependent	Model	Variables	Regression	r^2 and p
D_p	A-1	Total root	$D_p = 2.79 + 0.72$ total root	0.48 ***
	A-2	Total root, dead root	$D_p = 2.95 + 0.63$ total root+0.11 dead root	0.53 ***
D_{total}	B-1	Total root	$D_{total} = 3.03 + 0.67$ total root	0.42 ***
	B-2	Total root, dead root	$D_{total} = 3.20 + 0.58$ total root+0.12 dead root	0.49 ***

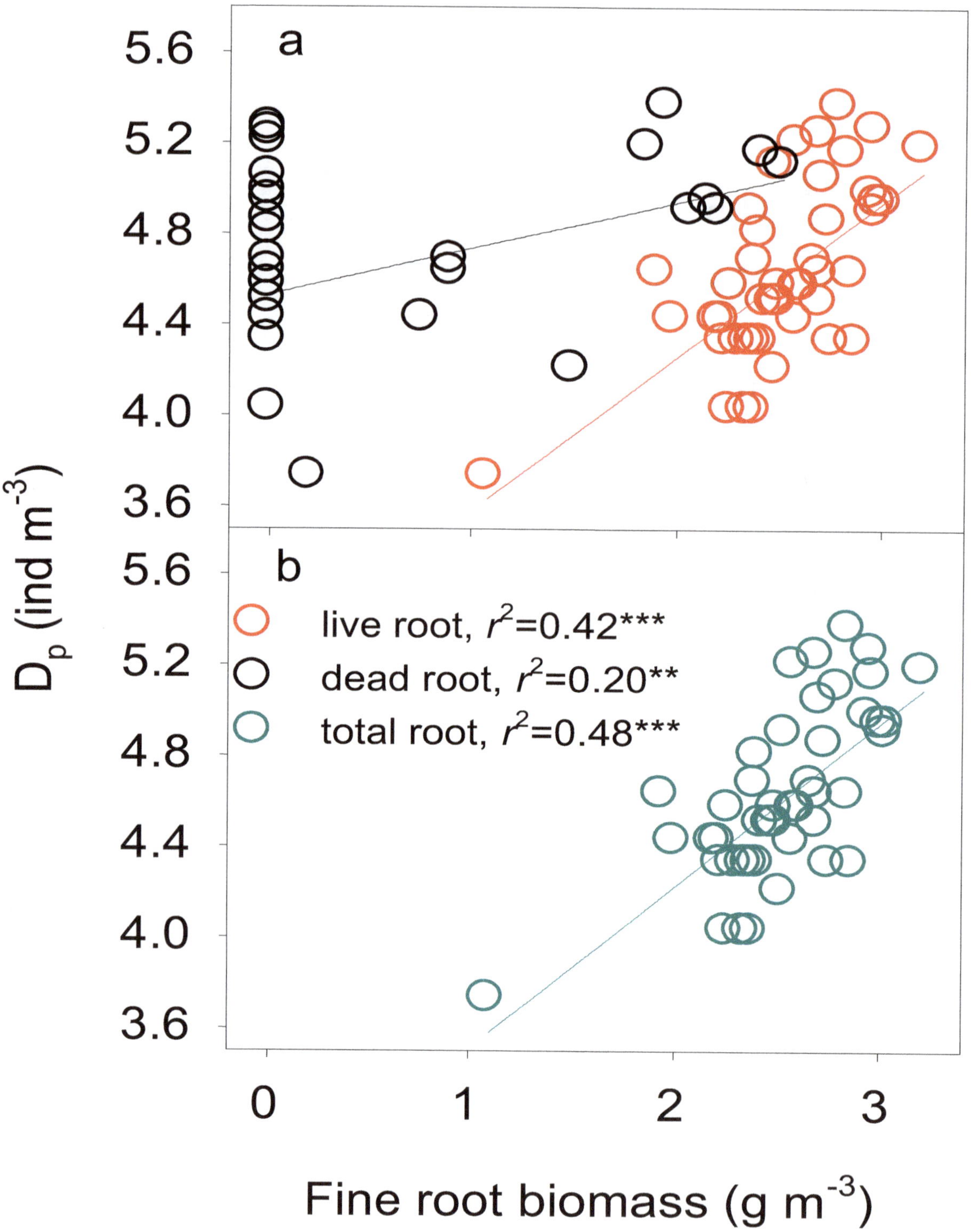

Figure 4. Log-log regulation of fine root biomass ((**a**) live and dead root biomass; (**b**) total root biomass) on the density of phytophagous soil fauna (D_p). **: $p < 0.01$, ***: $p < 0.001$.

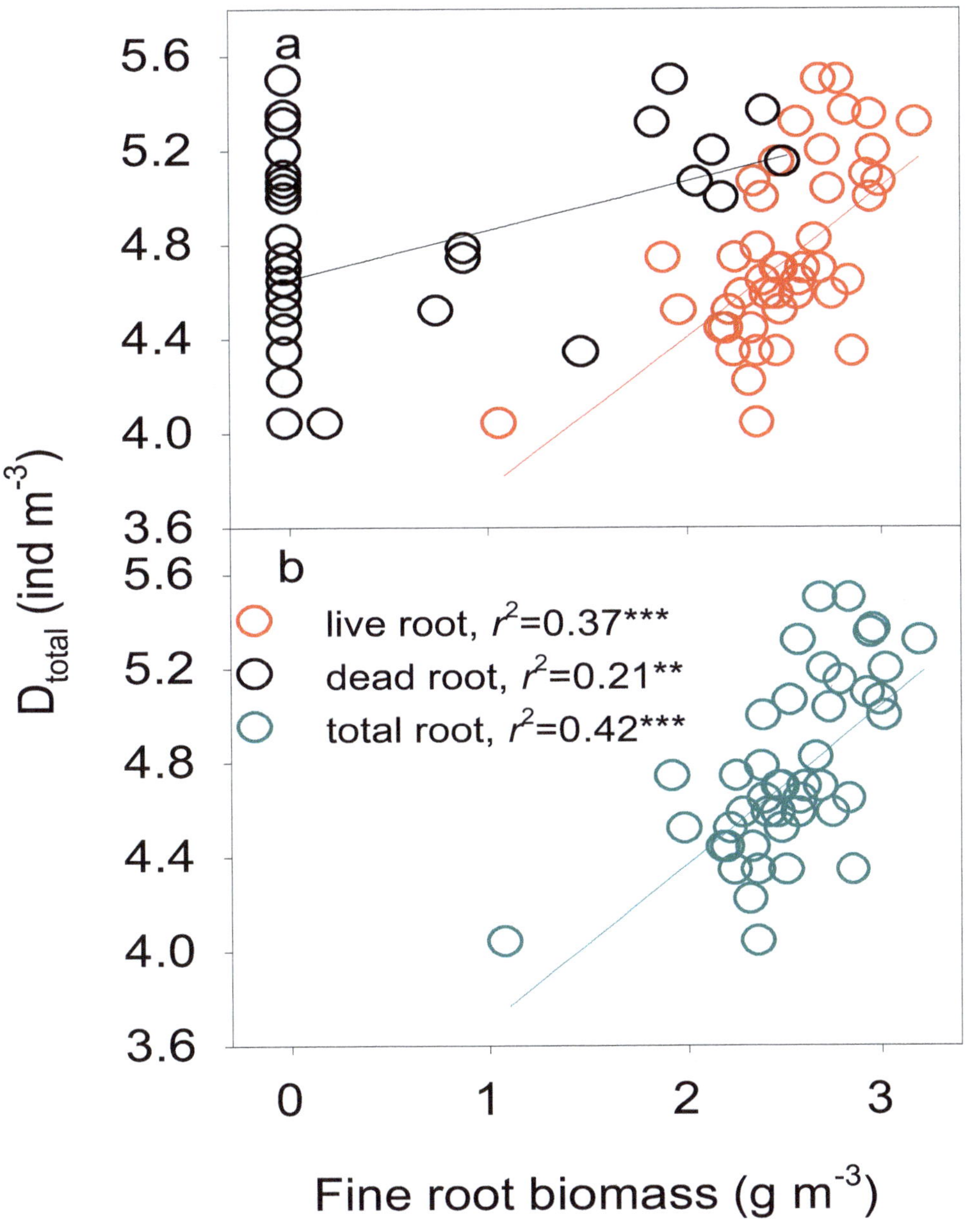

Figure 5. Log-log regulation of fine root biomass ((**a**) live and dead root biomass; (**b**) total root biomass) on the density of total soil fauna (D_{total}). **: $p < 0.01$, ***: $p < 0.001$.

4. Discussion

4.1. N-addition Impacts on Soil Fauna Density

Although N addition, up to a certain rate, could have positive impacts on soil fauna, high rates of N input appears to negatively affect the fauna community. The increases in D_{total} at low N addition rates are in line with previous findings, which show that N addition, to a certain extent, can be beneficial to soil fauna (e.g., [57–59]). The increases in D_{total} under N addition may have resulted from (1) increases in available N in the soil [60]; (2) increases in the food sources (soil bacteria and fungi) for the soil fauna [5,61]; (3) increases in rhizodeposition-C accessible to soil fauna [14,62]; and (4) more hospitable soil conditions for soil fauna (e.g., decreases in soil pH associated with N addition) [63]. Interestingly, D_{total} dramatically decreased from N_3 to N_4, exhibiting a negative response to high N addition. Decreases in the density of soil fauna at high levels of N addition may be attributed to at least two reasons. First, excessive acidic substances resulted from N addition may negatively affect soil fauna [64,65]. For instance, soil acidification [66] significantly restricts the number of soil nematodes and could thus decrease D_{total} [67,68]. Second, N may cause soil eutrophication due to the increase in nitrate (NO_3^-) and NH_4^+ concentrations in soil, and thereby has a negative influence on soil fauna [19,67]. Similar to the variations observed in D_{total}, D_p increased from N_0 to N_3 and decreased from N_3 to N_4. The decline in D_p is consistent with previous studies, which found that N addition that increases NH_4^+ concentration significantly decreases D_p, as the NH_4^+ suppresses phytophagous soil fauna [13,67,69]. Additionally, we found that D_p positively responds to low N addition rates, which is similar to the trend observed for D_{total}, probably due to the large proportion (~75% across all treatments) of phytophagous soil fauna to total soil fauna in our study. Since few studies have reported any positive impacts of N addition on D_p, the mechanisms remain to be explored.

A threshold effect of N addition on soil fauna is usually widely found [19,70]. In our study, we detected a threshold in the effects of N addition on soil fauna at approximately 150 kg N ha^{-1} $year^{-1}$ for both D_{total} and D_p. N addition within a certain concentration range enables competitive soil fauna to coexist with less competitive soil fauna, leading to an increase in density, while excessive N addition is a threat to the less competitive soil fauna, as it negatively alters the soil's physical and chemical properties [2]. When an ecosystem reaches N saturation, excessive exogenous N might increase leaching of some basic ions, such as calcium ion (Ca^{2+}), aluminum ion (Al^{3+}), and magnesium ion (Mg^{2+}), from the system [24], consequently destroying the structure and function of the ecosystem, causing soil acidification and nutrient imbalances, breaking intraspecific competition dynamics, and eroding the system's resistance to external disturbances [71]. Differences in N thresholds among ecosystems probably result from a quantitative variation in the primary available N. N addition may increase the quality and quantity of organic matter in young, developing ecosystems while having a toxic effect on the soil fauna in old ecosystems that reached N saturation from N addition [12]. Thus, the threshold effects of N deposition in different ecosystems and their potential impacts on soil fauna are worthy of further investigation from a community structure perspective.

4.2. N-addition Impacts on the Number of Groups and Diversity of Soil Fauna

N addition has either positive or negative effects on the number of groups and Shannon-Wiener diversity index of soil fauna in various ecosystems (e.g., [58,59,72,73]). However, a non-significant impact of N addition on those variables has been widely reported (e.g., [16,17,19]), supporting our findings in this study. The lack of N addition effects may be explained as follows: (1) N addition did not significantly alter the proportion of soil fauna in each group to the total number. Statistically, we found that the number of soil fauna in each group increased evenly with N addition (Table S1, Figures S5 and S6); (2) litter, as one of the main food sources for soil fauna [26], had no significant response to N addition either in terms of quantity (litter mass) or quality (C:N ratio) (all $p > 0.05$, Tables 1 and 2); (3) N addition had no significant influence on soil microclimate (soil temperature and moisture) (all $p > 0.05$, Table 1). The abrupt decreases in G_{total} and G_p at N_2 may have been related to a decrease in food resources, e.g., fine root biomass, at N_2. Additionally, the discrepancy in the effects of N addition on the number of groups and the diversity of soil fauna between our study and the previous studies may have resulted from the differences in soils, as our experimental site is located in an alkaline coastal area.

4.3. Increasing N-Addition Thresholds along Soil Profile

N addition did not affect the "surface aggregation" of soil fauna but increased the thresholds with increasing soil depth. The density and the number of groups of soil fauna usually decrease along soil profile [48,74], and this holds true under N addition in our study. Surface soil is more hospitable to soil fauna, as it always has a higher substrate content and air circulation rate [53,75]. Higher N addition thresholds for soil fauna may be attributed to the fact that topsoil could be more sensitive to environmental stress (e.g., N addition) than deeper soil layers [75]. Soil fauna move from the topsoil downwards to the deeper soil layers when high amounts of N are added to escape the depression of high N addition in the topsoil [76].

4.4. Regulation of Fine Root Biomass on Soil Fauna

Responses of soil fauna to N addition at the site level may be influenced by many factors, such as soil temperature and moisture [77], pH value [78], and resource supply [15]. Based on the multifactor regression analyses, we found that fine root biomass was the dominant factor controlling variations in the density of soil fauna for both D_{total} and D_P. Fine roots and their exudates may positively influence soil biota through stimulated hyphal growth with increasing quantities of fine root biomass [79]. A growing body of evidence suggests that soil food webs rely heavily on "root C" because fine roots can be a source of readily available C for both soil microorganisms and phytophagous soil fauna [80]. Likewise, many studies have shown that phytophagous soil fauna are largely controlled by root-derived nutrition [30,81], just as aboveground herbivore populations are strongly linked to net primary production [82].

5. Conclusions

Our study in poplar plantations on the east coast of China suggests that N addition could have substantial effects on soil fauna along the soil profile. Across the N addition treatments, low N addition rates had positive effects on soil fauna densities, likely due to improved physical and chemical soil conditions. With increasing N addition rates, responses of soil fauna community returned to the control level. Soil fauna responded differently to N addition in different soil layers, which may largely relate to changes in the soil substrate content with N addition. Multifactor linear regressions showed that fine root biomass mediated N induced changes in the soil fauna density. The global N deposition rate is projected to continuously increase in the future [3,83], which will undoubtedly affect soil fauna and further the biogeochemical cycles. Therefore, it is imperative to explore impacts of N addition on soil fauna communities to understand the way an ecosystem functions under the global N deposition.

Supplementary Materials

Figure S1: Variations in soil temperature (a) and moisture (b) from January to December along the nitrogen (N) addition gradient in 2016. N0, N1, N2, N3, N4 indicate different N additions (N0: 0 kg N ha^{-1} $year^{-1}$; N1: 50 kg N ha^{-1} $year^{-1}$; N2: 100 kg N ha^{-1} $year^{-1}$; N3: 150 kg N ha^{-1} $year^{-1}$; N4: 300 kg N ha^{-1} $year^{-1}$), Figure S2: The live (a), dead (b), and total (c) fine root biomass along the N addition gradient and the soil profile (red for 0–10 cm, green for 10–25 cm, and blue for 25–40 cm). See Figure S1 for abbreviations, Figure S3: Regulation of fine root biomass (a: live and dead root biomass; b: total root biomass) on the density of phytophagous soil fauna (Dp), without the very small points (circled in the panels). **: $p < 0.01$, ***: $P < 0.001$, Figure S4: Regulation of fine root biomass (a: live and dead root biomass; b: total root biomass) on the density of total soil fauna (Dtotal), without the very small points (circled in the panels). **: $p < 0.01$, ***: $p < 0.001$, Figure S5: The proportion of each individual soil fauna over total soil fauna along the nitrogen (N) addition gradient and the soil profile. See Figure S1 for abbreviations, Figure S6: The proportions of macrofauna, mesofauna, and microfauna over total soil fauna along the nitrogen (N) addition gradient and the soil profile. See Figure S1 for abbreviations. Table S1: Results of two-way ANOVA for the responses of each individual soil fauna's proportion over total soil fauna to nitrogen (N) addition (0, 50, 100, 150, and 300 kg N ha^{-1} $year^{-1}$) and soil depth (0–10, 10–25, and 25–40 cm). P levels are indicated beside the F values by the following: *: <0.05, **: <0.01, ***: <0.001, and no asterisks indicate $p > 0.05$.

Author Contributions: Conceptualization, H.R. and X.X.; Data curation, H.B., Q.G. and H.X.; Formal analysis, H.B., Q.G. and X.X.; Funding acquisition, H.R. and X.X.; Investigation, H.B., Q.G. and H.X.; Resources, C.S., H.R. and X.X.; Supervision, H.R. and X.X.; Validation, H.X.; Writing–original draft, H.B. and X.X.; Writing–review & editing, Q.L., X.C., Y.L., H.R. and X.X.

Acknowledgments: We thank many lab members for their help with the field work.

References

1. Penuelas, J.; Poulter, B.; Sardans, J.; Ciais, P.; van der Velde, M.; Bopp, L.; Boucher, O.; Godderis, Y.; Hinsinger, P.; Llusia, J.; et al. Human-induced nitrogen-phosphorus imbalances alter natural and managed ecosystems across the globe. *Nat. Commun.* **2013**, *4*, 2934. [CrossRef] [PubMed]
2. Niu, S.; Classen, A.T.; Dukes, J.S.; Kardol, P.; Liu, L.; Luo, Y.; Rustad, L.; Sun, J.; Tang, J.; Templer, P.H. Global patterns and substrate-based mechanisms of the terrestrial nitrogen cycle. *Ecol. Lett.* **2016**, *19*, 697–709. [CrossRef] [PubMed]
3. Schwede, D.B.; Simpson, D.; Tan, J.; Fu, J.S.; Dentener, F.; Du, E.; deVries, W. Spatial variation of modelled total, dry and wet nitrogen deposition to forests at global scale. *Environ. Pollut.* **2018**, *243*, 1287–1301. [CrossRef]
4. Xu, G.-L.; Mo, J.-M.; Zhou, G.-Y.; Fu, S.-L. Preliminary response of soil fauna to simulated N deposition in three typical subtropical forests. *Pedosphere* **2006**, *16*, 596–601. [CrossRef]
5. Sun, F.; Tariq, A.; Chen, H.; He, Q.; Guan, Y.; Pan, K.; Chen, S.; Li, J.; Zhao, C.; Wang, H.; et al. Effect of nitrogen and phosphorus application on agricultural soil food webs. *Arch. Agron. Soil Sci.* **2017**, *63*, 1176–1186. [CrossRef]
6. Xin, W.D.; Yin, X.Q.; Song, B. Contribution of soil fauna to litter decomposition in Songnen sandy lands in northeastern China. *J. Arid Environ.* **2012**, *77*, 90–95. [CrossRef]
7. Oliverio, A.M.; Gan, H.; Wickings, K.; Fierer, N. A DNA metabarcoding approach to characterize soil arthropod communities. *Soil. Biol. Biochem.* **2018**, *125*, 37–43. [CrossRef]
8. Li, X.; Yin, X.; Wang, Z.; Fan, W. Litter mass loss and nutrient release influenced by soil fauna of Betula ermanii forest floor of the Changbai Mountains, China. *Appl. Soil Ecol.* **2015**, *95*, 15–22. [CrossRef]
9. Lu, P.; Dai, N.H.; Sun, X.Y.; Zhang, G.H.; Xu, D.R.; Zhan, Y.H. Composition and structure of soil fauna community in the Dexing Copper Mine tailings pool after revegetation. *Turk. J. Zool.* **2018**, *42*, 307–315.
10. Lin, H.; He, Z.; Hao, J.; Tian, K.; Jia, X.; Kong, X.; Akbar, S.; Bei, Z.; Tian, X. Effect of N addition on home-field advantage of litter decomposition in subtropical forests. *For. Ecol. Manag.* **2017**, *398*, 216–225. [CrossRef]
11. Raub, F.; Scheuermann, L.; Hoefer, H.; Brandl, R. No bottom-up effects of food addition on predators in a tropical forest. *Basic Appl. Ecol.* **2014**, *15*, 59–65. [CrossRef]
12. Gan, H.; Zak, D.R.; Hunter, M.D. Chronic nitrogen deposition alters the structure and function of detrital food webs in a northern hardwood ecosystem. *Ecol. Appl.* **2013**, *23*, 1311–1321. [CrossRef] [PubMed]
13. Eisenhauer, N.; Cesarz, S.; Koller, R.; Worm, K.; Reich, P.B. Global change belowground: Impacts of elevated CO2, nitrogen, and summer drought on soil food webs and biodiversity. *Glob. Chang. Biol.* **2012**, *18*, 435–447. [CrossRef]
14. Preece, C.; Penuelas, J. Rhizodeposition under drought and consequences for soil communities and ecosystem resilience. *Plant Soil* **2016**, *409*, 1–17. [CrossRef]
15. Cole, L.; Buckland, S.M.; Bardgett, R.D. Influence of disturbance and nitrogen addition on plant and soil animal diversity in grassland. *Soil Biol. Biochem.* **2008**, *40*, 505–514. [CrossRef]
16. Jiang, M.; Wang, X.; Liusui, Y.; Sun, X.; Zhao, C.; Liu, H. Diversity and Abundance of Soil Animals as Influenced by Long-Term Fertilization in Grey Desert Soil, China. *Sustainability* **2015**, *7*, 10837–10853. [CrossRef]
17. Zhu, X.; Zhu, B. Diversity and abundance of soil fauna as influenced by long-term fertilization in cropland of purple soil, China. *Soil. Tillage Res.* **2015**, *146*, 39–46. [CrossRef]
18. Liu, X.; Duan, L.; Mo, J.; Du, E.; Shen, J.; Lu, X.; Zhang, Y.; Zhou, X.; He, C.; Zhang, F. Nitrogen deposition and its ecological impact in China: An overview. *Environ. Pollut.* **2011**, *159*, 2251–2264. [CrossRef]
19. Ochoa-Hueso, R.; Rocha, I.; Stevens, C.J.; Manrique, E.; Luciañez, M.J. Simulated nitrogen deposition affects soil fauna from a semiarid Mediterranean ecosystem in central Spain. *Biol. Fertil. Soils* **2014**, *50*, 191–196. [CrossRef]
20. Nkem, J.N.; Bruyn, L.A.L.D.; Hulugalle, N.R.; Grant, C.D. Changes in invertebrate populations over the growing cycle of an N-fertilised and unfertilised wheat crop in rotation with cotton in a grey Vertosol. *Appl. Soil Ecol.* **2002**, *20*, 69–74. [CrossRef]

21. Xu, G.L.; Schleppi, P.; Li, M.H.; Fu, S.L. Negative responses of Collembola in a forest soil (Alptal, Switzerland) under experimentally increased N deposition. *Environ. Pollut.* **2009**, *157*, 2030–2036. [CrossRef] [PubMed]
22. Yin, X.; Song, B.; Dong, W.; Xin, W.; Wang, Y. A review on the eco-geography of soil fauna in China. *J. Geogr. Sci.* **2010**, *20*, 333–346. [CrossRef]
23. Potapov, A.M.; Goncharov, A.A.; Semenina, E.E.; Korotkevich, A.Y.; Tsurikov, S.M.; Rozanova, O.L.; Anichkin, A.E.; Zuev, A.G.; Samoylova, E.S.; Semenyuk, I.I. Arthropods in the subsoil: Abundance and vertical distribution as related to soil organic matter, microbial biomass and plant roots. *Eur. J. Soil Biol.* **2017**, *82*, 88–97. [CrossRef]
24. Xu, G.L.; Mo, J.M.; Fu, S.L.; Gundersen, P.; Zhou, G.Y.; Xue, J.H. Response of soil fauna to simulated nitrogen deposition: A nursery experiment in subtropical China. *Acta Sci. Circumstantiae* **2007**, *19*, 603–609. [CrossRef]
25. Morrien, E. Understanding soil food web dynamics, how close do we get? *Soil Biol. Biochem.* **2016**, *102*, 10–13. [CrossRef]
26. Zieger, S.L.J. Trophic Structure of Soil Animal Food Webs of Deciduous Forests as Analyzed by Stable Isotope Labeling. Ph.D. Thesis, Universität Göttingen, Göttingen, Germany, 2016.
27. Leppälammi-Kujansuu, J.; Ostonen, I.; Strömgren, M.; Nilsson, L.O.; Kleja, D.B.; Sah, S.P.; Helmisaari, H.-S. Effects of long-term temperature and nutrient manipulation on Norway spruce fine roots and mycelia production. *Plant Soil* **2013**, *366*, 287–303. [CrossRef]
28. Li, W.; Jin, C.; Guan, D.; Wang, Q.; Wang, A.; Yuan, F.; Wu, J. The effects of simulated nitrogen deposition on plant root traits: A meta-analysis. *Soil Biol. Biochem.* **2015**, *82*, 112–118. [CrossRef]
29. Tsunoda, T.; Dam, N.M.V. Root chemical traits and their roles in belowground biotic interactions. *Pedobiologia* **2017**, *65*, 58–67. [CrossRef]
30. Eissfeller, V.; Beyer, F.; Valtanen, K.; Hertel, D.; Maraun, M.; Polle, A.; Scheu, S. Incorporation of plant carbon and microbial nitrogen into the rhizosphere food web of beech and ash. *Soil Biol. Biochem.* **2013**, *62*, 76–81. [CrossRef]
31. Sauvadet, M.; Chauvat, M.; Fanin, N.; Coulibaly, S.; Bertrand, I. Comparing the effects of litter quantity and quality on soil biota structure and functioning: Application to a cultivated soil in Northern France. *Appl. Soil Ecol.* **2016**, *107*, 261–271. [CrossRef]
32. Sayer, E.J.; Sutcliffe, L.M.E.; Ross, R.I.C.; Tanner, E.V.J. Arthropod Abundance and Diversity in a Lowland Tropical Forest Floor in Panama: The Role of Habitat Space vs. Nutrient Concentrations. *Biotropica* **2010**, *42*, 194–200. [CrossRef]
33. Ilieva-Makulec, K.; Olejniczak, I.; Szanser, M. Response of soil micro- and mesofauna to diversity and quality of plant litter. *Eur. J. Soil Biol.* **2006**, *42*, S244–S249. [CrossRef]
34. Gilbert, K.J.; Fahey, T.J.; Maerz, J.C.; Sherman, R.E.; Bohlen, P.; Dombroskie, J.J.; Groffman, P.M.; Yavitt, J.B. Exploring carbon flow through the root channel in a temperate forest soil food web. *Soil Biol. Biochem.* **2014**, *76*, 45–52. [CrossRef]
35. Melaniem, P.; Reinhard, L.; Stefan, S.; Mark, M. Compartmentalization of the soil animal food web as indicated by dual analysis of stable isotope ratios (15N/14N and 13C/12C). *Soil Biol. Biochem.* **2009**, *41*, 1221–1226.
36. Wang, S.; Tan, Y.; Fan, H.; Ruan, H.; Zheng, A. Responses of soil microarthropods to inorganic and organic fertilizers in a poplar plantation in a coastal area of eastern China. *Appl. Soil Ecol.* **2015**, *89*, 69–75. [CrossRef]
37. Liang, W.J.; Lou, Y.L.; Qi, L.; Shuang, Z.; Zhang, X.K.; Wang, J.K.; Coleman, D.; Fu, S.L.; Zou, X.M. Nematode faunal response to long-term application of nitrogen fertilizer and organic manure in Northeast China. *Soil Biol. Biochem.* **2009**, *41*, 883–890. [CrossRef]
38. Zhao, J.; Wang, F.; Li, J.; Zou, B.; Wang, X.; Li, Z.; Fu, S. Effects of experimental nitrogen and/or phosphorus additions on soil nematode communities in a secondary tropical forest. *Soil Biol. Biochem.* **2014**, *75*, 1–10. [CrossRef]
39. Boxman, A.W.; Blanck, K.; Brandrud, T.E.; Emmett, B.A.; Gundersen, P.; Hogervorst, R.F.; Kjonaas, O.J.; Persson, H.; Timmermann, V. Vegetation and soil biota response to experimentally-changed nitrogen inputs in coniferous forest ecosystems of the NITREX project. *For. Ecol. Manag.* **1998**, *101*, 65–79. [CrossRef]
40. WIKIPEDIA. Köppen Climate Classification. 1884. Available online: https://en.wikipedia.org/wiki/Köppen_climate_classification (accessed on 3 January 2019).

41. Wang, S.; Chen, H.Y.; Tan, Y.; Fan, H.; Ruan, H. Fertilizer regime impacts on abundance and diversity of soil fauna across a poplar plantation chronosequence in coastal Eastern China. *Sci. Rep.* **2016**, *6*, 20816. [CrossRef]
42. Zhu, J.; Wang, Q.; He, N.; Smith, M.D.; Elser, J.J.; Du, J.; Yuan, G.; Yu, G.; Yu, Q. Imbalanced atmospheric nitrogen and phosphorus depositions in China: Implications for nutrient limitation. *J. Geophys. Res. Biogeosci.* **2016**, *121*, 1605–1616. [CrossRef]
43. Ramnarine, R.; Voroney, R.P.; WagnerRiddle, C.; Dunfield, K.E. Carbonate removal by acid fumigation for measuring the δ13C of soil organic carbon. *Can. J. Soil Sci.* **2011**, *91*, 247–250. [CrossRef]
44. Wotherspoon, A.; Voroney, R.P.; Thevathasan, N.V.; Gordon, A.M. Comparison of Three Methods for Measurement of Soil Organic Carbon. *Commun. Soil Sci. Plant Anal.* **2015**, *46*, 362–374. [CrossRef]
45. Xu, X.; Niu, S.; Sherry, R.A.; Zhou, X.; Zhou, J.; Luo, Y. Interannual variability in responses of belowground net primary productivity (NPP) and NPP partitioning to long-term warming and clipping in a tallgrass prairie. *Glob. Chang. Biol.* **2012**, *18*, 1648–1656. [CrossRef]
46. Derner, J.D.; Briske, D.D. Does a tradeoff exist between morphological and physiological root plasticity? A comparison of grass growth forms. *Acta Oecol.* **1999**, *20*, 519–526. [CrossRef]
47. Gao, Y.Z.; Giese, M.; Lin, S.; Sattelmacher, B.; Zhao, Y.; Brueck, H. Belowground net primary productivity and biomass allocation of a grassland in Inner Mongolia is affected by grazing intensity. *Plant Soil* **2008**, *307*, 41–50. [CrossRef]
48. Yang, B.; Zhang, W.; Xu, H.; Wang, S.; Xu, X.; Fan, H.; Chen, H.Y.H.; Ruan, H. Effects of soil fauna on leaf litter decomposition under different land uses in eastern coast of China. *J. For. Res.* **2018**, *29*, 973–982. [CrossRef]
49. Wang, G.B.; Deng, F.F.; Xu, W.H.; Chen, H.Y.H.; Ruan, H.H. Poplar plantations in coastal China: Towards the identification of the best rotation age for optimal soil carbon sequestration. *Soil Use Manag.* **2016**, *32*, 303–310. [CrossRef]
50. Wallwork, J.A. The Distribution and Diversity of Soil Fauna. *Q. Rev. Biol.* **1977**, *52*, 319.
51. Wu, H.; Lu, M.; Lu, X.; Guan, Q.; He, X. Interactions between earthworms and mesofauna has no significant effect on emissions of CO_2 and N_2O from soil. *Soil Biol. Biochem.* **2015**, *88*, 294–297. [CrossRef]
52. Bokhorst, S.; Huiskes, A.; Convey, P.; Pmvan, B.; Aerts, R. Climate change effects on soil arthropod communities from the Falkland Islands and the Maritime Antarctic. *Soil Biol. Biochem.* **2008**, *40*, 1547–1556. [CrossRef]
53. Li, Y.; Chen, Y.; Xu, C.; Xu, H.; Zou, X.; Chen, H.Y.H.; Ruan, H. The abundance and community structure of soil arthropods in reclaimed coastal saline soil of managed poplar plantations. *Geoderma* **2018**, *327*, 130–137. [CrossRef]
54. Yin, W. *Subtropical Soil Animals in China (in Chinese)*; Science Press: Beijing, China, 2000.
55. Yin, W. *Chinese subtropical soil animals (in Chinese)*; Science Press: Beijing, China, 1992.
56. Whittaker, R.H. Evolution and Measurement of Species Diversity. *Taxon* **1972**, *21*, 213–251. [CrossRef]
57. Sjursen, H.; Michelsen, A.; Jonasson, S. Effects of long-term soil warming and fertilisation on microarthropod abundances in three sub-arctic ecosystems. *Appl. Soil Ecol.* **2005**, *30*, 148–161. [CrossRef]
58. Van, d.W.A.; Rhem, G.; Korevaar, H.; Schouten, A.J.; Gajmjagers, O.A.; Rutgers, M.; Mulder, C. Dissimilar response of plant and soil biota communities to long-term nutrient addition in grasslands. *Biol. Fertil. Soils* **2009**, *45*, 663–667.
59. Cusack, D.F.; Firestone, M.K. Changes in microbial community characteristics and soil organic matter with nitrogen additions in two tropical forests. *Ecology* **2011**, *92*, 621–632. [CrossRef] [PubMed]
60. Wall, D.H.; Bardgett, R.D.; Kelly, E. Biodiversity in the dark. *Nat. Geosci.* **2010**, *3*, 297–298. [CrossRef]
61. Kudrin, A.A.; Tsurikov, S.M.; Tiunov, A.V. Trophic position of microbivorous and predatory soil nematodes in a boreal forest as indicated by stable isotope analysis. *Soil Biol. Biochem.* **2015**, *86*, 193–200. [CrossRef]
62. Li, J.; Wen, Y.C.; Li, X.H.; Li, Y.T.; Yang, X.D.; Lin, Z.; Song, Z.Z.; Cooper, J.M.; Zhao, B.Q. Soil labile organic carbon fractions and soil organic carbon stocks as affected by long-term organic and mineral fertilization regimes in the North China Plain. *Soil Tillage Res.* **2018**, *175*, 281–290. [CrossRef]
63. Fenn, M.E.; Baron, J.S.; Allen, E.B.; Rueth, H.M.; Nydick, K.R.; Geiser, L.; Bowman, W.D.; Sickman, J.O.; Meixner, T.; Johnson, D.W. Ecological Effects of Nitrogen Deposition in the Western United States. *Bioscience* **2003**, *53*, 404–420. [CrossRef]
64. Fu, G.; Shen, Z.X. Response of alpine soils to nitrogen addition on the Tibetan Plateau: A meta-analysis. *Appl. Soil Ecol.* **2017**, *114*, 99–104. [CrossRef]

65. Lu, X.; Mo, J.; Gilliam, F.S.; Zhou, G.; Fang, Y. Effects of experimental nitrogen additions on plant diversity in an old-growth tropical forest. *Glob. Chang. Biol.* **2010**, *16*, 2688–2700. [CrossRef]
66. Xu, C.; Geng, Q.; Guo, L.; Li, Q.; Cheng, X.; Luo, Y.; Ruan, H.; Xu, X. Variation in Soil Carbon Content in Response to Long-Term Nitrogen Addition in Poplar Plantations (*Populus Deltoids*) on the East Coast of China. in preparation.
67. Wei, C.; Zheng, H.; Li, Q.; Lü, X.; Yu, Q.; Zhang, H.; Chen, Q.; He, N.; Kardol, P.; Liang, W. Nitrogen Addition Regulates Soil Nematode Community Composition through Ammonium Suppression. *PLoS ONE* **2012**, *7*, e43384. [CrossRef] [PubMed]
68. Guo, J.H.; Liu, X.J.; Zhang, Y.; Shen, J.L.; Han, W.X.; Zhang, W.F.; Christie, P.; Goulding, K.W.T.; Vitousek, P.M.; Zhang, F.S. Significant Acidification in Major Chinese Croplands. *Science* **2010**, *327*, 1008–1010. [CrossRef] [PubMed]
69. Collange, B.; Navarrete, M.; Peyre, G.; Mateille, T.; Tchamitchian, M. Root-knot nematode (Meloidogyne) management in vegetable crop production: The challenge of an agronomic system analysis. *Crop Prot.* **2011**, *30*, 1251–1262. [CrossRef]
70. Pinho, P.; Theobald, M.R.; Dias, T.; Tang, Y.S.; Cruz, C.; Martinsloução, M.A.; Máguas, C.; Sutton, M.; Branquinho, C. Critical loads of nitrogen deposition and critical levels of atmospheric ammonia for semi-natural Mediterranean evergreen woodlands. *Biogeosciences* **2012**, *9*, 1205–1215. [CrossRef]
71. Yang, J.-Y.; Fan, J. Review of study on mineralization, saturation and cycle of Nitrogen in forest ecosystems. *J. For. Res.* **2003**, *14*, 239–243.
72. Van Diepen, L.T.A.; Lilleskov, E.A.; Pregitzer, K.S.; Miller, R.M. Simulated Nitrogen Deposition Causes a Decline of Intra- and Extraradical Abundance of Arbuscular Mycorrhizal Fungi and Changes in Microbial Community Structure in Northern Hardwood Forests. *Ecosystems* **2010**, *13*, 683–695. [CrossRef]
73. Högberg, M.N.; Briones, M.J.I.; Keel, S.G.; Metcalfe, D.B.; Campbell, C.; Midwood, A.J.; Thornton, B.; Hurry, V.; Linder, S.; Näsholm, T. Quantification of effects of season and nitrogen supply on tree below-ground carbon transfer to ectomycorrhizal fungi and other soil organisms in a boreal pine forest. *New Phytol.* **2010**, *187*, 485–493. [CrossRef]
74. Doblasmiranda, E.; Sánchezpiñero, F.; Gonzálezmegías, A. Vertical distribution of soil macrofauna in an arid ecosystem: Are litter and belowground compartmentalized habitats? *Pedobiologia* **2009**, *52*, 361–373. [CrossRef]
75. Wei, H.; Liu, W.; Zhang, J.; Qin, Z. Effects of simulated acid rain on soil fauna community composition and their ecological niches. *Environ. Pollut.* **2016**, *220*, 460–468. [CrossRef]
76. Rusek, J.; Marshall, V.G. Impacts of Airborne Pollutants on Soil Fauna. *Annu. Rev. Ecol. Syst.* **2000**, *31*, 395–423. [CrossRef]
77. Wang, S.; Ruan, H. Effects of soil mesofauna and microclimate on nitrogen dynamics in leaf litter decomposition along an elevation gradient. *Afr. J. Biotechnol.* **2011**, *10*, 6732–6742.
78. Li, Q.; Bai, H.; Liang, W.; Xia, J.; Wan, S.; Wh, V.D.P. Nitrogen addition and warming independently influence the belowground micro-food web in a temperate steppe. *PLoS ONE* **2013**, *8*, e60441. [CrossRef] [PubMed]
79. Hishi, T.; Fujimaki, R.; McGonigle, T.P.; Takeda, H. Relationships among fine roots, fungal hyphae and soil microarthropods among different soil microhabitats in a temperate coniferous forest of Chamaecyparis obtusa. *Eur. J. Soil Biol.* **2008**, *44*, 473–477. [CrossRef]
80. Bradford, M.A. Re-visioning soil food webs. *Soil Biol. Biochem.* **2016**, *102*, 1–3. [CrossRef]
81. Pollierer, M.M.; Langel, R.; Körner, C.; Maraun, M.; Scheu, S. The underestimated importance of belowground carbon input for forest soil animal food webs. *Ecol. Lett.* **2007**, *10*, 729–736. [CrossRef] [PubMed]
82. Mcnaughton, S.J.; Oesterheld, M.; Frank, D.A.; Williams, K.J. Ecosystem-level patterns of primary productivity and herbivory in terrestrial habitats. *Nature* **1989**, *341*, 142–144. [CrossRef]
83. Du, E.; de Vries, W. Nitrogen-induced new net primary production and carbon sequestration in global forests. *Environ. Pollut.* **2018**, *242*, 1476–1487. [CrossRef]

11

Evaluation of Whole Tree Growth Increment Derived from Tree-Ring Series for Use in Assessments of Changes in Forest Productivity across Various Spatial Scales

Juha M. Metsaranta and Jagtar S. Bhatti *

Natural Resources Canada, Canadian Forest Service, Northern Forestry Centre, 5320 122 Street, Edmonton, AB T6H 3S5, Canada; juha.metsaranta@canada.ca
* Correspondence: jagtar.bhatti@canada.ca

Academic Editors: Scott X. Chang and Xiangyang Sun

Abstract: The inherent predictability of inter-annual variation in forest productivity remains unknown. Available field-based data sources for understanding this variability differ in their spatial resolution, temporal resolution, and typical units of measure. Nearly all other tree and forest characteristics are in practice derived from measurements of diameter at breast height (DBH). Therefore, diameter increment reconstructed annually from tree-ring data can be used to estimate annual growth increments of wood volume, but the accuracy and precision of these estimates requires assessment. Annual growth estimates for n = 170 trees sampled for whole stem analysis from five tree species (jack pine, lodgepole pine, black spruce, white spruce, and trembling aspen) in Western Canada were compared against increments derived from breast height measurements only. Inter-annual variability of breast height and whole tree growth increments was highly correlated for most trees. Relative errors varied by species, diameter class, and the equation used to estimate volume (regional vs. national). A simple example of the possible effect of this error when propagated to the stand level is provided.

Keywords: tree rings; volume increment; height increment; stem analysis

---

1. Introduction

Changes in climate and other environmental conditions can cause annual forest growth to deviate from its long-term historic average, with significant impacts on large-scale estimates of carbon stock changes and forest dynamics. Both spatial and inter-annual variability of tree growth and forest productivity are fundamental for the assessment of future forest management options as well as the annual reporting of forest carbon balances, but the inherent predictability of this inter-annual variation remains unknown [1]. Available data for understanding inter-annual variation differ in their temporal resolution, spatial coverage, and typical units of measure [2,3]. For example, eddy covariance sites provide data on ecosystem carbon fluxes with high temporal resolution. These data can be used to assess site-specific factors affecting inter-annual variation in fluxes, but the measurements are limited to a few sites over a limited number of recent years. Larger amounts of data may be obtained from forest inventory plots, where data such as the diameter, height, vital status, and species of trees are periodically recorded. Allometric relationships can be used to convert measured tree characteristics into stock and stock change estimates for various tree- and stand-level metrics such as wood volume or biomass. This approach provides large-scale estimates on changes resulting from both growth and mortality, allowing the assessment of long-term trends over landscapes [4]. However, the temporal resolution of these data is limited, typically only 5 to 10 years, so they may not provide the annual scale data required for analyzing tree response to annual climate fluctuations. Recent studies have noted the potential utility of tree ring data in terrestrial carbon cycle research [5], as it can lengthen the available time series of annually resolved estimates of ecosystem production or growth increment [3,6–8]. In these applications, ring width measurements are instead used to generate various tree or forest level estimates of annual growth increment. This is possible because past diameter at breast height (DBH), the most basic measurement of trees from which nearly all other tree and forest characteristics are derived, can be reconstructed annually from ring-width measurements. Therefore, tree-ring data could potentially form the basis for assessing stand-level changes in forest productivity, as well as in the development and evaluation of data products and ecosystem models [3,9,10].

An issue remains in that relatively little effort has been put into evaluating the accuracy of estimates derived from breast height, relative to actual whole tree growth increments [11–13]. For example, stem growth at breast height can over- or underestimate whole-tree annual increments under warm and dry climate [13–15]. Presently, the best available method for obtaining annual growth increment data is full stem analysis [16–18], which involves felling and sectioning the main stem at regular intervals. Annual wood volume and height increments can be calculated from ring-widths measured and cross-dated on these sections. Stem analysis is labour-intensive, so large-scale application for obtaining annual growth increment data is likely cost-prohibitive. However, it can provide a useful estimate of true annual growth increment, against which alternatives derived from data at breast height only, which is more widely available, can be compared. In this paper, we evaluate annual growth increment estimates of wood volume (m^3) derived from past diameters reconstructed from breast height ring widths compared to the same estimates obtained from full stem analysis data, when both estimates are available for the same tree. We did this using data from five species common in the Western Canadian boreal forest: lodgepole pine (*Pinus contorta* Dougl. *var. latifolia* Engelm.) jack pine (*Pinus banksiana* Lamb.), black spruce (*Picea mariana* (Mill.) B.S.P.), white spruce (*Picea glauca* (Moench) Voss.), and trembling aspen (*Poplus tremuloides* Michx.). The purpose of this evaluation is to determine the potential error associated with measures of volume growth increment, derived from tree ring measurements at breast height only, in terms of the actual value of the growth increment, as well as its pattern of inter-annual variation. We also apply our estimates of error to an example stand where tree-ring data have previously been used to estimate inter-annual variation in growth [2,3], to estimate the potential stand level magnitude of this error.

2. Materials and Methods

2.1. Stem Analysis Data Collection and Measurement

Full stem analysis data were obtained from 173 trees sampled at various points in time from 1994 to 2010 in the Western Canadian boreal forest. Samples were collected in four jurisdictions: Alberta, Saskatchewan, Manitoba, and the Northwest Territories, and for five species, lodgepole pine (n = 30) jack pine (n = 15), black spruce (n = 68), white spruce (n = 28), and trembling aspen (n = 31). The characteristics of the sample trees (at time of sampling) are further detailed in Table 1. These stem analysis data were acquired from several projects. Some were collected during the 1994 field campaign of the Boreal Ecosystem-Atmosphere Study (BOREAS) campaign from locations near Prince Albert, Saskatchewan, and Thompson, Manitoba. Sites, stands, soils, and the detailed stem analysis data and procedures are described elsewhere [19,20].

Briefly, tree height and DBH were measured on each sample tree prior to felling. Disks were cut at heights of 0.3 m, 1.0 m, 1.3 m, 2.0 m, and at 1 m intervals to the top of the stem. Fresh diameters were measured in the field for each disk. In the lab, each disk was dried in a 70 °C oven for three days, after which diameters and bark thickness were re-measured. For each disk, two radii were selected, 180° apart and avoiding compression wood. These radii were then sampled, X-rayed, and analyzed for ring width using DendroScan [21]. Additional samples were collected in 2005 and 2006 as a part of a study validating tree biomass models in the Canadian province of Alberta. The field methods were similar to those for the BOREAS project, with the exception that disks were sampled at slightly different heights: 0.3 m, 1.3 m, 2.8 m, and then at intervals of 2.5 m to the top of the stem. Disks were dried as for the BOREAS data. However, rather than processing for X-ray analysis with DendroScan, the samples were sanded with progressively finer sandpaper, scanned at 1600 dpi, and ring widths were then measured using CDendro (Cybis Elektronik & Data, Stockholm, Sweden). An additional 21 stem analyses were conducted during the summer of 2010 as part of a study validating tree biomass models for the Liard Valley, NWT, Canada. The disks on these trees were collected at heights of 0.3 m, 1.3 m, 5.0 m, and every 4.0 m to the top of the stem, and processed in the lab in the same way as the 2005–2006 samples. For all samples, ring widths were cross-dated between all the disks in each tree to ensure correct ring dating, and corrected by proportion to the mean fresh outside bark diameter of each disk.

Table 1. Summary of characteristics of trees sampled for stem analysis in this study.

Species	DBH (cm)	Height (m)	Age (Years)	Year Sampled (Year (*n*))
White spruce (SW) (*Picea glauca*)	23.0 (6.2, 51.4)	18.7 (6.7, 33.7)	86 (15, 226)	1994 (5) 2005 (4) 2006 (11) 2010 (8)
Black spruce (SB) (*Picea mariana*)	11.8 (5.0, 24.6)	10.8 (6.4, 20.9)	89 (18, 222)	1994 (26) 2005 (19) 2006 (16) 2010 (7)
Trembling aspen (TA) (*Populus tremuloides*)	21.4 (5.3, 54.8)	17.7 (6.5, 35.6)	69 (19, 176)	1994 (9) 2005 (15) 2010 (7)
Jack pine (PJ) (*Pinus banksiana*)	13.1 (7.0, 18.6)	12.3 (8.5, 15.8)	64 (52, 76)	1994 (15)
Lodgepole pine (PL) (*Pinus contorta*)	15.3 (4.2, 27.0)	14.7 (5.8, 28.1)	77 (14, 153)	2005 (24) 2006 (6)

For DBH, height, and age, the value in the table represents the mean of all sampled trees, and the values in parentheses are the minimum and maximum values for that species. For year sampled, values in parentheses are the number of trees of that species sampled in that year.

2.2. *Stem Analysis Growth Estimates*

The ring-width measurements for each disk section for each tree were input into a spreadsheet program that calculated annual estimates of tree height (m) and stem volume (m^3). Historical tree height (m) was interpolated assuming equal annual height growth within a log section, and assuming that each disk cuts through the middle of a growth cycle, and was referred to as stem analysis height (HTs). Volume was calculated from ring width measurements on each of the disk sections and summed for the entire tree for each year, and referred to as stem analysis volume (Vs). The volume for each log section was calculated using Smalian's formula, $V = (L(Ab + At)/2)$, for the top section from the volume of a paraboloid, $V = (LAb)/2$, and for the stump section from the volume of a cylinder $V = LAb$, where V is the volume of the section (cm^3), L is the length of the section (cm), Ab is the cross-sectional area of the base of the section (cm^2), and At is the cross-sectional area of the top of the section (cm^2). Note that, while the stem analysis-based measurements were taken as the true standard against which alternatives derived from the breast height sample were compared, they are not themselves without uncertainty. For example, estimates derived from stem analysis will differ depending on the method used to interpolate height between cross sections [18], which standard model is used to determine log volume [22,23], the number of cross-cuts sampled, and the number of radii measured along each sample [12], and because trees are not perfectly round in cross section [24,25].

2.3. Model-Based Growth Estimates

The breast height (1.3 m) sample taken from each tree was used as a reconstructed estimate of the past breast height diameter (DBH) of each tree, from which we then calculated annual increments of wood volume for comparison to the estimates derived from stem analysis. We describe the methods used for each of these in the following sections, and an additional summary is provided in Table 2.

Table 2. Summary of the volume and height estimation methodologies used in the comparisons.

Metric	Comparison	Description	References for Equations and Parameters
(A) Volume	(1) National DBH	(1.1) Volume is estimated from DBH only, using a national equation	[26]
	(2) National DBH and HT	(2.1) Volume is estimated from DBH and uncorrected height (HTr), using a national equation.	[26]; heights as in B1.1 or B1.2
		(2.2) Volume is estimated from DBH and corrected height (HTc) using a national volume equation	
	(3) Regional DBH and HT	(3.1) Volume is estimated from DBH and uncorrected height (HTr), using a regional equation.	[27] for the equation formulation, Regional parameters from published sources for Saskatchewan [28], Manitoba [29], Alberta [30], and the Northwest Territories [30]. Heights estimated as in B1.1 or B1.2
		(3.2) Volume is estimated from DBH and corrected height (HTc) using a regional equation	
(B) Tree height	(1) HT DBH model	(1.1) Heights estimated from diameters reconstructed on the breast height sample are compared to heights interpolated between stem analysis sections.	(1.1 and 1.2) Provincial parameter sets for trees in Manitoba [29]. Alberta parameter sets for other provinces (NWT, SK, and AB) [31]
		(1.2) Heights estimated as above, but a correction factor is calculated from the difference between measured and predicted height at time of sampling, and applied to the rest of the height time series.	

2.3.1. Height Estimation

Height was estimated from DBH using regional height–diameter equations, which use the Chapman-Richards function:

$$H = 1.3 + a\left(1 - e^{-b \cdot D_{ob}}\right)^{c} \quad (1)$$

where H is the tree height (m), D_{ob} is the outside bark diameter (DBH, cm), and a, b, and c are the parameters obtained from published sources for Alberta ([31], Table S1) and Manitoba ([29], Table S1). Trees in Saskatchewan and the Northwest Territories also used the Alberta parameters, as regional parameters were not available for these jurisdictions. We calculated two height increment series from the annually reconstructed DBH, (1) a raw height series (HTr), which used the height estimate directly from Equation (1), as well as (2) a corrected height series (HTc), where a ratio between the final predicted height from the raw height series and the measured tree height was used as a correction factor and applied to the raw height series to scale past predicted heights so that the final height in the corrected series was equal to the measured height at the time the tree was sampled.

2.3.2. Volume Estimation

Estimates of volume increment were obtained from the reconstructed DBH, both alone and in combination with height estimated using Equation (1), both in raw (HTr) and corrected form (HTc). We first used a national scale taper equation that uses DBH only as a predictor (VNd) [26]:

$$d^2{}_{ijkm} = DBH^2{}_{ijk}\frac{\beta_0 DBH^{\beta_1+\delta_{i1}+\delta_{ij1}} - h_{ijkm}}{\beta_0 DBH^{\beta_1+\delta_{i1}+\delta_{ij1}} - 1.3}\left(\frac{h_{ijkm}}{1.3}\right)^{2-(\beta_2+\delta_{i2}+\delta_{ij2}+\delta_{ijk})} + \varepsilon_{ijkm} \quad (2)$$

where d is the diameter at different cross-section heights (h), indices i, j, k, and m respectively refer to province, plot, tree, and cross section, β_0, β_1, and β_2 are fixed-effect parameters, and δ_i, δ_{ij}, and δ_{ijk} are random effects associated with the province, plot, and tree, respectively [26]. We also tested two methods that use both DBH and H as predictors. The first was also a national level taper equation (VNdh) [26]:

$$d^2{}_{ijkm} = DBH^2{}_{ijk}\frac{H_{ijk} - h_{ijkm}}{H_{ijk} - 1.3}\left(\frac{h_{ijkm}}{1.3}\right)^{2-(\beta_2+\delta_{i2}+\delta_{ij2}+\delta_{ijk})} + \varepsilon_{ijkm} \quad (3)$$

where d is the diameter at different cross-section heights (h), indices i, j, k, and m respectively refer to province, plot, tree, and cross section, β_2 is a fixed-effect parameter, and δ_i, δ_{ij}, and δ_{ijk}, are random effects associated with the province, plot, and tree, respectively [26] (Table S2). The second was a different taper equation (VRdh) [27], for which regional parameter estimates were generally available [28–30]:

$$d_i = a_0 DBH^{a1} a2^{D_{ob}} X_i^{b1z_i{}^2 + b2\ln(z_i+0.001) + b3\sqrt{z_i} + b4e^{z_i} + b5(DBH/H)} \quad (4)$$

where d is the inside bark diameter at height i along the stem (cm), DBH is the diameter at breast height (cm) of the tree outside bark, H is the total height (m), and X_i is $\left(1-\sqrt{h_i/H}\right)/\left(1-\sqrt{p}\right)$, where h_i is the cross-section height i, p is the relative height of the inflection point, typically assumed to be 0.25, and z_i is the relative height (h_i/H) [27]. Parameters a_0, a_1, a_2, b_1, b_2, b_3, b_4, and b_5 were obtained from published sources Saskatchewan [28], Manitoba [29], and Alberta [30] (Table S3). Regional parameters were not available for the Northwest Territories; therefore, for trees in that jurisdiction, the Alberta parameters were used. Estimates of volume for each of the methods tested were obtained by numerical integration, and the volume increment from the difference in volume between subsequent years.

2.4. Comparison of Estimates

For volume increment, we conducted five comparisons, representing combinations of different volume equations (VNd, VNdh, and VRdh) and height estimation options (HTr (VNdhr or VRDhr) or HTc (VNdhc or VRdhc)), which were in each case assessed against stem analysis volume (Vs), which was considered true. Details on these comparisons are provided in Table 2. Differences were assessed using relative error,

$$100(\hat{Y}_i - Y_i)/Y_i \quad (5)$$

where $\hat{Y}_i$ is the estimated growth metric (height or volume) determined from the reconstructed DBH only, and Y_i is the same metric determined from the stem analysis for the ith year. For analysis, we grouped relative errors by species and 5 cm DBH class, and used the distribution of errors in each class as an indicator of the magnitude and potential significance of differences relative to the estimates derived from stem analysis. When the value zero (indicating no difference between the breast height and whole tree-derived volume increment) was below the 2.5th percentile of all differences in a class, we considered the estimates to be biased high. When zero was between the 2.5th and 25th percentiles, we considered them borderline high, between the 25th and 75th percentiles unbiased, between the 75th and 97.5th percentiles borderline low, and greater than the 97.5th percentile, low.

2.5. Inter-Annual Variation

An additional aim was to determine if annual increment values derived from stem analysis and their alternative were statistically in phase. To do this, we compared a first-differenced time series of each comparison variable and assessed the value of the cross correlation at lag zero, as well as determined the fraction of cases in which the cross correlation peaked at lag zero, relative to shifting the comparison time series forwards or backwards by up to five years.

2.6. Stand-Level Example Application

Errors at the tree-level were also propagated to the stand level for an example fixed area plot in Saskatchewan, Canada, for which samples for dendrochronological analysis were collected at breast height from all live and dead (standing and fallen) trees present at the time of sampling, and that has previously been used in assessments of mortality, competition dynamics, and ecosystem production [2,3]. The observed distribution of error for the regional volume model for jack pine using DBH and HTc as input was propagated using $n = 1000$ Monte Carlo simulations, under a simple assumption of an independence of errors among trees and size classes.

3. Results

3.1. Comparison of Estimates

Relative errors varied by species, diameter class, the equation used to estimate volume (regional vs. national), and whether or not the equation used only DBH or both DBH and HT as predictors (Figure 1). Estimates derived from breast height only were only biased high or low in relatively few species, diameter class, and volume equation categories ($n = 13/175$, 7%). Most (52%) showed no bias, and the rest (41%) were borderline high or low (Figure 1). The average range of errors across classes was ±50%, with some small differences among species. In general, the widest distribution of errors for any species or diameter class was for small trees (<10 cm DBH), and in particular for the national model that used only DBH as a predictor. For models that used both DBH and HT, the use of HTc rather than HTr, or a regional rather than a national model had only a marginal influence on the distribution of relative errors in any species or diameter class. The number of annual increment observations available for comparison by species and diameter class is provided in Table S4.

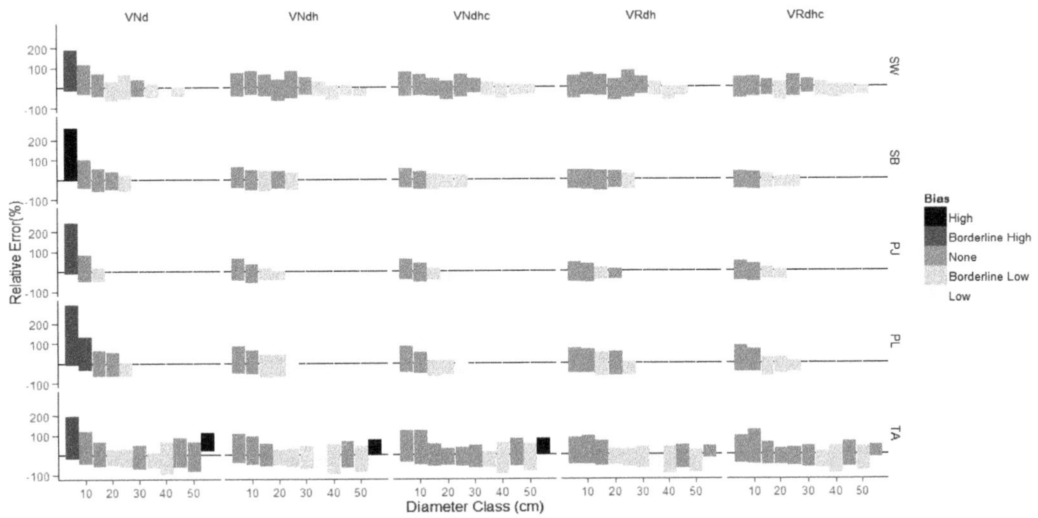

Figure 1. Relative error in estimation of volume increment from DBH reconstructed from tree-ring data, in comparison to volume increment derived from whole stem analysis for the same trees. Errors are plotted by species (rows) and 5 cm DBH class. The bars represent the range of the 2.5th and 97.5th percentiles of the relative errors in each class. Bars are plotted in different shades as a function of the percentile location of zero with the distribution of errors in a diameter class, and interpreted as described in the text. The columns represent different potential methods for estimating volume increment, national equations using DBH only (VNd), national equations using DBH and raw HT (VNdh), national equations using DBH and corrected height (VNdhc), regional equations using DBH and raw HT (VRdh), and regional equations using DBH and corrected height (VRdhc).

3.2. Inter-Annual Variation

For the assessment of the cross-correlation of inter annual variability between breast height and stem analysis-derived growth metrics, we show results for HTr only, as results for HTc are virtually identical. For volume increment, the cross correlation was maximized at lag zero for 94% of cases for estimates derived from DBH only using the national volume model (median 0.83 and 95% were between 0.20 and 0.99), and for 99% of samples for estimates derived from both DBH and HT for the national (median 0.92, 95% between 0.56 and 0.99), and regional models (median 0.93, 95% between 0.63 and 0.99).

3.3. Example Application

Results for the propagation of errors to a stand level example are provided in Figure 2. This particular example used a regional volume for jack pine in Saskatchewan and corrected estimate of height increment. Estimates of volume increment derived from breast height were within the propagated error range for ages less than 30 years. For ages greater than 30 years, the ranges fell below the point estimate derived from the breast height sample, indicating that it was biased to a high degree relative to whole tree increments. It is likely that many trees passed from a size class where the error distribution was unbiased (VRdhc for PJ, Figure 1), to one where the bias was borderline low. Relative to the median of the error simulations, the range of values typically spanned ±5%.

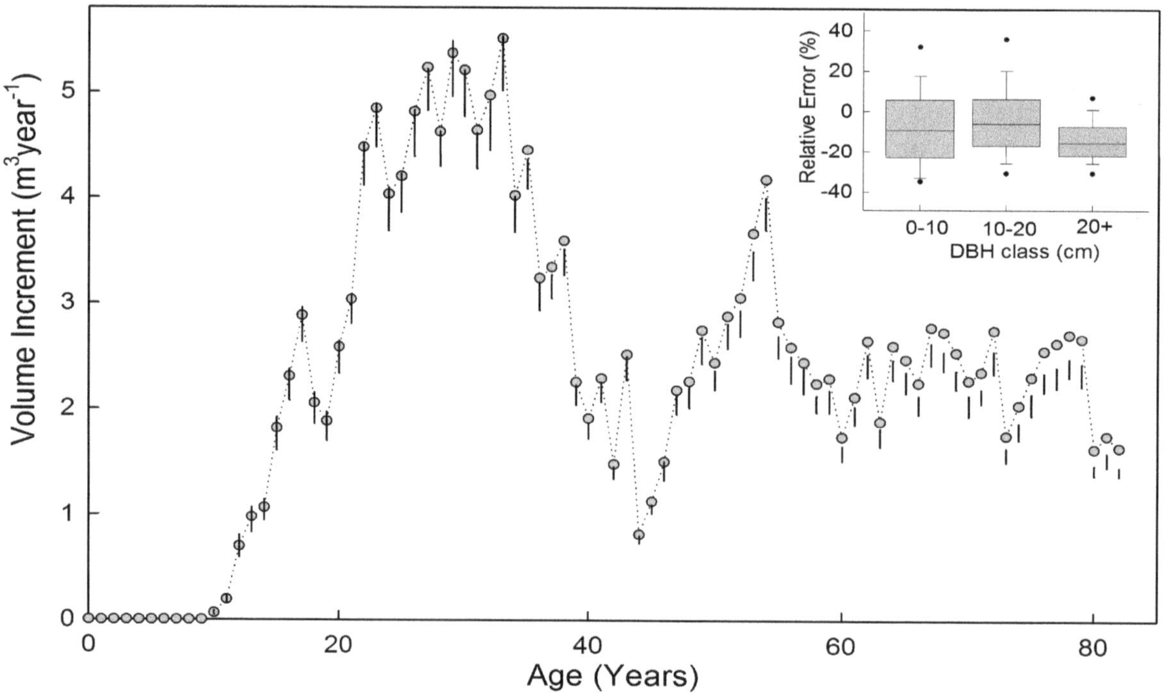

Figure 2. Time series of above ground volume increment ($m^3 \cdot ha^{-1} \cdot year^{-1}$) with stand age for a 900 m^2 fixed area jack pine plot in Saskatchewan, Canada, for which all live and dead trees present in the stand at age 82 were sampled at breast height for ring width measurement. The line plot represents volume increment derived from these measurements using a regional volume model using corrected height as input. Error bars represent the 2.5th and 97.5th percentiles of stand level estimates of volume increment, derived from diameter class-based relative errors (inset graph) derived from the observed difference in breast height and whole tree-derived growth increment for this species and volume estimation method.

4. Discussion

In this paper, we evaluated annual growth increment estimates of wood volume (m^3) derived from past DBH reconstructed from ring widths and compared these to estimates obtained from full stem analysis data, when both estimates are available for the same tree. This was done using data from n = 170 trees for five species common in the Western Canadian boreal forest: lodgepole pine, jack pine, black spruce, white spruce, and trembling aspen. The main results were that relative errors varied by species, diameter class, and the equation used to estimate volume and that the direction of inter-annual variability for whole tree and breast height estimates of volume increment were highly correlated for nearly all trees. In general, when estimates of volume increment are derived from models that use both DBH and HT as predictors, the total range of values spanned by the relative errors is narrower, relative to volume increment derived from a model using DBH only. The range of observed relative errors for national models was similar to that of regional models, at least for the species tested here. When tree-level errors were propagated to estimates of error for volume increment at the stand level, the overall range of errors was narrower, typically $\pm$5% for our example jack pine stand, at least for our simple procedure that assumed the observed errors were independent. The overall range of errors is less than what is currently assumed to be the uncertainty of growth increments applied by Canada's National Forest Carbon Monitoring, Accounting, and Reporting system to estimate the carbon balance of Canada's managed forest [32], but it should be kept in mind that additional error would arise from model and measurement errors that were not accounted for, and at young ages from trees that have died and subsequently decayed such that they could not be detected when the stand was sampled [33]. In this analysis, breast height estimates were consistently high relative to whole tree estimates after age 30–40, but this is likely a result specific to this particular combination of tree species and volume estimation method.

Few other studies have attempted to compare the relative accuracy of different methods of measuring the annual incremental growth of trees. Similar to our analysis, standardized ring-width growth series at breast height and standardized whole-tree growth series of volume increment were highly correlated for high-elevation red spruce (*Picea rubens*) forests [11]. For European beech (*Fagus sylvatica*), the breast-height growth series were also highly correlated to whole-tree volume or biomass-increment series [13], but the breast-height series showed higher sensitivity to weather variation. An analysis of the relative accuracy of various full stem analysis sampling protocols for measuring volume increment found that it was possible to increase precision by increasing the density of sections sampled along the stem, and by measuring more radii at each sample [12]. Our results are essentially in line with these previous findings, in that we found that the direction of inter-annual variation in volume increment derived from breast height only was highly correlated with whole tree increment obtained by stem analysis. Our observation that using models with a height predictor generally improved the estimation of volume increment is interesting because this occurs even though total height estimates derived from stem analysis are calculated as the mean height increment between cross-cut segments. Interpolation between cross-cut segments means that stem analysis samples do not actually measure the inter-annual variability of height growth. There is no method presently available for retrospectively obtaining this information, except for some species that consistently grow a single annual whorl of branches. Ground-based measurements of height cannot resolve annual height increments for tall, slow growing trees [2,34], though height growth can be prospectively monitored with some accuracy by tree climbing [35].

Although it likely would have been of interest, we could not conduct a comparison for biomass increment because stem analysis does not provide a direct estimate of biomass that could be considered true. We could have compared estimates generated using reconstructed DBH only with estimates generated using DBH and HT, as alternative models that could have been applied do exist [36,37]. However, this analysis would have been a model inter-comparison exercise that compares the relative difference between estimates of biomass increment using DBH and HT relative to the use of DBH only, not an accuracy assessment directly. The estimation of biomass from tree level allometric equations remains challenging, and the current state of work in this area shows that changes in allometric equations can result in large variations in estimated biomass [38].

Some further challenges need to be considered when inferring above-ground forest increment from radial tree growth. Stand structure and density along with necessary biometric and metadata would be useful to upscale tree biomass estimates to the site level [39,40] and to efficiently select sample trees for estimating climate relationships in an unbiased manner [41]. An alternative, such as was used in our example application where tree level errors were propagated to the stand level, is to sample all live and dead trees in fixed area plots to reconstruct the growth of the whole stand [2,3], a method that has shown promise if the aim is to successfully detect climate and global change-related trends from tree-ring data [42], but remains challenged by the inability to sample trees that have died in the past and subsequently decayed [33]. In addition, replacing the relatively basic equations used in this study with more sophisticated methods that are sensitive to the effect of stand conditions on height–diameter relationships [43], tree volume [44,45], or biomass [46] should also improve the estimates of annual growth increment that could be obtained at both the tree and forest levels, for species where such models are available.

5. Conclusions

The ability to estimate annual growth increment over a long time period at a large spatial scale from field-based data would be a tremendous asset. Such data can be obtained by tree-ring analysis, which is increasingly being used in forestry applications to address questions around the effect of forestry practices, climate, insect outbreaks, global change, and other disturbances in commercially viable, closed-canopy forests as well as a wide range of unmanaged forest and woodland types. Traditionally, tree-ring data have been collected from old trees at climate-sensitive sites for reconstructing past climatic variation, and expressed in unitless ring-width indices. Therefore, observed causes and trends may not be applicable to more productive and managed forests. We have shown that translating these instead into tree- and forest-level estimates of annual growth increment in units that are of greater interest in forestry applications can provide reasonable estimates relative to alternative data on whole tree growth obtained by stem analysis.

Acknowledgments: The authors would like to thank Shongming Huang and Yonghe Wang for comments on a previous version of the manuscript, and Thierry Varem-Sanders for extensive assistance with sample processing and data analysis. Funding for this study was provided in part by Canadian Forest Service and program of the Federal Panel on Energy Research and Development (PERD).

Author Contributions: Juha Metsaranta and Jagtar Bhatti both have contributed substantially to idea, data collection, data analysis and finally the manuscript writing.

References

1. Luo, Y.; Keenan, T.F.; Smith, M. Predictability of the terrestrial carbon cycle. *Glob. Chang. Biol.* **2015**, *21*, 1737–1751. [CrossRef] [PubMed]
2. Metsaranta, J.M.; Lieffers, V.J. Using dendrochronology to obtain annual data for modelling stand development: A supplement to permanent sample plots. *Forestry* **2009**, *82*, 163–173. [CrossRef]
3. Metsaranta, J.M.; Kurz, W.A. Inter-annual variability of ecosystem production in boreal jack pine forests (1975–2004) estimated from tree-ring data using CBM-CFS3. *Ecol. Mod.* **2012**, *224*, 111–123. [CrossRef]
4. Hember, R.A.; Kurz, W.A.; Metsaranta, J.M.; Black, T.A.; Coops, N.C.; Guy, R.D. Accelerated regrowth of temperate-maritime forests due to environmental change. *Glob. Chang. Biol.* **2012**, *18*, 2026–2040. [CrossRef]
5. Babst, F.; Alexander, M.R.; Szejner, P.; Bouriaud, O.; Klesse, S.; Roden, J.; Ciais, P.; Poulter, B.; Frank, D.; Moore, D.J.P.; et al. A tree-ring perspective on the terrestrial carbon cycle. *Oecologia* **2014**, *176*, 307–322. [CrossRef] [PubMed]

6. Graumlich, L.J.; Brubaker, L.B.; Grier, C.C. Long-term trends in forest net primary productivity: Cascade Mountains, Washington. *Ecology* **1989**, *70*, 405–410. [CrossRef]
7. Babst, F.; Bouriaud, O.; Alexander, R.; Trouet, V.; Frank, D. Toward consistent measurement of carbon accumulation: A multi-site assessment of biomass and basal area increment across Europe. *Dendrochronologia* **2014**, *32*, 153–161. [CrossRef]
8. Fang, O.; Wang, Y.; Shao, X. The effect of climate on the net primary production (NPP) of *Pinus koraiensis* in the Changbai Mountains over the past 50 years. *Trees* **2016**, *30*, 281–294. [CrossRef]
9. Bunn, A.G.; Hughes, M.K.; Kirdyanov, A.V.; Losleben, M.; Shishov, V.V.; Berner, L.T.; Oltchev, A.; Vaganov, E.A. Comparing forest measurements from tree rings and a space based index of vegetation activity in Siberia. *Environ. Res. Lett.* **2013**, *8*, 035034. [CrossRef]
10. Babst, F.; Bouriaud, O.; Papale, D.; Gielen, B.; Jansenns, I.A.; Nikinmaa, E.; Ibrom, A.; Wu, J.; Bernhofer, C.; Köstner, B.; et al. Above-ground woody carbon sequestration measured from tree rings is coherent with net ecosystem productivity at five eddy-covariance sites. *New Phytol.* **2014**, *201*, 1289–1303. [CrossRef] [PubMed]
11. LeBlanc, D.C. Relationship between breast-height and whole-stem growth indices for red spruce on Whiteface Mountain, New York. *Can. J. For. Res.* **1990**, *20*, 1399–1407. [CrossRef]
12. Newton, P.F. A stem analysis computational algorithm for estimating volume growth and its empirical evaluation under various sampling strategies. *Comput. Electron. Agric.* **2004**, *44*, 21–31. [CrossRef]
13. Bouriaud, O.; Breda, N.; Dupouey, J.-L.; Granier, A. Is ring width a reliable proxy for stem-biomass increment? A case study in European beech. *Can. J. For. Res.* **2005**, *35*, 2920–2933. [CrossRef]
14. Chhin, S.; Hogg, E.H.; Lieffers, V.J.; Huang, S. Growth–climate relationships vary with height along the stem in lodgepole pine. *Tree Phys.* **2010**, *30*, 335–345. [CrossRef] [PubMed]
15. Van der Maaten-Theunissen, M.; Bouriaud, O. Climate-growth relationships at different stem heights in silver fir and Norway spruce. *Can. J. For. Res.* **2012**, *42*, 958–969. [CrossRef]
16. Duff, G.H.; Nolan, N.J. Growth and morphogenesis in the Canadian forest species: I. The controls of cambial and apical activity in *Pinus resinosa*. *Can. J. Bot.* **1957**, *31*, 471–513. [CrossRef]
17. Shea, S.R.; Armson, K.A. Stem analysis of jack pine (*Pinus banksiana* Lamb.): Techniques and concepts. *Can. J. For. Res.* **1972**, *2*, 392–406. [CrossRef]
18. Dyer, M.E.; Bailey, R.L. A test of six methods for estimating true height from stem analysis data. *For. Sci.* **1987**, *33*, 3–13.
19. Halliwell, D.H.; Apps, M.J. *BOReal Ecosystem-Atmosphere Study (BOREAS) Biometry and Auxiliary Sites: Overstory and Understory Data*; Natural Resources Canada, Canadian Forest Service, Northern Forestry Centre: Edmonton, AB, Canada, 1997.
20. Varem-Sanders, T.M.L.; Cambpell, I.D. *BOReal Ecosystem-Atmosphere Study (BOREAS) Biometry and Auxiliary Sites: X-ray Densitometry of Tree Allometry Samples*; Natural Resources Canada, Canadian Forest Service, Northern Forestry Centre: Edmonton, AB, Canada, 1998.
21. Varem-Sanders, T.M.L.; Cambell, I.D. *DendroScan: A Tree-Ring Width and Density Measurement System*; UBC Press: Vancouver, BC, Canada, 1996.
22. Martin, A.J. Testing volume equation accuracy with water displacement techniques. *For. Sci.* **1984**, *30*, 41–50.
23. Figueiredo-Filho, A.; Schaaf, L.B. Comparison between predicted volumes estimated by taper equations and true volumes obtained by the water displacement technique (xylometer). *Can. J. For. Res.* **1999**, *29*, 451–461. [CrossRef]
24. Biging, G.S.; Wensel, L.C. The effect of eccentricity on the estimation of basal area and basal area increment of coniferous trees. *For. Sci.* **1988**, *34*, 621–633.
25. Bakker, J.D. A new, proportional method for reconstructing historical tree diameters. *Can. J. For. Res.* **2005**, *35*, 2515–2520. [CrossRef]
26. Ung, C.H.; Guo, X.J.; Fortin, M. Canadian national taper models. *For. Chron.* **2014**, *89*, 211–224. [CrossRef]
27. Kozak, A. A variable-exponent taper equation. *Can. J. For. Res.* **1988**, *18*, 1362–1368. [CrossRef]
28. Gál, J.; Bella, I.E. *New Stem Taper Functions for 12 Saskatchewan Timber Species*; Information Report NOR-X-338; Natural Resources Canada, Canadian Forest Service, Northern Forestry Centre: Edmonton, AB, Canada, 1994.
29. Klos, R. Ecologically Based Taper Equations for Major Tree Species in Manitoba. Master's Thesis, Lakehead University, Thunder Bay, ON, Canada, 2004.

30. Huang, S. *Ecologically Based Individual Tree Volume Estimation for Major Alberta Tree Species Report #1 Individual Tree Volume Estimation Procedures for Alberta: Methods of Formulation and Statistical Foundations*; Alberta Sustainable Resource Development, Public Lands and Forests Division: Edmonton, AB, Canada, 1994.
31. Huang, S. *Ecologically Based Individual Tree Volume Estimation for Major Alberta Tree Species Report #2 Ecologically Based Individual Tree Height-Diameter Models for Major Alberta Tree Species*; Alberta Sustainable Resource Development, Public Lands and Forests Division: Edmonton, AB, Canada, 1994.
32. Metsaranta, J.M.; Shaw, C.H.; Kurz, W.A.; Boisvenue, C.; Morken, S. Uncertainty of inventory-based estimates of the carbon dynamics of Canada's managed forest (1990–2014). *Can. J. For. Res.* **2016**. in review.
33. Metsaranta, J.M.; Lieffers, V.J.; Wein, R.W. Dendrochronological reconstruction of jack pine snag and downed log dynamics in Saskatchewan and Manitoba, Canada. *For. Ecol. Manag.* **2008**, *255*, 1262–1270. [CrossRef]
34. Hasenauer, H.; Monserud, R.A. Biased predictions for tree height increment models developed from smoothed 'data'. *Ecol. Mod.* **1997**, *98*, 13–22. [CrossRef]
35. Sumida, A.; Miyaura, T.; Torii, H. Relationships of tree height and diameter at breast height revisited: Analysis of stem growth using 20-year data of an even-aged *Chamaecyparis obtuse* stand. *Tree Phys.* **2013**, *33*, 106–118. [CrossRef] [PubMed]
36. Lambert, M.C.; Ung, C.H.; Raulier, F. Canadian national tree aboveground biomass equations. *Can. J. For. Res.* **2005**, *35*, 1996–2018. [CrossRef]
37. Ung, C.H.; Bernier, P.; Guo, X.J. Canadian national biomass equations: New parameter estimates that include British Columbia data. *Can. J. For. Res.* **2008**, *38*, 1123–1132. [CrossRef]
38. Weiskittel, A.R.; MacFarlane, D.W.; Radtke, P.J.; Affleck, D.L.R.; Temesgen, H.; Woodall, C.W.; Westfall, J.A.; Coulston, J.W. A call to improve methods for estimating tree biomass for regional and national assessments. *J. For.* **2015**, *113*, 414–424. [CrossRef]
39. Garcia, O. Sampling for tree-ring analysis. In Presented at Integrating Forest Information Over Space and Time, Canberra, Australia, 13–17 January 1992; pp. 110–128.
40. Osawa, A.; Abaimov, A.P.; Kajimoto, T. Feasibility of estimating total stem volume and aboveground biomass from measurement on the largest trees in even-aged pure stands. *Can. J. For. Res.* **2001**, *31*, 2042–2048. [CrossRef]
41. Mérian, P.; Bert, D.; Lebourgeois, F. An approach for quantifying and correcting sample size-related bias in population estimates of climate-tree growth relationships. *For. Sci.* **2013**, *59*, 444–452. [CrossRef]
42. Nehrbass-Ahles, C.; Babst, F.; Klesse, S.; Nöttzli, M.; Bouriaud, O.; Neukom, R.; Dobbertin, M.; Frank, D. The influence of sampling design on tree-ring based quantification of forest growth. *Glob. Chang. Biol.* **2014**, *20*, 2867–2885. [CrossRef] [PubMed]
43. Sharma, M.; Zhang, S.Y. Height-diameter models using stand characteristics for *Pinus banksiana* and *Picea mariana*. *Scand. J. For. Res.* **2004**, *19*, 442–451. [CrossRef]
44. Sharma, M.; Zhang, S.Y. Variable-exponent taper equations for jack pine, black spruce and balsam fir in eastern Canada. *For. Ecol. Manag.* **2004**, *198*, 39–53. [CrossRef]
45. Sharma, M.; Parton, J. Modelling stand density effects on taper for jack pine and black spruce plantations using dimensional analysis. *For. Sci.* **2009**, *55*, 268–282.
46. Almedag, I.S.; Stiell, W.M. Spacing and Age Effects on Biomass Production in Red Pine Plantations. *For. Chron.* **1982**, *58*, 220–224. [CrossRef]

12

Decreased Temperature with Increasing Elevation Decreases the End-Season Leaf-to-Wood Reallocation of Resources in Deciduous *Betula ermanii* Cham. Trees

Yu Cong [1], Mai-He Li [1,2], Kai Liu [1], Yong-Cai Dang [1], Hu-Dong Han [1] and Hong S. He [1,3,*]

1 School of Geographical Sciences, Northeast Normal University, Changchun 130024, China; congy345@nenu.edu.cn (Y.C.); maihe.li@wsl.ch (M.-H.L.); liuk368@nenu.edu.cn (K.L.); dangyc818@nenu.edu.cn (Y.-C.D.); hanhd254@nenu.edu.cn (H.-D.H.)

2 Swiss Federal Institute for Forest, Snow and Landscape Research WSL, Zuercherstrasse 111, CH-8903 Birmensdorf, Switzerland

3 School of Natural Resources, University of Missouri, Columbia, MO 65211, USA

* Correspondence: HeH@missouri.edu

Abstract: Global air temperature has increased and continues to increase, especially in high latitude and high altitude areas, which may affect plant resource physiology and thus plant growth and productivity. The resource remobilization efficiency of plants in response to global warming is, however, still poorly understood. We thus assessed end-season resource remobilization from leaves to woody tissues in deciduous *Betula ermanii* Cham. trees grown along an elevational gradient ranging from 1700 m to 2187 m a.s.l. on Changbai Mountain, northeastern China. We hypothesized that end-season resource remobilization efficiency from leaves to storage tissues increases with increasing elevation or decreasing temperature. To test this hypothesis, concentrations of non-structural carbohydrates (NSCs), nitrogen (N), phosphorus (P), and potassium (K) during peak shoot growth (July) were compared with those at the end of growing season (September on Changbai Mt.) for each tissue type. To avoid leaf phenological effects on parameters, fallen leaves were collected at the end-season. Except for July-shoot NSC and July-leaf K, tissue concentrations of NSC, N, P, and K did not decrease with increasing elevation for both July and September. We found that the end-season leaf-to-wood reallocation efficiency decreased with increasing elevation. This lower reallocation efficiency may result in resource limitation in high-elevation trees. Future warming may promote leaf-to-wood resource reallocation, leading to upward shift of forests to higher elevations. The NSC, N, P, and K accumulated in stems and roots but not in shoots, especially in trees grown close to or at their upper limit, indicating that stems and roots of deciduous trees are the most important storage tissues over winter. Our results contribute to better understand the resource-related ecophysiological mechanisms for treeline formation, and vice versa, to better predict forest dynamics at high elevations in response to global warming. Our study provides resource-related ecophysiological knowledge for developing management strategies for high elevation forests in a rapidly warming world.

Keywords: Alpine treeline; Nitrogen; Non-structural carbohydrates; Phosphorus; Potassium; Remobilization; Storage; Upper limits

---

1. Introduction

Global air temperature has increased and continues to increase, especially in high latitude and high altitude areas. Climate change such as global warming will inevitably influence tree growth rate [1] via changing resource availability and use efficiency [2], and probably also resource allocation and reallocation. Resources can be recycled through remobilization from senescing leaves to storage tissues in deciduous trees which enables plants to reuse these resources for regrowth after dormancy [3,4], and for defense and reproduction [5]. Plants remobilize carbon components and nutrients through internal resources cycling from senescing tissues to maximize the resource use efficiency and to minimize the costs [5]. Thus, remobilization is an internal conservation process which can contribute a substantial annual nutrient supply to increase the resource use efficiency [6]. Owing to the importance of resource resorption for plants, differences in patterns of resource remobilization have been exhaustively studied at the intra- and interspecific levels [3,7,8]. The main recycled carbon components in trees are non-structural carbohydrates (NSCs), which are mainly stored in ligneous tissues, particularly in stems and roots [5,8]. Carbohydrate storage is particularly important for deciduous trees that lack photosynthesizing tissues in early spring [9,10]. As a result of greater asynchrony of supply and demand undergone by deciduous species, more carbon compounds are stored in deciduous than in evergreen species [5]. Woody roots or the stems seem to play a crucial role in resource storage of deciduous species [11,12]. Compared to evergreen species that can directly retain resources in over-wintering leaves, rather than translocating them to stems and roots [4], deciduous species recycle resources back to storage tissues before leaf abscission [5]. The resources withdrawn from senescent tissues can be depleted during spring for new growth [13].

Environmental factors have strong impacts on resource remobilization during leaf senescence, because leaf senescence relies most on photoperiod, temperature, precipitation and nutrient availability. For example, resource remobilization efficiency is higher when leaves senesce prematurely as a consequence of low nutrient availability [14–16], indicating an adaptation to infertile habitats [17]. Similarly, recent studies suggested that climatic warming would increase the proficiency of nutrient resorption due to delayed leaf senescence [18]. However, González-Zurdo et al. (2015) studied leaf nitrogen resorption of three evergreen tree species, and concluded that N resorption efficiency was lower when winter temperature was low [19]. Nutrient remobilization efficiency can be lower under water stress, because drought advances leaf abscission [18,20,21]. Current rapid global warming, for example, may result in less water availability in water-limited areas, and warming, in combination with warming-induced changes in water availability, will influence plant ecophysiology such as resource remobilization. Using elevational gradients as a proxy of decreasing temperature, Kutbay and Ok (2003) found that foliar N and P resorption efficiencies did not change in evergreen *Juniperus oxycedrus* L. subsp. *macrocarpa* (Sibth. & Sm.) Ball trees [22]. However, it is unclear how resource remobilization efficiency in deciduous trees responds to low temperature at the alpine treeline and, contrastingly, to future global warming.

We, therefore, studied the effects of elevation on end-season remobilization of carbon and nutrients in deciduous *Betula ermanii* Cham. (Erman's birch) trees grown along an elevational gradient from lower elevation (1700 m a.s.l.) up to the upper limit (2200 m a.s.l.) on Changbai Mountain. We measured tissue NSC, N, P and K concentrations both at the peak shoot growth period and at the end of growing season, and calculated the resource remobilization efficiency of trees along the elevational gradient, to test our hypothesis that the end-season resource remobilization efficiency from leaves to storage tissues in deciduous *B. ermanii* trees increases with increasing elevations, to increase resource use efficiency of trees in a harsh environment.

2. Materials and Methods

2.1. Study Area

The study was conducted in the Natural Reserve, Changbai Mountain (41°59′ N, 127°59′–128° E) in northeastern China. The area was not disturbed because of its remoteness and high elevation. The area has a temperate continental climate [23] with mean growing season (late May to late September) temperature of 5.87 °C (ranging from 3.37 to 8.82 °C) [24]. The annual precipitation ranges from 700 to 1400 mm (data from Tian-Chi Meteorological Station located at 2623 m a.s.l.) [25] and the annual frost-free period is about 65 to 70 days [26]. Soils on the Changbai Mountain are classified as mountain soddy forest soil [25]. The Changbai Mountain has vertical spectra with four vegetation belts [27]: mixed coniferous broad leaved forests distributed from 740 to 1100 m a.s.l., coniferous forests from 1100 to 1700 m a.s.l., birches (*Betula ermanii*) distributed from 1700 to 2200 m a.s.l., with the birch alpine treeline at 2030 m a.s.l., and tundra above 2000 m a.s.l. [22,26].

2.2. Field Sampling

Birch (*B. ermanii*) forests are distributed between 1700 and 2200 m a.s.l. (upper limit) on the Changbai Mt. The *Betula* elevational gradient can be divided by the alpine treeline (2030 m a.s.l.) into two parts, i.e., below the alpine treeline with trees of > 2 m in height and above the alpine treeline with trees of < 2 m in height. We used this elevational gradient as a proxy of environmental change, mainly air warming [1,28]. To compare patterns of remobilization efficiency across environmental gradients from low elevations to high elevations, samples were taken at the peak-growing season and at the end of growing season. We selected end-July (23 July 2017) to collect samples, since for that time the difference in leaf phenology along the elevational gradient was negligible [29] and the physiological processes such as photosynthesis and growth are most active [30]. The end-season samples were collected from the higher elevations downslope to the lower elevations between September 5 and 9, when the nighttime temperature at an elevation dropped to below 0 °C according to a linear model using climate data from Tian-Chi Meteorological Station located at 2623 m a.s.l. and from the Meteorological Station (738 m a.s.l.) of Changbai Mountain Forest Ecosystem Research facility. Five healthy *B. ermanii* trees ($n = 5$) with similar age, height and diameter at breast height (DBH) or basal diameter (2107 and 2187 m a.s.l.) were selected at each of the six elevations, i.e., 1700, 1800, 1900, 2027, 2107 and 2187 m a.s.l. (Table 1). Using a temperature lapse rate of 0.6 °C decrease with 100 m increase in elevation, the annual mean air temperature is 3 °C lower at 2187 m than at 1700 m a.s.l. One-year-old shoots (with bark), main stem sapwood (with bark) and fine roots (< 0.5 cm in diameter, with bark) were collected from each selected tree. For leaf samples, mature green leaves were collected on 23 July (mid-season samples), whereas only freshly fallen leaves were collected from the forest floor close to each sample tree (end-season samples), to minimize difference in leaf phenology among elevations. Also, to minimize the effects of diurnal temperature variation and sunlight on samples, all samples were taken around noon and immediately stored in a cool box [31]. The samples were killed in a microwave oven at 600 W for 40 s to minimize the physiological activities, dried to a constant weight at 65 °C and then ground to fine powder to pass through a 1-mm sieve for analysis. We measured the height and DBH of each tree at the first sampling date (July) (Table 1). For juvenile trees with DBH of < 5 cm, basal diameter of trees (about 1 cm above ground surface) was measured in July. As limitations have been imposed on tree-harvesting within the Changbai Mountain Nature Reserve, we, therefore, only measured the biomass of leaves, shoots, stems, and roots of 5 average trees at the two highest elevations (2107 and 2187 m a.s.l.) in July 2017.

Table 1. Characteristics of the plots and the sampling trees *Betula ermanii* Cham. (mean ± 1 standard deviation; $n = 5$ trees) on the Changbai Mountain (Jilin, NE China).

Site No.	Elevation (m a.s.l.)	Average DBH (cm)	Average Height (m)	Slope Exposure
1	2187	1.0 ± 0.1 [a]	0.9 ± 0.1	West
2	2107	1.1 ± 0.1 [a]	1.5 ± 0.1	West
3	2027	4.7 ± 0.4	4.2 ± 0.1	West
4	1900	9.4 ± 1.0	9.3 ± 0.4	West
5	1800	9.2 ± 0.7	9.2 ± 0.5	West
6	1700	10.9 ± 0.7	11.2 ± 0.8	West

[a] basal diameter (about 1 cm above ground surface).

2.3. Analysis of NSC

Plant tissue (30 mg) was placed into a 10-mL centrifuge tube and mixed with 5 mL of 80% ethanol. The mixture was incubated at 80 °C in a water shaker (SHA-C, Jintan Jingda Instrument Manufacturing Co., Ltd., Jintan, China) for 30 min, cooled to ambient temperature, and thereafter centrifuged at 4000 rpm for 10 min. The sediments were re-extracted twice with 80% ethanol (Sinopharm Chemical Reagent Co., Ltd., Shanghai, China) to extract the soluble sugars [29]. The ethanol-insoluble pellets were used for starch extraction, and the combined supernatants were retained for soluble sugars analysis by the anthrone method. Glucose was used as a standard. Starch was extracted from the ethanol-insoluble pellets after placing in water at 80 °C to remove the ethanol by evaporation. The ethanol-insoluble residues were boiled with 2 mL of distilled water for 15 min. After cooling to room temperature, 2 mL of 9.2M $HClO_4$ (Sinopharm Chemical Reagent Co., Ltd., Shanghai, China) was added to hydrolyze the starch for 15 min, 4 mL distilled water was added and mixed thereafter, and then the mixture was centrifuged at 4000 rpm for 10 min. Subsequently, the solid residues were added with 2 mL of 4.6M $HClO_4$ for one more extraction. Soluble sugars and starch concentrations were both measured at 620 nm using a spectrophotometer (TU-1810, Beijing Purkinje General Instrument Co., Ltd., Beijing, China) [32]. The NSC, soluble sugars and starch concentrations were expressed on a dry matter basis (% d.m.).

2.4. Analysis of Nitrogen, Phosphorus and Potassium

Plant material (0.1 g) was digested in concentrated sulphuric acid and hydrogen peroxide [33]. The digested solutions were used for the determination of nitrogen, phosphorus and potassium. Nitrogen (N) concentration was determined by the idophenol blue colorimetric method on an automatic chemical analyzer (SmartChem 140, AMS-Alliance Instruments, Rome, Italy). Phosphorus (P) was colorimetrically determined by the ammonium molybdate-ascorbic acid method [34] on a spectrophotometer (TU-1810, Beijing Purkinje General Instrument Co., Ltd., Beijing, China). Potassium (K) concentration was measured with a flame photometer (FP6410, Shanghai Precise Scientific Instrument Co., Ltd., Shanghai, China).

2.5. Methods for Evaluating Resource Remobilization

Remobilization efficiency (*R*) was calculated based on differences in concentrations of mobile carbohydrates and nutrients in tissues collected at the peak shoot growth period and at the end season, using the Equation (1):

$$R\% = 100\% \times (C_s - C_j)/C_j = (C_s/C_j - 1) \times 100\% \quad (1)$$

where C_s (September) and C_j (July) represent concentrations of NSC, N, P or K within each tissue type at the end-season (September) and at the peak shoot growth period (July), respectively. A negative *R*-value ($C_s < C_j$) indicates resource reallocation from that tissue, whereas a positive *R*-value ($C_s > C_j$) reflects resource accumulation in that tissue at the end-season. A more negative *R*-value indicates higher remobilization efficiency, whereas a more positive *R*-value suggests higher accumulation efficiency.

2.6. Data Analysis

Data (NSC, N, P and K) were confirmed for normality (Kolmogorov-Smirnov test) before statistical analysis. Two-way analyses of variance (ANOVAs) were used to test the effects of elevation, time, and their interactions on the concentrations of NSC, N, P and K in each tissue. Two-way ANOVAs were performed with elevation and tissue type as factors, with random selection of trees as random factor to identify the trends in the remobilization efficiency of NSC, N, P, and K. Differences in the parameters studied among elevations were tested for significance using Tukey's HSD test at $\alpha = 0.05$. Because we had equal sample size, Duncan's test at $\alpha = 0.05$ was used to analyze the difference for remobilization means next to each adjacent tree among the elevations.

3. Results

3.1. NSC Concentration

Elevation did not affect leaf NSC but significantly affected the NSC concentrations in shoots, stem sapwood, and roots (Table 2). Tissue NSC concentrations varied significantly with time (season), and time interacted with elevation to influence the NSC concentrations in aboveground tissues (leaves, shoots, and stem sapwood) (Table 1). Leaf NSC concentration was higher in July than in September, whereas NSC concentrations in sink tissues (shoots, stem, and roots) were higher in September than in July (Figure 1a–h). Root NSC concentrations seemed to be stable across elevations in both July and September (Figure 1d,h). Apart from this, clear elevational trends in NSC were found only in July shoots where the NSC concentration significantly decreased with increasing elevation (Figure 1b), and in September stem sapwood where the NSC concentrations were significantly higher at higher elevations (Figure 1g).

Table 2. Results of two-way nested ANOVAs with elevation and time as fixed factors. *F* and *p* values are given. NSC = Non-structural carbohydrates.

	NSC		Nitrogen		Phosphorus		Potassium	
	F	*p*	*F*	*p*	*F*	*p*	*F*	*p*
				Leaves				
Elevations (E)	0.42	0.829	20.15	**<0.001**	25.57	**<0.001**	7.21	**<0.001**
Time (T)	27.80	**<0.001**	478.52	**<0.001**	554.96	**<0.001**	125.46	**<0.001**
E × T	14.82	**<0.001**	7.19	**<0.001**	11.61	**<0.001**	2.74	**0.043**
				1-year-old-shoots				
Elevations (E)	8.25	**<0.001**	2.11	0.098	3.96	**0.009**	8.42	**<0.001**
Time (T)	253.11	**<0.001**	301.44	**<0.001**	222.21	**<0.001**	465.96	**<0.001**
E × T	5.77	**0.001**	5.77	**0.001**	3.53	**0.016**	9.37	**<0.001**
				Stem sapwood				
Elevations (E)	23.45	**<0.001**	206.63	**<0.001**	215.03	**<0.001**	22.26	**<0.001**
Time (T)	36.69	**<0.001**	306.85	**<0.001**	113.94	**<0.001**	28.97	**<0.001**
E × T	11.69	**<0.001**	101.70	**<0.001**	49.27	**<0.001**	28.55	**<0.001**
				Fine roots				
Elevations (E)	3.63	**0.014**	7.13	**<0.001**	10.60	**<0.001**	5.88	**0.001**
Time (T)	232.16	**<0.001**	84.80	**<0.001**	72.13	**<0.001**	37.25	**<0.001**
E × T	1.50	0.228	2.28	0.079	1.21	0.333	0.62	0.688

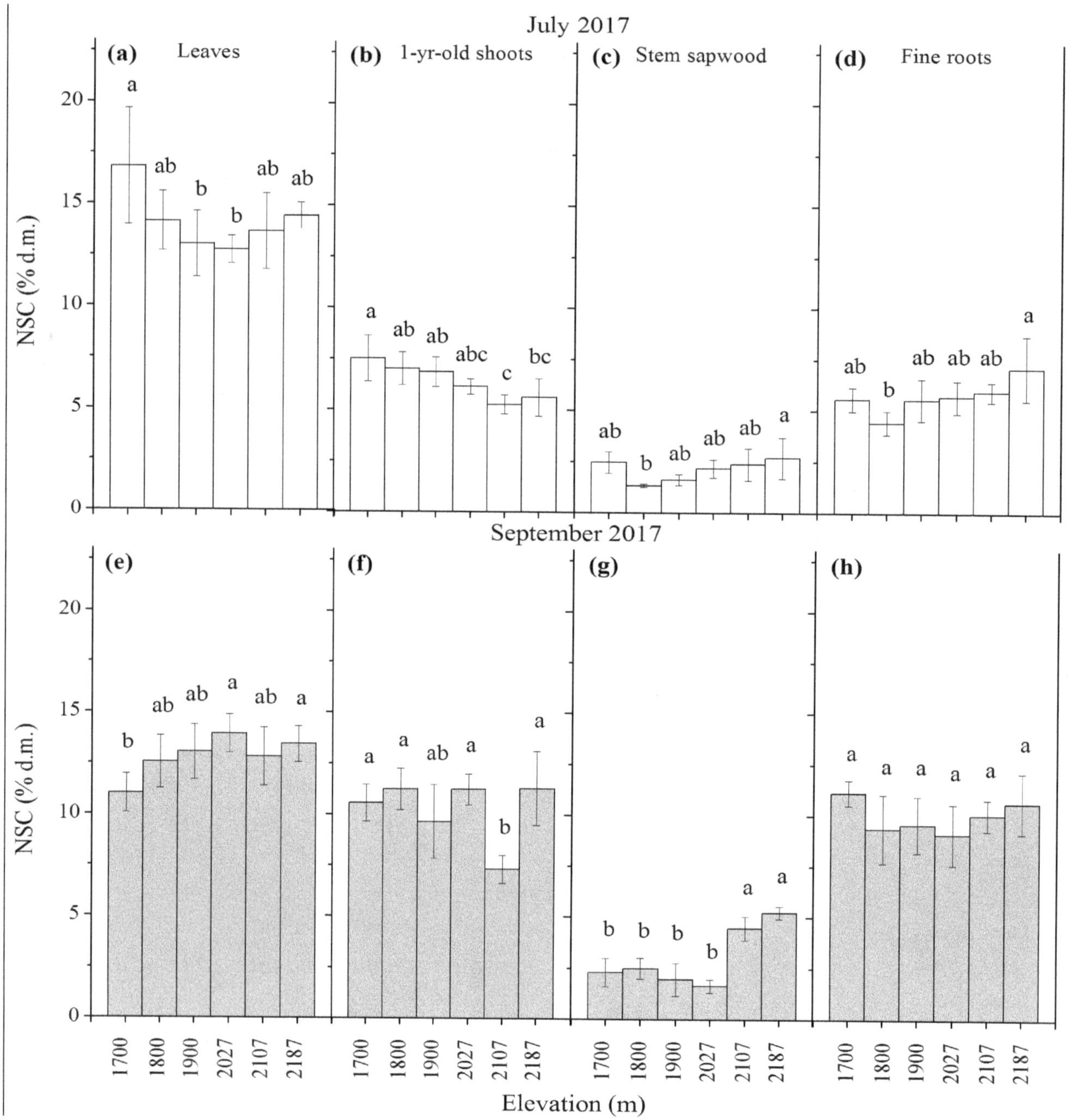

Figure 1. Seasonal tissues concentration (Mean ± 1SD; % of dry matter) changes in non-structural carbohydrate (NSC) compounds of *Betula ermanii* Cham. trees along elevational gradients in Changbai Mountain ($n = 5$ for each elevational site and tissue type). Different letters display significant differences at the 0.05 level among elevations as determined by Tukey's HSD test.

3.2. N Concentration

Elevation significantly affected the N concentrations in leaves, stem sapwood, and roots but not in shoots (Table 2). Tissue N concentrations changed significantly with time (season), and the interaction between time and elevation was significant for N concentrations in aboveground tissues (leaves, shoots, and stem sapwood) (Table 2). Stem N concentration was higher in September than in July, whereas N concentrations in other tissues (leaves, shoots, and roots) were higher in July than in September (Figure 2a–h). In both July and September, trees at higher elevations had higher N concentrations in tissues, except the N concentrations in shoots and roots did not change with elevation in July (Figure 2a–h).

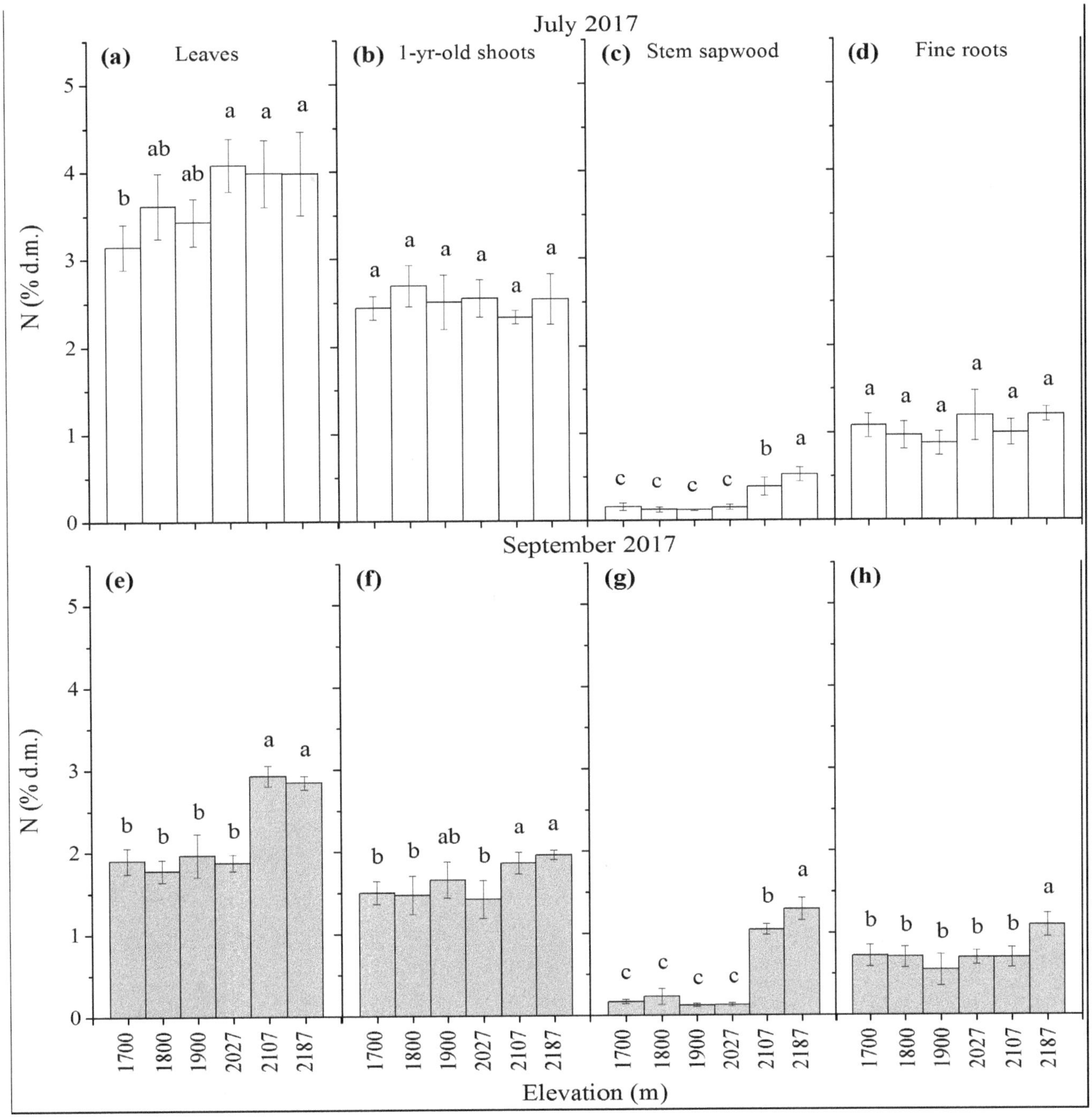

Figure 2. Seasonal tissues concentration (Mean ± 1SD; % of dry matter) changes in total nitrogen (N) of *Betula ermanii* trees along elevational gradients in Changbai Mountain (n = 5 for each elevational site and tissue type). Different letters display significant differences at the 0.05 level among elevations as determined by Tukey's HSD test.

3.3. P Concentration

Both elevation and time (season) significantly affected the tissue P concentrations, and the elevation × time interaction was significant for P concentrations in aboveground tissues (leaves, shoots, and stem sapwood) (Table 2). Leaf and shoot P concentrations were higher in July than in September, whereas stem and root P concentrations were higher in September than in July (Figure 3a–h). Tissue P concentrations were higher in trees at higher elevations in both July and September, except for P in leaves and shoots in July (Figure 3a–h).

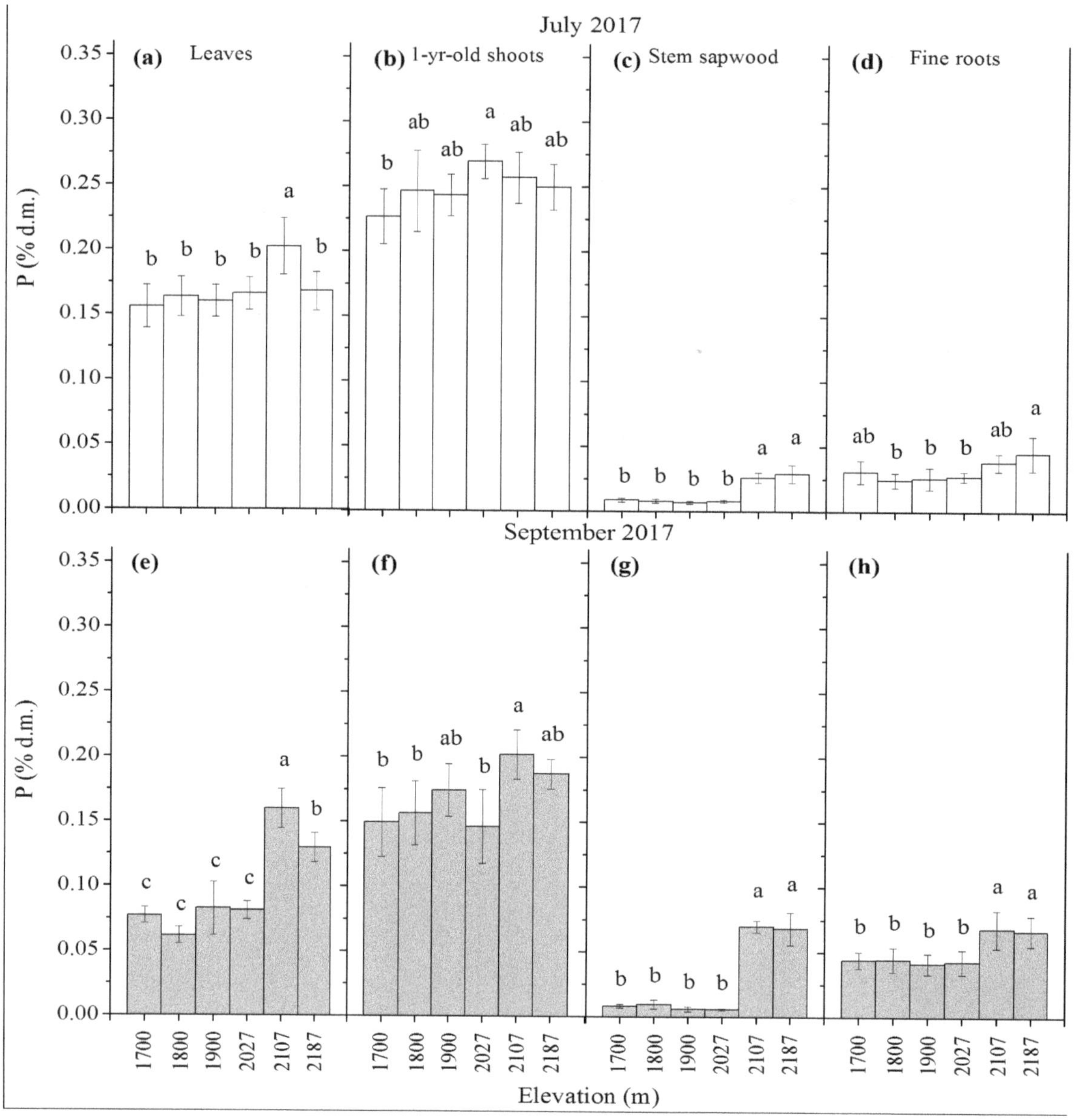

Figure 3. Seasonal tissues concentration (Mean ± 1SD; % of dry matter) changes in total phosphorus (P) of *Betula ermanii* trees along elevational gradients in Changbai Mountain (n = 5 for each elevational site and tissue type). Different letters display significant differences at the 0.05 level among elevations as determined by Tukey's HSD test.

3.4. K Concentration

Tissue K concentrations were significantly affected by both elevation and time (season) (Table 2). The interaction between elevation and time was significant for K concentrations in leaves, shoots and stem sapwood (Table 2). Leaf and shoot K concentrations were higher in July than in September, whereas stem and root K concentrations were higher in September than in July (Figure 4a–h). Stem and root K concentrations in July, as well as leaf, shoot and root K in September remained stable among trees at different elevations (Figure 4c,d–f,h). Apart from those, leaf K concentrations significantly decreased but shoot K significantly increased with increasing elevation in July (Figure 4a,b), and stem sapwood had higher K concentrations in trees close to their upper limit in September (Figure 4g).

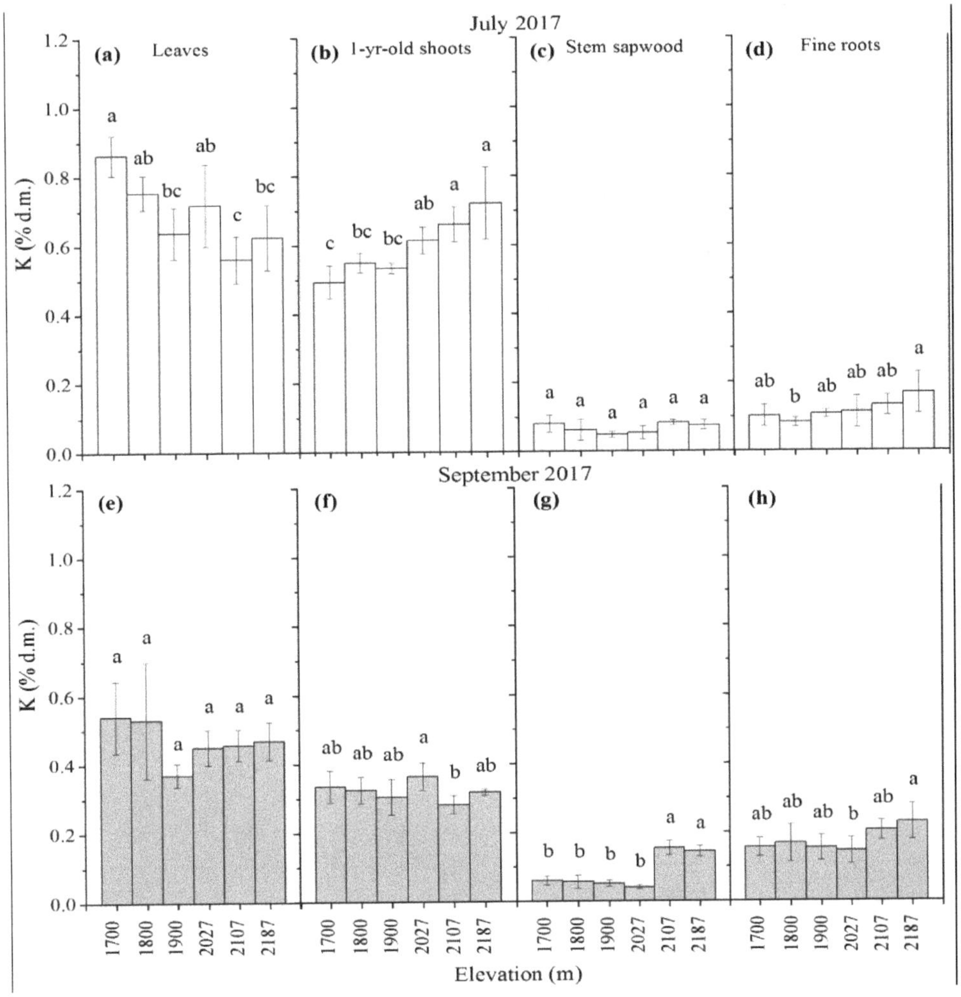

Figure 4. Seasonal tissues concentration (Mean ± 1SD; % of dry matter) changes in total potassium (K) of *Betula ermanii* trees along elevational gradients in Changbai Mountain (n = 5 for each elevational site and tissue type). Different letters display significant differences at the 0.05 level among elevations as determined by Tukey's HSD test.

3.5. Elevational Effects on Resource Remobilization

Both elevation and tissue type, and their interaction had significant effects on the remobilization efficiency of NSC, N, P and K (Table 3). At the end of the growing season, leaves reallocated NSC, N, P, and K to other tissues (Figure 5a,e,i,m), and the remobilization efficiency tended to decrease with increasing elevation, and even for an NSC accumulation at 1900 and 2027 m a.s.l. (Figure 5a). Shoots accumulated NSC but remobilized N, P, and K (Figure 5b,f,j,n), and the NSC accumulation (Figure 5b) and K reallocation (Figure 5n) tended to increase but N reallocation (Figure 5f) seemed to decrease with increasing elevation. Stems in trees close to the upper limit accumulated NSC, N, P and K (Figure 5c,g,k,o). Roots accumulated NSC, P and K, but reallocated N to other tissues (Figure 5d,h,l,p). The root NSC accumulation efficiency decreased with increasing elevation (Figure 5d). At the two highest elevations (i.e., 2107 and 2187 m a.s.l.), for example, both the amount of resource reallocation (less negative values at 2187 m compared to 2107 m) and accumulation (positive values) decreased with increasing elevation (Table 4). The mean reallocation efficiency of resources from leaves to other tissues was significantly lower in trees above the alpine treeline than in trees below the treeline (Table 5), and the mean efficiency across the entire transect was 40% (N), 42% (P), 31% (K), and 8% (NSC) (Table 5).

Table 3. Results of two-way nested ANOVAs with elevation and tissue types as fixed factors. The *F* and *p* values are given. R refers to remobilization rate.

	R NSC		R Nitrogen		R Phosphorus		R Potassium	
	F	*p*	*F*	*p*	*F*	*p*	*F*	*p*
Elevations (E)	8.67	**<0.001**	17.20	**<0.001**	10.72	**<0.001**	3.49	**0.016**
Tissue types (T)	287.57	**<0.001**	42.36	**<0.001**	79.04	**<0.001**	84.44	**<0.001**
E × T	5.73	**<0.001**	2.50	**0.005**	3.22	**<0.001**	2.50	**0.005**

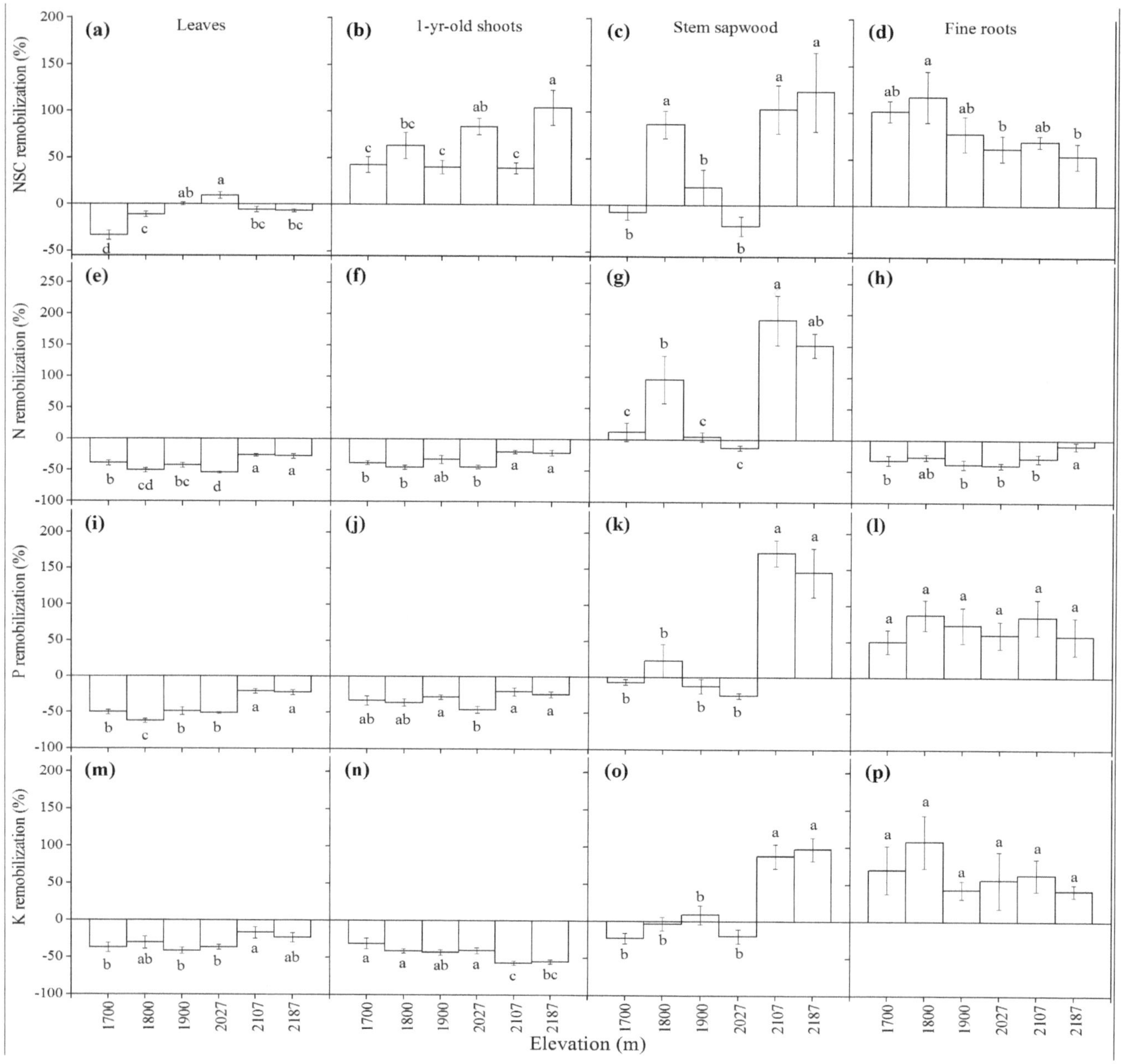

Figure 5. Carbohydrates, N, P and K remobilization efficiency (Mean ± 1SE) (R%) in *Betula ermanii* trees along the altitudinal gradients in Changbai Mountain (n = 5 for each elevational site and tissue type). Different letters display significant differences at the 0.05 level among elevations as determined by Duncan's test (A negative *R*-value indicates resource translocation, whereas a positive *R*-value reflects resource accumulation).

Table 4. End-season resource reallocation (mean value ± 1 standard deviation; n = 5 trees; calculated using biomass × reallocation rate) for *Betula ermanii* trees at 2187 m (upper limit of distribution) and 2107 m a.s.l. on Changbai Mt. Different letters display significant differences at the 0.05 level among elevations as determined by Tukey's HSD test. Negative values (dark-highlighted) indicate resource "loss" due to resource reallocation, and positive values indicate resource "gain" because of accumulation.

Elevation (m a.s.l.)	NSC (g)	N (g)	P (g)	K (g)
		Leaves		
2187	**−1.4 ± 0.4^a**	**−5.8 ± 1.5^b**	**−4.3 ± 0.8^b**	**−5.3 ± 1.9^a**
2107	**−3.0 ± 1.3^a**	**−12.9 ± 2.7^a**	**−10.5 ± 2.8^a**	**−4.7 ± 3.6^a**
		1-year-old shoots		
2187	31.1 ± 8.3^a	**−7.0 ± 2.3^a**	**−7.6 ± 2.2^a**	**−17.4 ± 4.0^a**
2107	26.6 ± 9.9^a	**−17.1 ± 7.0^a**	**−18.2 ± 8.1^a**	**−43.7 ± 16.3^a**
		Stem sapwood		
2187	63.0 ± 21.2^b	77.5 ± 10.8^b	76.7 ± 20.1^b	49.8 ± 7.6^b
2107	240.4 ± 72.6^a	391.5 ± 65.1^a	369.6 ± 50.5^a	193.4 ± 43.3^a
		Fine roots		
2187	22.0 ± 6.7^b	**−4.1 ± 2.2^b**	22.3 ± 8.8^a	17.0 ± 4.7^a
2107	85.1 ± 15.1^a	**−36.7 ± 13.1^a**	102.7 ± 34.7^a	59.3 ± 26.5^a

Table 5. Mean end-season reallocation rate (%) of resources (NSC, N, P, K) from leaves to other tissues in *Betula ermanii* trees grown at higher elevations (2107 and 2187 m a.s.l.) and lower elevations (1700, 1800, 1900, and 2027 m), separated by the alpine treeline located at 2030 m a.s.l. on the Changbai Mt., northeastern China.

	Trees across the Entire Transect	Trees below the Alpine Treeline	Trees above the Alpine Treeline
Non-structural carbohydrates (NSC)	8	9	6
Nitrogen (N)	40	47	27
Phosphorus (P)	42	52	22
Potassium (K)	31	36	20

4. Discussion

4.1. Tissue- and Resource-Dependent Reallocation or Accumulation

At the end of season, NSC reallocated from leaves to woody tissues (Figure 5a–d) and N accumulated only in stems (Figure 5g), whereas P and K seemed to store in stems and roots (Figure 5k,l,o,p), indicating tissue- and resource-dependent reallocation or accumulation (Table 4). *B. ermanii* leaves reallocated NSC to other tissues (Figure 5a; Table 4), leading to decreases in leaf NSC (Figure 1e vs. Figure 1a) but increases in NSC concentrations in shoots (Figure 1f vs. Figure 1b), stem sapwood (Figure 1g vs. Figure 1c), and roots (Figure 1h vs. Figure 1d), and thus an NSC accumulation in those storage tissues (Figure 5b–d). In line with previous studies, our results confirmed that leaf carbon compounds in deciduous trees were reallocated and stored in ligneous tissues, particularly in stems and roots [5,8]. Leaf carbon components recycled through remobilization from leaves to storage tissues before leaf senescence are particularly important for regrowth of new leaves in a leafless state of deciduous trees in early spring [5,9,10]. We found that the NSC remobilization efficiency of *B. ermanii* leaves ranged from 6% at high elevation to 33% at low elevation (Figure 5a). Chapin et al. (1990) proposed a carbon remobilization efficiency of ~10% based on an assumption that the loss of leaf weight was a direct result of carbon remobilization [5]. Similarly, Eckstein et al. (1998) proposed a carbon remobilization efficiency of 6%–13% for deciduous species [35].

Nitrogen (N) is the most important stored compound [36], paralleling to carbohydrates in supporting growth and reproduction [37]. Our results indicated that all tissue types decreased in N concentration during leaf senescence except stems (Figure 5e–h), suggesting that stems participated in the N accumulation rather than N translocation. Our results agreed with previous studies, showing that N remobilized from senescing tissues (e.g., leaves and shoots) for storage prior to leaf abscission [3,37,38]. As previously reported for *Betula pendula* Roth., the deciduous broadleaf tree tended to store N in roots and stem [11,12]. Our findings differ from coniferous evergreen species, which store N in the youngest age class of needle [11,13]. N stored in specific tissues with an individual species, dependent on leaf habit [37]. In addition, the mean leaf N remobilization efficiency of 40% (Table 5) was lower than that value reported by Aerts (1996) for deciduous shrubs and trees (mean 54%) [3]. In another study, Vergutz et al. (2012) showed that, based on a leaf mass loss correction, average nutrient resorption was 62% for N during senescence [39].

Apart from mobile carbohydrates and nitrogen (N), plants also store and remobilize macronutrients such as phosphorus (P) and potassium (K) [38,40]. By the process of leaf senescence, P and K is reallocated from leaves and shoots to and stored in roots and stems (Figure 5i–p). The mean efficiency of P remobilized from senescing leaves was 42% (Table 5), which is lower than that reported by Aerts (1996) for deciduous shrubs and trees (50% for N) [3]. Vergutz et al. (2012) even suggested that, in general, leaf average P resorption value was 65%, after correction for leaf mass loss [39]. Similar results were found for K. As K occurs in ionic form but not in macromolecule form [18], K is thus highly mobile and has been transferred to storage tissues (i.e., stems and roots) before leaf abscission. However, our remobilization efficiency from senescing leaves of 31% (Table 5) was much lower than published resorption efficiency. Chapin et al. (1990) and Vergutz et al. (2012) proposed 50 and 70%, respectively [5,39].

As previously reported for deciduous species, carbohydrate remobilization took place in the leaves, indicating that carbon components recycled from senescing leaves (Figure 5a) are used as a carbon source for new growth [5,41]. But unlike carbohydrates, nitrogen was recycled from senescing leaves (Figure 5e), shoots (Figure 5f) and roots (Figure 5h) to support new leaf and shoot growth, showing that senescing tissues play an important role as potential supplier of nitrogen. Besides, our findings were in agreement with other studies that demonstrate N storage in roots and stems in broad-leaved deciduous species [11,12]. P and K were reallocated from leaves and shoots to stems and roots at the end-season (Figure 5i–p), although these patterns differed from those of NSC and N mentioned above (Table 4). In *B. ermanii* trees, 6%–33% of the leaf mobile carbohydrates was recycled at the end-season. However, about half the N and P was remobilized from senescing leaves before leaves were shed, and 36% of the leaf K was resorbed during leaf senescence. Our results support the fact that resource remobilization is composed of a large storage of nutrients but a small storage of carbohydrates [5].

4.2. Elevational Effects on Resource Remobilization

In the present study, we calculated the remobilization efficiency using the resource concentration within the same tissue type measured at two time points (end-season vs. peak growth season), which may provide insights into resource remobilization processes during leaf senescence. This calculation revealed that the end-season remobilization efficiency of NSC, N, P, and K from leaves to storage tissues tended to decrease with increasing elevation (Figure 5a,e,i,m), which does not support our hypothesis that expected an increase in remobilization efficiency at higher elevations, especially at the upper limit. In our previous study [25], we calculated the remobilization efficiency using the end-season difference in the resource concentrations between leaves (source) and shoots (sink), and found that the remobilization efficiency from leaves to shoots increased with increasing elevation. Using that method [25], we re-calculated the end-season remobilization efficiency from leaves to shoots for the present study and found that NSC remobilization efficiency had a bimodal curve of R% with a bigger remobilization efficiency (smaller R% value) at the upper distribution limit (Figure 6a). Leaf to shoot K reallocation efficiency did not show any response to elevational gradients (Figure 6d), and leaf to shoot N and P reallocation efficiency tended to be lower above the treeline compared to below the treeline (Figure 6b,c). This comparison suggests that a standard method for calculating the remobilization efficiency and thus for understanding the reallocation processes of resources is urgently needed. Previous studies proposed that resource remobilization efficiency can be higher at low nutrient availability or in low temperature condition [14,16], representing an adaptation to stress environment [17]. However, Kutbay and Ok (2003) have demonstrated that the absolute and proportional N and P resorption efficiency did not change significantly along elevational gradients [22,29,31,42]. Many reports have revealed the importance of temperature on regulating leaf senescence and fall [18]. For example, leaf senescence and fall were delayed, in response to warmer temperature [43–45]. The timing or phenology of leaf abscission is considered as one of the reasons leading to altered resorption patterns of nutrients. To minimize such leaf phenological effects on resource remobilization, we did not collect end-season leaves from trees, instead, we collected the fallen leaves as end-season samples in the present study. However, *B. ermanii* in low temperature at the upper limit may have less time to complete nutrient resorption from leaves, which may result in a lower resorption of resources [18], as discussed in our previous study [25].

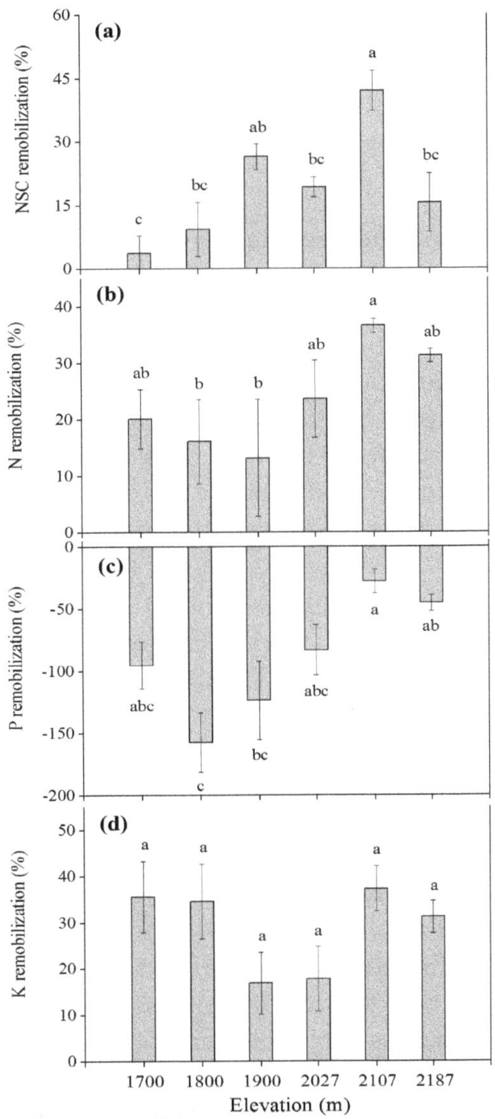

Figure 6. Carbohydrates, N, P and K remobilization efficiency (Mean ± 1SE) (R%) in *Betula ermanii* trees along the altitudinal gradients in Changbai Mountain ($n = 5$ for each elevational site). R% was calculated using $R\% = ((C_l - C_s)/C_l) \times 100\% = (1 - C_s/C_l) \times 100\%$, where C_l and C_s represent concentrations in leaves and shoots at the end of growing season, respectively. Different letters display significant differences at the 0.05 level among elevations as determined by Tukey's HSD test (The larger the R-value, the lower the remobilization efficiency is).

Stem and shoot NSC accumulation increased but root NSC accumulation decreased with increasing elevation (Figure 5b–d). Given that the NSC accumulation increased in stems and shoots up to the upper limit, this process may act to resist the harsh cold alpine environment for survival in winter [46], because soluble sugars participate in cell osmotic regulation and prevent intracellular ice formation by decreasing the freezing point of the cytoplasm [47]. Moreover, accumulation of soluble sugars would prime stems for recovery after the stress is alleviated, and stem is an important organ not only for long-distance carbon transport, but also for the regulation of the tree's carbon balance [48].

Similarly, our results showed that N, P, and K were also accumulated in stems, especially in trees grown close to or at their upper limit (Figure 5g,k,o). Oleksyn et al. (1998) and Körner (1999) have reported that nutrient accumulation and conservation are adaptive responses that enhance metabolic activity and growth rates in harsh cold environment [49,50]. In such situations, the tree would internally store more recycling nutrients to recover from harsh environment because of losses in respiration. The reserve storage is a risk-aversion function to minimize risk of catastrophic loss [5]. The greater the risk (high frequency or large losses), the more the tree should invest in internally stored reserves.

5. Conclusions

Many studies have stated that the end-season resource remobilization is strategic since it supports growth at the beginning of bud break in deciduous species [18,49,50]. Except for July-shoot NSC and July-leaf K, tissue concentrations of NSC, N, P, and K did not decrease with increasing elevation for both July and September. Our results partly supported our hypothesis that resources (i.e., mobile carbohydrates, N, P, and K) were reallocated from leaves to storage tissues at the end of growing season in deciduous *B. ermanii* trees. In an economic view, shoots should be an important storage tissue for resources to reduce the transport costs at both the end-season and early growing season. However, our study indicates that stems and roots of deciduous *B. ermanii* trees at high elevations are the most important storage tissues for resources over winter.

Inconsistent with our initial hypothesis, we found trees had lower end-season reallocation efficiency of NSC, N, P, and K from leaves to woody tissues at the higher elevations above the alpine treeline compared to lower elevations (Table 5; Figure 5a,e,i,m), and this may result in a resource limitation in high-elevation trees and further limit tree growth at high elevations. Our results contribute to better understand the resource-related ecophysiological mechanisms for treeline formation, and vice versa, to better predict treeline dynamics in response to global warming. Our study provides resource-related ecophysiological knowledge for developing management strategies for high elevation forests in a rapidly warming world. To more closely explore carbon and nutrient remobilization mechanisms, further studies are needed to estimate the pool size of resources remobilized (concentration × biomass, see Table 4) [51] and their stoichiometry [52]. Furthermore, stable isotope labeling experiments would be taken into account (e.g., ^{13}C, ^{15}N) to accurately evaluate carbon and N remobilization from senescing tissues, and to determine which tissues contribute to the most internal resource remobilization [13].

Author Contributions: Conceptualization, Y.C., M.-H.L. and H.-S.H.; methodology, M.-H.L. and H.-S.H.; software, Y.C.; investigation, K.L., Y.-C.D., H.-D.H.; data analysis, Y.C. and M.-H.L.; writing—original draft preparation, Y.C. and M.-H.L.; writing—review and editing, M.-H.L. and H.-S.H.; funding acquisition, M.-H.L. and H.-S.H.

Acknowledgments: We sincerely thank Danae Yu and two anonymous referees for valuable comments on our manuscript, Xinhua Zhou for her assistance in the laboratory and Samuel Allen for English proofreading to improve our manuscript.

References

1. Li, M.; Yang, J. Effects of elevation and microsite on growth of *Pinus cembra* in the subalpine zone of the Austrian Alps. *Annu. For. Sci.* **2004**, *61*, 319–325. [CrossRef]
2. Bhattacharya, A. *Changing Climate and Resource use Efficiency in Plants*; Academic Press: London, UK, 2018.
3. Aerts, R. Nutrient resorption from senescing leaves of perennials: Are there general patterns? *J. Ecol.* **1996**, *84*, 597–608. [CrossRef]
4. Cherbuy, B.; Joffre, R.; Gillon, D.; Rambal, S. Internal remobilization of carbohydrates, lipids, nitrogen and phosphorus in the Mediterranean evergreen oak *Quercus ilex*. *Tree Physiol.* **2001**, *21*, 9–17. [CrossRef] [PubMed]
5. Chapin, F.; Schulze, E.; Mooney, H. The ecology and economics of storage in plants. *Annu. Rev. Ecol. Syst.* **1990**, *21*, 423–447. [CrossRef]
6. Li, Y.; Lan, G.; Xia, Y. Rubber Trees demonstrate a clear retranslocation under seasonal drought and cold stresses. *Front. Plant Sci.* **2016**, *7*. [CrossRef] [PubMed]
7. Chapin, F.; Kedrowski, R.A. Seasonal changes in nitrogen and phosphorus fractions and autumn retranslocation in evergreen and deciduous taiga trees. *Ecology* **1983**, *64*, 376–391. [CrossRef]
8. Millard, P. Ecophysiology of the internal cycling of nitrogen for tree growth. *Zeitschrift für Pflanzenernährung Bodenkunde* **1996**, *159*, 1–10. [CrossRef]
9. Marchi, S.; Sebastiani, L.; Gucci, R.; Tognetti, R. Changes in sink-source relationships during shoot development in olive. *J. Am. Soc. Hortic. Sci.* **2005**, *130*, 631–637. [CrossRef]
10. Marchi, S.; Sebastiani, L.; Gucci, R.; Tognetti, R. Sink-source transition in peach leaves during shoot development. *J. Am. Soc. Hortic. Sci.* **2005**, *130*, 928–935. [CrossRef]
11. Millard, P.; Hester, A.; Wendler, R.; Baillie, G. Interspecific defoliation responses of trees depend on sites of winter nitrogen storage. *Funct. Ecol.* **2001**, *15*, 535–543. [CrossRef]
12. Millard, P.; Wendler, R.; Hepburn, A.; Smith, A. Variations in the amino acid composition of xylem sap of *Betula pendula* Roth. trees due to remobilization of stored N in the spring. *Plant Cell Environ.* **1998**, *21*, 715–722. [CrossRef]
13. Millard, P.; Proe, M.F. Storage and internal cycling of nitrogen in relation to seasonal growth of *Sitka spruce*. *Tree Physiol.* **1992**, *10*, 33–43. [CrossRef] [PubMed]
14. Bridgham, S.D.; Pastor, J.; McClaugherty, C.A.; Richardson, C.J. Nutrient-use efficiency: A litterfall index, a model, and a test along a nutrient-availability gradient in North Carolina peatlands. *Am. Nat.* **1995**, *145*, 1–21. [CrossRef]
15. Maillard, A.; Diquélou, S.; Billard, V.; Laîné, P.; Garnica, M.; Prudent, M.; Garcia-Mina, J.-M.; Yvin, J.-C.; Ourry, A. Leaf mineral nutrient remobilization during leaf senescence and modulation by nutrient deficiency. *Front. Plant Sci.* **2015**, *6*, 317. [CrossRef] [PubMed]
16. Pugnaire, F.I.; Chapin, F.S. Environmental and physiological factors governing nutrient resorption efficiency in barley. *Oecologia* **1992**, *90*, 120–126. [CrossRef] [PubMed]
17. Dissanayaka, D.M.S.B.; Maruyama, H.; Nishida, S.; Tawaraya, K.; Wasaki, J. Landrace of japonica rice, Akamai exhibits enhanced root growth and efficient leaf phosphorus remobilization in response to limited phosphorus availability. *Plant Soil* **2017**, *414*, 327–338. [CrossRef]
18. Estiarte, M.; Peñuelas, J. Alteration of the phenology of leaf senescence and fall in winter deciduous species by climate change: Effects on nutrient proficiency. *Glob. Chang. Biol.* **2015**, *21*, 1005–1017. [CrossRef]
19. González-Zurdo, P.; Escudero, A.; Mediavilla, S. N resorption efficiency and proficiency in response to winter cold in three evergreen species. *Plant Soil* **2015**, *394*, 87–98. [CrossRef]
20. Killingbeck, K.T. Nutrient Resorption. In *Plant Cell Death Processes*; Academic Press: San Diego, CA, USA, 2004; pp. 215–226.
21. Etienne, P.; Diquelou, S.; Prudent, M.; Salon, C.; Maillard, A.; Ourry, A. Macro and micronutrient storage in plants and their remobilization when facing scarcity: the case of drought. *Agriculture* **2018**, *8*, 14. [CrossRef]
22. Kutbay, H.; Ok, T. Foliar N and P resorption and nutrient levels along an elevational gradient in *Juniperus oxycedrus* L. subsp. *macrocarpa* (Sibth. & Sm.) Ball. *Ann. For. Sci.* **2003**, *60*, 449–454.
23. Du, H.; Liu, J.; Li, M.-H.; Büntgen, U.; Yang, Y.; Wang, L.; Wu, Z.; He, H.S. Warming-induced upward migration of the alpine treeline in the Changbai Mountains, northeast China. *Glob. Chang. Biol.* **2018**, *24*, 1256–1266. [CrossRef] [PubMed]

24. Zong, S.; He, H.; Liu, K.; Du, H.; Wu, Z.; Zhao, Y.; Jin, H. Typhoon diverged forest succession from natural trajectory in the treeline ecotone of the Changbai Mountains, Northeast China. *For. Ecol. Manag.* **2018**, *407*, 75–83. [CrossRef]
25. Cong, Y.; Wang, A.; He, H.; Yu, F.; Tognetti, R.; Cherubini, P.; Wang, X.; Li, M.H. Evergreen *Quercus aquifolioides* remobilizes more soluble carbon components but less N and P from leaves to shoots than deciduous *Betula ermanii* at the end-season. *iForest* **2018**, *11*, 517–525. [CrossRef]
26. Yu, D.; Wang, Q.; Liu, J.; Zhou, W.; Qi, L.; Wang, X.; Zhou, L.; Dai, L. Formation mechanisms of the alpine Erman's birch (*Betula ermanii*) treeline on Changbai Mountain in Northeast China. *Trees* **2014**, *28*, 935–947. [CrossRef]
27. Liu, Q.-J.; Li, X.-R.; Ma, Z.-Q.; Takeuchi, N. Monitoring forest dynamics using satellite imagery—A case study in the natural reserve of Changbai Mountain in China. *For. Ecol. Manag.* **2005**, *210*, 25–37. [CrossRef]
28. Li, M.; Yang, J.; Kräuchi, N. Growth responses of *Picea abies* and *Larix decidua* to elevation in subalpine areas of Tyrol, Austria. *Can. J. For. Res.* **2003**, *33*, 653–662. [CrossRef]
29. Li, M.; Jiang, Y.; Wang, A.; Li, X.; Zhu, W.; Yan, C.-F.; Du, Z.; Shi, Z.; Lei, J.; Schönbeck, L.; et al. Active summer carbon storage for winter persistence in trees at the cold alpine treeline. *Tree Physiol.* **2018**, *38*, 1345–1355. [CrossRef]
30. Yamaguchi, D.P.; Nakaji, T.; Hiura, T.; Hikosaka, K. Effect of seasonal change and experimental warming on the temperature dependence of photosynthesis in the canopy leaves of *Quercus serrata*. *Tree Physiol.* **2016**, *36*, 1283–1295. [CrossRef]
31. Li, M.; Xiao, W.-F.; Shi, P.; Wang, S.-G.; Zhong, Y.-D.; Liu, X.-L.; Wang, X.-D.; Cai, X.-H.; Shi, Z.-M. Nitrogen and carbon source–sink relationships in trees at the Himalayan treelines compared with lower elevations. *Plant Cell Environ.* **2008**, *31*, 1377–1387. [CrossRef]
32. Wang, X.; Xu, Z.; Yan, C.; Luo, W.; Wang, R.; Han, X.; Jiang, Y.; Li, M.-H. Responses and sensitivity of N, P and mobile carbohydrates of dominant species to increased water, N and P availability in semi-arid grasslands in northern China. *J. Plant Ecol.* **2017**, *10*, 486–496. [CrossRef]
33. Parkinson, J.A.; Allen, S.E. A wet oxidation procedure suitable for the determination of nitrogen and mineral nutrients in biological material. *Commun. Soil Sci. Plant Anal.* **1975**, *6*, 1–11. [CrossRef]
34. Murphy, J.; Riley, J.P. A modified single solution method for the determination of phosphate in natural waters. *Anal. Chim. Acta* **1962**, *27*, 31–36. [CrossRef]
35. Eckstein, R.L.; Karlsson, P.S.; Weih, W. The significance of resorption of leaf resources for shoot growth in evergreen and deciduous woody plants from a subarctic environment. *Oikos* **1998**, *81*, 567–575. [CrossRef]
36. Millard, P.; Sommerkorn, M.; Grelet, G.A. Environmental change and carbon limitation in trees: a biochemical, ecophysiological and ecosystem appraisal. *New Phytol.* **2007**, *175*, 11–28. [CrossRef] [PubMed]
37. Millard, P.; Grelet, G.-A. Nitrogen storage and remobilization by trees: ecophysiological relevance in a changing world. *Tree Physiol.* **2010**, *30*, 1083–1095. [CrossRef] [PubMed]
38. Nambiar, E.K.S.; Fife, D.N. Nutrient retranslocation in temperate conifers. *Tree Physiol.* **1991**, *9*, 185–207. [CrossRef]
39. Vergutz, L.; Manzoni, S.; Porporato, A.; Novais, R.F.; Jackson, R.B. Global resorption efficiencies and concentrations of carbon and nutrients in leaves of terrestrial plants. *Ecol. Monogr.* **2012**, *82*, 205–220. [CrossRef]
40. Villar-Salvador, P.; Uscola, M.; Jacobs, D.F. The role of stored carbohydrates and nitrogen in the growth and stress tolerance of planted forest trees. *New For.* **2015**, *46*, 813–839. [CrossRef]
41. Gessler, A.; Treydte, K. The fate and age of carbon—Insights into the storage and remobilization dynamics in trees. *New Phytol.* **2016**, *209*, 1338–1340. [CrossRef]
42. Körner, C. The nutritional status of plants from high altitudes. A worldwide comparison. *Oecologia* **1989**, *81*, 379–391. [CrossRef]
43. Chung, H.; Muraoka, H.; Nakamura, M.; Han, S.; Muller, O.; Son, Y. Experimental warming studies on tree species and forest ecosystems: A literature review. *J. Plant Res.* **2013**, *126*, 447–460. [CrossRef] [PubMed]
44. Gunderson, C.A.; Edwards, N.T.; Walker, A.V.; O'Hara, K.H.; Campion, C.M.; Hanson, P.J. Forest phenology and a warmer climate—growing season extension in relation to climatic provenance. *Glob. Chang. Biol.* **2012**, *18*, 2008–2025. [CrossRef]
45. Xu, Z.; Hu, T.; Zhang, Y. Effects of experimental warming on phenology, growth and gas exchange of treeline birch (*Betula utilis*) saplings, Eastern Tibetan Plateau, China. *Eur. J. For. Res.* **2012**, *131*, 811–819. [CrossRef]

46. Sheen, J.; Zhou, L.; Jang, J.-C. Sugars as signaling molecules. *Curr. Opin. Plant Biol.* **1999**, *2*, 410–418. [CrossRef]
47. Morin, X.; Améglio, T.; Ahas, R.; Kurz-Besson, C.; Lanta, V.; Lebourgeois, F.; Miglietta, F.; Chuine, I. Variation in cold hardiness and carbohydrate concentration from dormancy induction to bud burst among provenances of three European oak species. *Tree Physiol.* **2007**, *27*, 817–825. [CrossRef] [PubMed]
48. Furze, M.E.; Trumbore, S.; Hartmann, H. Detours on the phloem sugar highway: Stem carbon storage and remobilization. *Curr. Opin. Plant Biol.* **2018**, *43*, 89–95. [CrossRef] [PubMed]
49. Oleksyn, J.; Modrzýnski, J.; Tjoelker, M.G.; Z·ytkowiak, R.; Reich, P.B.; Karolewski, P. Growth and physiology of *Picea abies* populations from elevational transects: common garden evidence for altitudinal ecotypes and cold adaptation. *Funct. Ecol.* **1998**, *12*, 573–590. [CrossRef]
50. Körner, C. *Alpine Plant Life: Functional Plant Ecology of High Mountain Ecosystems*; Springer: Berlin, Genmary, 1999.
51. Li, M.; Kräuchi, N.; Dobbertin, M. Biomass distribution of different-aged needles in young and old Pinus cembra trees at highland and lowland sites. *Trees* **2006**, *20*, 611–618. [CrossRef]
52. Wang, A.; Wang, X.; Tognetti, R.; Lei, J.-P.; Pan, H.-L.; Liu, X.-L.; Jiang, Y.; Wang, X.-Y.; He, P.; Yu, F.-H.; et al. Elevation alters carbon and nutrient concentrations and stoichiometry in *Quercus aquifolioides* in southwestern China. *Sci. Total Environ.* **2018**, *622*, 1463–1475. [CrossRef] [PubMed]

13

Growth Characteristics of Ectomycorrhizal Seedlings of *Quercus glauca*, *Quercus salicina*, *Quercus myrsinaefolia* and *Castanopsis cuspidata* Planted in Calcareous Soil

Masazumi Kayama [1,*,†] and Takashi Yamanaka [2]

1 Kyushu Research Center, Forestry and Forest Products Research Institute, Kumamoto 860-0862, Japan
2 Forestry and Forest Products Research Institute, Tsukuba 305-8687, Japan; yamanaka@affrc.go.jp
* Correspondence: kayama@affrc.go.jp
† Present address: Japan International Research Center for Agricultural Sciences, Ibaraki 305-8686, Japan.

Academic Editors: Scott X. Chang and Xiangyang Sun

Abstract: To verify the acclimation capacity of evergreen Fagaceae species on calcareous soil, we compared ecophysiological traits between *Quercus glauca* Thunb., *Q. salicina* Blume, *Q. myrsinaefolia* Blume, and *Castanopsis cuspidata* (Thunb.) Schottky as typical woody species from southwestern Japan. We also examined the inoculation effects of the ectomycorrhizal (ECM) fungi *Astraeus hygrometricus* and *Scleroderma citrinum*, and planted seedlings in calcareous soil collected from a limestone quarry. We measured growth, ectomycorrhizal colonization, photosynthetic rate, and concentrations of nutrients in plant organs for *A. hygrometricus*-inoculated, *S. citrinum*-inoculated, and non-ECM seedlings. Six months after planting on calcareous soil, seedlings of the three *Quercus* species inoculated with *A. hygrometricus* were larger than non-ECM seedlings, especially *Q. salicina*, which showed the greatest increase in dry mass. The dry mass of *C. cuspidata* seedlings was inferior to that of the three *Quercus* species. In the nutrient-uptake analysis, phosphorus, manganese, and iron uptakes were suppressed in calcareous soil for each Fagaceae species. However, seedlings of Fagaceae species that showed better growth had increased concentrations of phosphorus in roots. We concluded that seedlings of *Q. salicina* and *Q. glauca* inoculated with *A. hygrometricus* were best suited to calcareous soil and were considered as useful species for the reforestation in limestone quarries.

Keywords: oak; ectomycorrhizal fungi; limestone; photosynthetic capacity; nutrient physiology

---

1. Introduction

Limestone occurs throughout East Asia [1]. Calcareous soil, which is formed by weathering of limestone, shows the pH range 7.5–8.5 owing to the buffering effects of the large amounts of calcium carbonate [2,3]. Under such conditions, the uptake of micronutrients (e.g., iron (Fe), manganese (Mn), copper (Cu) and zinc (Zn)), which exhibit low availability at high pH, is suppressed [3,4]. Additionally, carbonate in calcareous soil causes lime-induced chlorosis as a consequence of Fe deficiency [2,3]. Moreover, phosphorus (P) is bound by calcium (Ca), and the uptake of P is also suppressed [3,5]. Accordingly, some plant species exhibit poor growth on calcareous soils [5–8]. Limestone has been quarried at many sites worldwide [9], but once quarrying has finished, natural revegetation of such sites is slow [10,11]. A limestone quarry site in Japan is currently being rehabilitated to a forest stand [12]. The calcareous soil at the quarry has a high pH and large amounts of Ca [12]. Seed from a range of plant species from an adjacent natural vegetation site was dispersed across the quarry site; however, growth of germinated plants was slow [13]. Thus, limestone quarry sites require artificial reforestation for environmental conservation of such sites.

Some plant species have high adaptability to calcareous soil because these species can obtain micronutrients and P from such soil [7]. Evergreen broad-leaved forests are distributed across warm-temperature zones in southern parts of China and southwestern Japan [14–16], and several *Quercus* species are distributed widely across limestone regions [17–19]. In particular, *Q. glauca* Thunb. is abundant on calcareous soils in Japan and China [17–23], but, in contrast, the genus *Castanopsis* is not distributed generally on such soils [18,20,24]. However, the distributions of *Quercus* and *Castanopsis* in East Asia have not been widely examined for their ability to grow on calcareous soil.

Fagaceae species can form a symbiotic interaction with ectomycorrhizal (ECM) fungi [25,26]. On calcareous soil, exudation of organic acids from ECM fungi and the dissolution of insoluble P is important for the uptake of P [27,28]. In the case of Fe, exudation of organic acids, such as citric acid, results in the chelation of Fe, leading to more available for plant uptake [29]. The communities of ECM fungi in calcareous soil can differ from those in non-calcareous soil [30,31], although cosmopolitan ECM fungi, such as genera *Astraeus* and *Scleroderma*, are distributed on calcareous soil [32,33].

The removal of limestone at a quarry site disturbs the topsoil. The abundance of the species of ECM fungi was poor at a disturbed site compared with an undisturbed site [34]. It is predicted that when non-ECM seedlings of Fagaceae species are planted into limestone, colonization of ECM fungi may be poor. When woody species inoculated with ECM fungi are planted on calcareous soil, several species exhibit tolerance to these nutrient-limiting soil conditions [33]. The beneficial effects of inoculation of ECM fungi for plant growth on sites with a poor diversity of ECM fungi have been shown previously [35]. However, few studies have examined the inoculation of seedlings with ECM fungi prior to growing in calcareous soil. We expected that evergreen Fagaceae species inoculated with ECM fungi may be effective for growth on limestone quarry sites. The aim of our research was to verify the acclimation capacity of *Quercus* and *Castanopsis* on calcareous soil. Moreover, we examined the inoculation effects of ECM fungi and compared these between *Quercus* and *Castanopsis* species. We examined the ecophysiological traits of seedlings of *Quercus* and *Castanopsis*, specifically (1) the growth characteristics of the seedlings; (2) symbiosis of ECM fungi on short roots; and (3) concentrations of elements in plant organs. These parameters were compared between ECM-inoculated and non-inoculated seedlings. Our findings may be applied in the artificial reforestation of limestone quarries using Fagaceae seedlings inoculated with ECM fungi.

2. Materials and Methods

2.1. Plants and ECM Fungi

We used *Quercus glauca* Thunb., *Quercus salicina* Blume, *Quercus myrsinaefolia* Blume, and *Castanopsis cuspidata* (Thunb.) Schottky. The three *Quercus* species are distributed across limestone regions in southwestern Japan, whereas *C. cuspidata* is not present in these regions [17–19,21]. Acorns of the four Fagaceae species were collected from an evergreen broad-leaved forest located in the central part of Kyushu, Southwestern Japan in December 2007 for *Q. glauca* and *Q. salicina*, and in December 2008 for *Q. myrsinaefolia* and *C. cuspidata*. To eliminate differences in genetic characteristics caused by edaphic factors, the forest-collected acorns of the four Fagaceae species were from an area with non-calcareous soil. The acorns were stored in a refrigerator at 4 °C prior to sowing.

Information about species of ECM fungi distributed in calcareous soil in Japan has rarely been reported [36]. However, reports from other countries indicate that *Astraeus hygrometricus* (Pers.) Morgan and *Scleroderma citrinum* Pers. can form symbiotic interactions in calcareous soil [31,32]. In addition, *A. hygrometricus* and *S. citrinum* are typical ECM fungi from Fagaceae forests in Japan [37,38], and we found previously that these species could form symbiotic relationships with Fagaceae species [39,40]. Thus, we chose *A. hygrometricus* and *S. citrinum* for the inoculation of Fagaceae seedlings.

In July 2008, fruiting bodies of *A. hygrometricus* and *S. citrinum* were collected from a slope with exposed mineral soil on the edge of a secondary evergreen, broad-leaved forest. This site was the same as that for the collection of acorns. Several fragments (3–7 mm) were cut off from the fruiting bodies and placed on agar medium. The medium consisted of 10 $g{\cdot}L^{-1}$ glucose, 10 $g{\cdot}L^{-1}$ malt extract, and 1 $g{\cdot}L^{-1}$ bactopeptone [41]. Fragments of *A. hygrometricus* and *S. citrinum* were incubated at 25 °C for 2 weeks in darkness. We confirmed the developed mycelial colony of *A. hygrometricus* and *S. citrinum* based on descriptions from previous literatures [38,41]. We used isolated mycelia for inoculation.

2.2. Preparation of Pot Experiments

Acorns of Fagaceae species were germinated in June 2008 for *Q. glauca* and *Q. salicina*, and in June 2009 for *Q. myrsinaefolia* and *C. cuspidata*. The acorns were surface sterilized with 30% H_2O_2 for 20 min, and rinsed 4 times with sterilized distilled water. Acorns were placed on a sterilized perlite and akadama (red ball earth, originated from subsoil in Kanto region, Japan) soil (1:1) in a free-draining container (width × length × depth = 31 cm × 21 cm × 8 cm). After germination, inoculation of ECM was performed in August of the same year.

Isolated mycelia of *A. hygrometricus* and *S. citrinum* were cultured in 40 mL of Ohta liquid medium [42] in 100 mL flasks at 24 °C for 2 weeks in darkness. The incubated mycelia of *A. hygrometricus* and *S. citrinum* were aseptically fragmented using a stirrer (DX, As One Co., Osaka, Japan) at 7000 rpm for 20 s. The suspension of hyphal fragments was poured into a 50 mL plastic tube and centrifuged at 2320 g for 10 min to collect the hyphal fragments, which were then mixed with sterilized distilled water and centrifuged at 2320 g for 10 min. After the supernatant was removed, the procedure was repeated. The hyphal fragments were resuspended in sterilized distilled water, and 48 seedlings each of *Q. glauca*, *Q. salicina*, *Q. myrsinaefolia*, and *C. cuspidata* were inoculated with *A. hygrometricus* or *S. citrinum*. Prior to the inoculation, seedlings were removed from their containers, and the soil was washed from their roots with sterilized distilled water. On the exposed roots of seedlings, 1 mL of hyphal suspension, which contained 1 mg of hyphal fragments, was added. In addition, 24 seedlings each of *Q. glauca*, *Q. salicina*, *Q. myrsinaefolia*, and *C. cuspidata* were prepared similarly but without inoculation of ECM fungi (non-ECM controls).

All seedlings were planted into Ray Leach Containers (164 mL, SC10, Stuewe and Sons, Tangent, OR, USA) filled with stabilized perlite: akadama soil (1:1). The pH of the perlite: akadama soil was 6.1. The seedlings were grown in an enclosed, air-conditioned (25.0–35.0 °C) greenhouse at the Forestry and Forest Products Research Institute (FFPRI), Tsukuba until April 2009 for *Q. glauca* and *Q. salicina.*, and until April 2010 for *Q. myrsinaefolia* and *C. cuspidata*. The greenhouse was sealed to prevent the seedlings from becoming contaminated by airborne spores or hyphal fragments of mycorrhizal fungi. No fertilizers were applied to the seedlings, and each pot was irrigated with 20 mL of distilled water per day. We also confirmed the contamination of ECM root visually, and the control seedlings contaminated with ECM on their roots were eliminated from the trial.

2.3. Soil Collection and Analysis

In January 2009, we collected calcareous soil from the B horizon of a limestone quarry (Todaka Mining Co., Tsukumi, Oita Pref., Japan, 33°03′ N, 131°48′ E, 580 m a.s.l.). The color of the soil was dark red (5YR 3/6), and it showed high viscosity owing to a high content of clay [43]. Collected soil was sieved (2 mm) at the same site and collected in 32 sandbags. After transportation, soil was kept in a storehouse in Kyushu Research Center, FFPRI.

The pot experiments were carried out by dividing them into two groups, in May 2009 for *Q. glauca* and *Q. salicina* and in May 2010 for *Q. myrsinaefolia* and *C. cuspidata*. Before planting for each year, the collected soil was placed in autoclave bags and sterilized using a wet cycle of an autoclave at 120 °C for 20 min. (KTS-2346B, Alp Co., Tokyo, Japan) to remove all other mycorrhizal species. After sterilization, five soil samples were collected from five bags for each year.

We measured the soil properties including pH and the concentration of carbon (C), nitrogen (N), exchanged P, base cations (Ca, magnesium (Mg), potassium (K) and sodium (Na)), Fe, and Mn. Soil pH was measured using a pH meter (SG2, Mettler Toledo, Zürich, Switzerland); 10 g of soil was mixed with 25 mL of distilled water and shaken for 1 h before the reading was taken [44].

After measuring the soil pH, samples were dried at 105 °C for 24 h for further analysis. Dried soil samples were used to determine the concentration of C and N using a NC analyzer (Sumigraph NC-220F, Sumika Chemical Analysis Service, Tokyo, Japan). Exchanged P was extracted by the Olsen method [45] by shaking for 1 h. P concentration in the soil solution was determined by the molybdenum blue method [46] using a spectrophotometer (UV-2500PC, Shimadzu, Kyoto, Japan).

The concentrations of exchanged base cations and Mn were determined; 2.5 g of dry soil was mixed with 50 mL of 1 M ammonium acetate solution and shaken for 1 h prior to analysis [45]. Solutions for analysis of exchanged Fe were obtained by adding 50 mL of 1 M KCl to 2.5 g of dry soil and shaking for 1 h [45]. The base cations, Fe, and Mn concentrations in the soil solution were analyzed using an inductivity coupled plasma analyser (ICPE-9000, Shimadzu, Kyoto, Japan).

2.4. Establishment of the Pot Experiment

A total of 48 mycorrhizal seedlings and 24 non-mycorrhizal seedlings of the four Fagaceae species were transported to Kyushu Research Center, FFPRI in May. A total of 24 mycorrhizal seedlings and 12 non-mycorrhizal seedlings each of *Q. glauca*, *Q. salicina*, *Q. myrsinaefolia*, and *C. cuspidata* were transplanted into free-draining pots (depth: 30 cm, diameter: 15 cm, volume: 3.7 L) filled with calcareous soil. The remaining seedlings, i.e., 24 mycorrhizal seedlings and 12 non-mycorrhizal seedlings of the four Fafaceae species, were sampled at this time. Fagaceae seedlings were raised for 6 months in a naturally illuminated phytotron (Koito Co., Tokyo, Japan) at Kyushu Research Center, FFPRI. The phytotron had three rooms, and seedlings of *Q. glauca*, *Q. salicina*, *Q. myrsinaefolia*, and *C. cuspidata* were inoculated with *A. hygrometricus* (abbreviated to AH), *S. citrinum* (SC), and non-ECM (NE), the seedlings were then put into the three rooms. Four seedlings of each species inoculated with AH, SC and NE were placed in each room.

After establishment, each pot was irrigated with 60 mL of distilled water per day. The average daily photosynthetically active radiation (PAR) during the experimental period was 39.6 $mol{\cdot}m^{-2}{\cdot}day^{-1}$ in 2009, and 37.5 $mol{\cdot}m^{-2}{\cdot}day^{-1}$ in 2010, calculated using data on solar radiation from the Japan Meteorological Agency [47] and the ratio of solar radiation to PAR [48]. Air temperature in the phytotron rooms was controlled at 30 °C (day) and 25 °C (night) from July to September, and 25 °C (day) and 20 °C (night) from May to June and in October. These temperatures were based on air temperature in Kumamoto city, Japan [47]. Relative humidity was controlled at 75%. During the experiment, some contamination by airborne spores was noted. The position of seedlings was changed at regular intervals (every month), and the phytotron rooms were disinfected using ethanol at this time.

2.5. Measurement of Seedling Growth

To determine the growth characteristics of seedlings of *Q. glauca*, *Q. salicina*, *Q. myrsinaefolia*, and *C. cuspidata*, we measured the dry mass of leaves, stems and branches, and roots. Twelve AH, SC, and NE seedlings each for the four tree species were harvested in May (first sampling) and November (second sampling). The roots of harvested seedlings were washed twice with tap water to remove soil, and then washed with distilled water. The washed seedlings were divided into leaves, stems and branches, and root components. Each component was put into its own envelope and oven-dried at 60 °C for 4 days. The dry mass of each component was determined. From the values of dry mass of each component at the first and second samplings, we calculated relative dry mass as follows:

$$\text{Dry mass at second sampling}/\text{Dry mass at first sampling} \times 100 \tag{1}$$

For the relative dry mass, 100 was set as the value at the first sampling. We also measured specific leaf area (SLA [7]) for each species. We used the SLA value to convert the concentration of nutrients into area based values.

2.6. Measurement of the Rate of Colonization of ECM Fungi

To measure the proportion of roots with ECM fungal colonization, we selected five large clusters of roots from each of seedling, and over 500 short roots (<5 mm in length) that diverged from them. This proportion was determined before drying. For assessment of ECM fungi, roots of seedlings were harvested at the first and second samplings and carefully washed free of soil under gently flowing water. Cluster roots were soaked in distilled water and observed at 10–40× magnification using a stereomicroscope (STZ-40TBIT, Shimadzu, Kyoto, Japan). ECM short roots of *A. hygrometricus* and *S. citrinum* were confirmed by the color, form, and size of the ECM fungi, the presence of rhizomorphs and the absence of root hairs [38,40,49]. The numbers of colonized and uncolonized short roots were counted, and we calculated the percentage of ECM fungal colonization from the proportion of colonized short roots to the total number of short roots [50]. We also examined the relationship between growth (total dry mass) and ECM fungal colonization.

2.7. Photosynthetic Capacity

The photosynthetic rate at light saturation (P_{sat}) was measured in all seedlings in October. Third leaves counted from the top of the seedling were used for the measurement of P_{sat} and stomatal conductance (g_s). Measurements were made using a portable gas analyzer (LI-6400, LI-COR Biosciences, Lincoln, NE, USA) under steady-state conditions at an ambient CO_2 concentration of 37.0 Pa. The LED light source was adjusted to a saturation light level of 1700 $\mu mol \cdot m^{-2} \cdot s^{-1}$·PPF.

2.8. Analysis of Element Concentrations in Plants

We measured the concentrations of N, P, K, Ca, Mg, Fe, and Mn in leaves and roots at the first and second samplings. Dried samples were ground to a fine powder using a sample mill (WB-1; Osaka Chemical Co., Osaka, Japan). The concentration of N was determined using a CN CORDER analyzer (MT-600, Yanako New Science Inc., Kyoto, Japan). The remaining samples were digested by the HNO_3-HCl-H_2O_2 method [51]. The concentration of P was determined by the molybdenum blue method using a spectrophotometer. Concentrations of K, Ca, Mg, Fe, and Mn were analyzed using an atomic absorption spectrophotometer (Z-2310, Hitachi High-Technologies Co., Tokyo, Japan).

The concentrations of each nutrient at the first and second samplings were calculated as relative nutrient concentration, as follows:

$$\text{Concentration at second sampling/Concentration at first sampling} \times 100 \quad (2)$$

For the relative concentrations, 100 was set as the value from the first sampling. Based on relative nutrient concentration and relative dry mass, we examined vector analysis in accordance with Valentine and Allen [52] and Scagel [53]. Changes in dry mass and the concentration of nutrients were plotted for AH, SC, and NE seedlings of the four Fagaceae species. The interpretation of vector shifts, as described in Valentine and Allen [52], are shown in Figure 1 as a reference.

We also examined the effects of nutrients on photosynthetic rate. In general, photosynthetic rate is closely related to N [7,54] and P [55,56]. Correlation analysis was performed between P_{sat} and the concentrations of N and P in leaves at the second sampling. For the concentrations of N and P, we used the SLA area-based values.

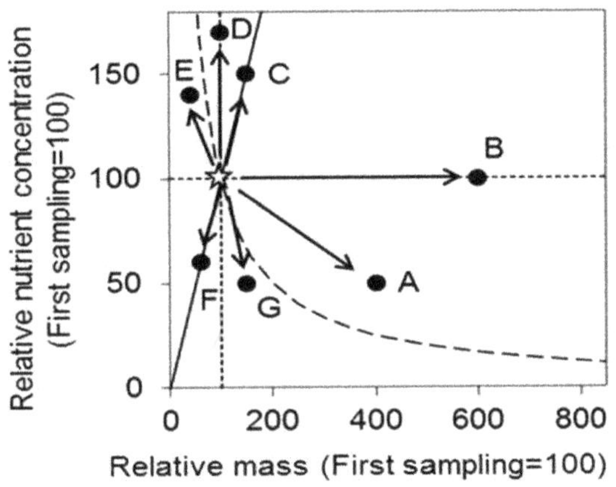

Vector shift	Concentration	Content	Mass	Interpretation
A	Decrease	Increase	Increase	Nutrient non-limiting **Dilution** due to growth
B	Unchanged	Increase	Increase	Nutrient non-limiting **Sufficiency** on growth
C	Increase	Increase	Increase	**Enhancement** of nutrient uptake, related to change in mass
D	Increase	Increase	Unchanged	**Luxury consumption** Little effect on growth
E	Increase	Decrease	Decrease	**Toxicity** of elevated nutrient Decrease of growth
F	Decrease	Decrease	Decrease	**Nutrient deficiency** due to low availability
G	Decrease	Decrease	Increase	**Efficient use** of very little nutrient

Figure 1. Graphical representation of interpretation comparing changes in dry mass and concentrations of nutrients grown in different media types. The content isograms represent combinations of dry mass and concentration required for constant content per unit dry mass. The arrows indicate the direction in which each interpretation holds (adapted from Valentine and Allen [48] and Scagel [49]). The curve indicated by the broken line shows the border to estimate increase or decrease of content. A plot located above the curve shows increase of content.

2.9. Statistical Analysis

All parameters were analysed for normality and homoscedasticity using the Shapiro–Wilk test with Sigma Plot 12.3 (Systat Software Inc., San Jose, CA, USA,). Some parameters did not show normality and homoscedasticity; therefore, we used nonparametric tests by Kyplot 3.0 (Kyens Lab. Inc., Tokyo, Japan). The data of soil chemical properties were compared between 2009 and 2010 using Mann-Whitney U test. Dry mass of the whole plant and each organ, ECM fungal colonization, P_{sat} and g_s, and concentrations of elements in leaves and roots were examined using a Steel–Dwass test. The mean values of total dry mass and ECM fungal colonization for *Q. glauca*, *Q. salicina*, *Q. myrsinaefolia*, and *C. cuspidata* were compared among the first and second samplings of AH, SC, and NE seedlings. The values of dry mass of each organ, P_{sat}, g_s and concentrations of elements in leaves and roots of the four Fagaceae species were compared among AH, SC and NE seedlings. Different letters are used to indicate statistically significant differences at $p < 0.05$. The mean values followed by the same letter showed no significant differences.

To examine the relationship between total dry mass and ECM fungal colonization, we conducted a Spearman's rank correlation. Additionally, we examined the relationship between P_{sat} and nutrients in leaves by the same method. For the data from the second sampling, we also compared each parameter of the seedlings from each of the three phytotron rooms by a Kruskal-Wallis test. There were no significant differences among the three rooms.

3. Results

3.1. Soil Chemical Properties

Table 1 lists the chemical properties of the calcareous soil. The pH of the calcareous soil was almost neutral (7.01). The Ca concentration was high. Comparing the two experimental years for calcareous soil, there was no significant difference.

Table 1. Chemical properties of calcareous soil used in this experiment (mean ± SD, $n = 5$).

Year	pH	C ($g \cdot kg^{-1}$)	N ($g \cdot kg^{-1}$)	P [1] ($mg \cdot kg^{-1}$)
2009	7.00 ± 0.03	26.5 ± 2.6	0.962 ± 0.082	59.2 ± 11.3
2010	7.02 ± 0.02	28.6 ± 1.1	0.935 ± 0.089	52.1 ± 10.6
	Ca [2] ($g \cdot kg^{-1}$)	Mg [2] ($mg \cdot kg^{-1}$)	K [2] ($mg \cdot kg^{-1}$)	Na [2] ($mg \cdot kg^{-1}$)
2009	2.64 ± 0.05	62.9 ± 1.3	26.0 ± 1.8	117 ± 6
2010	2.56 ± 0.05	62.1 ± 1.6	25.8 ± 0.8	115 ± 1
	Fe [3] ($mg \cdot kg^{-1}$)	Mn [2] ($mg \cdot kg^{-1}$)		
2009	8.34 ± 0.58	4.93 ± 0.59		
2010	9.52 ± 5.14	4.81 ± 0.68		

Mean values of each parameter were analyzed by Mann-Whitney U test (no significance; $p > 0.05$). [1] Extracted by the Olsen method; [2] Extracted in 1 M ammonium acetate; [3] Extracted in 1 M potassium chloride.

3.2. Growth Characteristics

At the first sampling stage before the experiment, the total dry mass of three *Quercus* species showed little difference among AH, SC, and NE seedlings (Figure 2). For *C. cuspidata*; the total dry mass of AH and SC seedlings showed significant increases compared with NE seedlings.

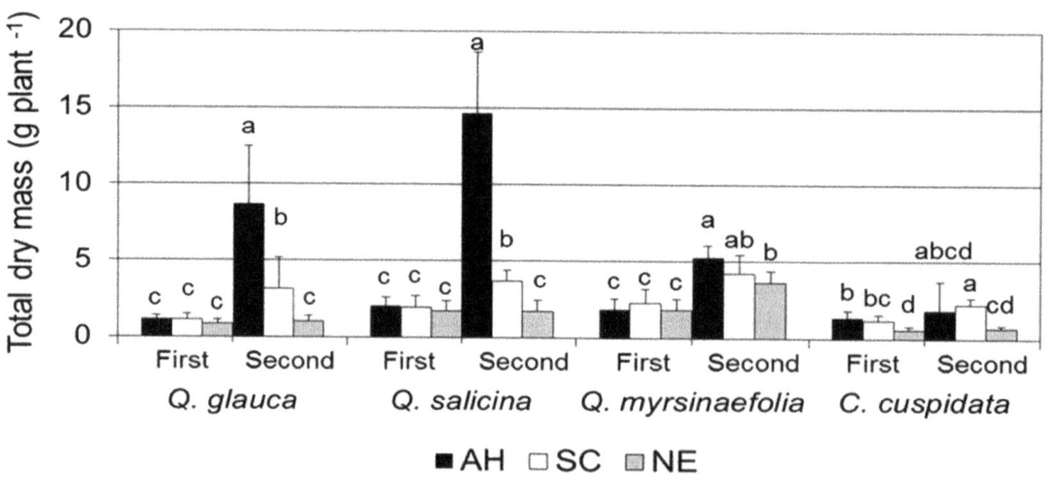

Figure 2. Total dry mass for seedlings of four Fagaceae species grown in a calcareous soil ($g \cdot plant^{-1}$, mean ± SD, $n = 12$). Different letters indicate significant effects according to a Steel-Dwass test ($p < 0.05$). Mean values followed by the same letter showed no significant difference ($p > 0.05$). AH, seedlings inoculated with *Astraeus hygrometricus*; SC, seedlings inoculated with *Scleroderma citrinum*; NE, non-ectomycorrhizal seedlings.

At the second sampling, the total dry mass was significantly increased in AH and SC seedlings of *Q. glauca*, *Q. salicina*, and *Q. myrsinaefolia*. For NE seedlings of *Q. myrsinaefolia*, the total dry mass increased significantly. In contrast, other NE seedlings did not show a significant increase in the total dry mass in *Q. glauca*, *Q. salicina*, and *C. cuspidata*. AH seedlings of *C. cuspidata* did not show significant increases in the total dry mass.

Comparing AH, SC, and NE seedlings at the second sampling, the total dry mass and dry mass of each organ were significantly higher in AH seedlings of *Q. glauca*, *Q. salicina*, and *Q. myrsinaefolia* than those in NE seedlings (Figures 2 and 3). Additionally, SC seedlings had significantly higher dry mass for each organ compared with NE seedlings of *Q. glauca*, *Q. salicina*, and *C. cuspidata*. For the SC seedlings of *Q. myrsinaefolia* and AH seedlings of *C. cuspidata*, the dry mass of each organ was not higher than that of NE seedlings. Comparing the AH seedlings of the four Fagaceae species, dry mass was highest for *Q. salicina* and lowest for *C. cuspidata*.

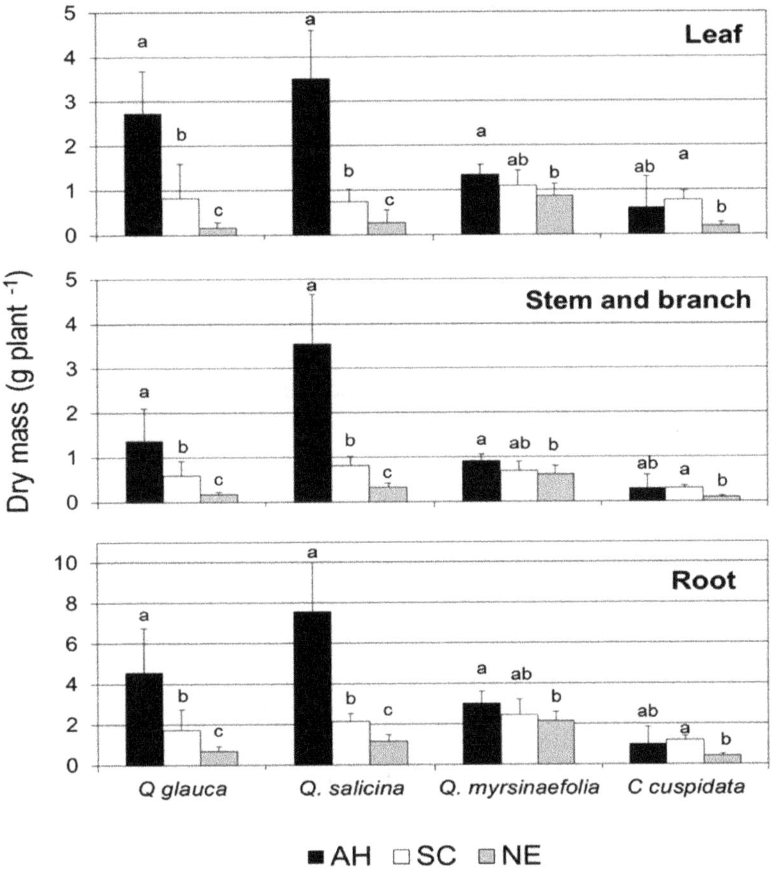

Figure 3. Dry mass of each organ (leaf, stem and branch, and root) for seedlings of four Fagaceae species grown in a calcareous soil at the second sampling (g·plant^{-1}, mean ± SD, $n = 12$). Different letters indicate significant effects according to a Steel–Dwass test ($p < 0.05$). Mean values followed by the same letter showed no significant difference ($p > 0.05$).

SLA of the four Fagaceae species was 85.2 cm^2·g^{-1} for *Q. glauca*, 95.6 cm^2·g^{-1} for *Q. salicina*, 87.7 cm^2· g^{-1} for *Q. myrsinaefolia*, and 162.2 cm^2·g^{-1} for *C. cuspidata*. There were no significant difference among AH, SC, and NE seedlings for each species.

3.3. The Percentage of ECM Colonization

At the first sampling, the percentages of ECM colonization in roots of AH seedlings were 69% for *Q. glauca* and *C. cuspidata*, 83% for *Q. salicina*, and 55% for *Q. myrsinaefolia* (Figure 4). In SC seedlings, these values were significantly lower than those of AH seedlings except for *Q. myrsinaefolia*. The percentage of ECM colonization of SC seedlings was around 40% for *Q. glauca*, *Q. myrsinaefolia*, and *C. cuspidate*, whereas this value was 54% for *Q. salicina*.

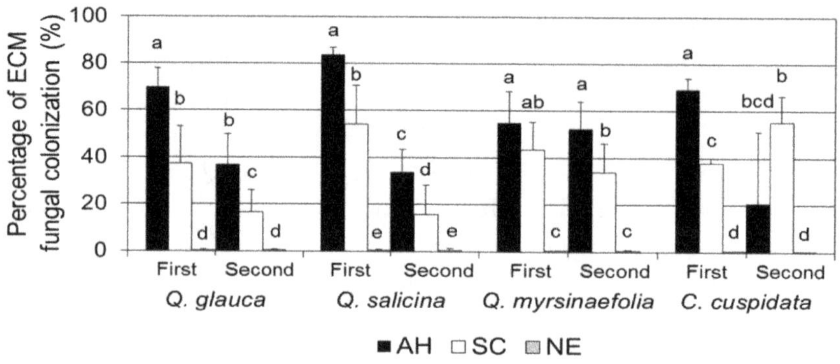

Figure 4. Percentage of ectomycorrhizal fungal colonization for seedlings of four Fagaceae species grown in a calcareous soil (%, mean ± SD, $n = 12$). Different letters indicate significant effects according to a Steel–Dwass test ($p < 0.05$). Mean values followed by the same letter showed no significant difference ($p > 0.05$).

After the second sampling, the percentages of ECM colonization of AH seedlings decreased significantly for *Q. glauca*, *Q. salicina*, and *C. cuspidata*. In addition, the percentages of ECM colonization decreased significantly for SC seedlings of *Q. glauca* and *Q. salicina*. The percentages of ECM colonization of the three *Quercus* species were significantly higher for AH seedlings than those of SC seedlings. In contrast, the percentage of ECM colonization of SC seedlings increased significantly to 55% for *C. cuspidata*. In half of the AH seedlings of *C. cuspidata*, the percentage of ECM colonization decreased to <1%. For *Q. myrsinaefolia*, the percentage of ECM colonization of AH and SC seedlings did not change after the experiment.

In NE seedlings of the four Fagaceae species, the percentage of ECM colonization was <1% at the two sampling points. In addition, these values of NE seedlings were significantly lower than those of AH and SC seedlings.

In terms of the relationship between total dry mass and the percentage of ECM colonization, we found a significant positive correlation for each Fagaceae species at the second sampling (Table 2, $p < 0.001$). Moreover, *Q. glauca* and *C. cuspidata* showed a significant positive difference at the first sampling ($p < 0.05$).

Table 2. Spearman's correlation coefficients (R) for seedlings of four Fagaceae species grown in a calcareous soil ($n = 12$).

	Q. glauca		*Q. salicina*		*Q. myrsinaefolia*		*C. cuspidata*	
	R	*p*	*R*	*p*	*R*	*p*	*R*	*p*
	0.826	<0.001 *	0.876	<0.001 *	0.638	<0.001 *	0.776	<0.001 *
Elements	*Q. glauca*		*Q. salicina*		*Q. myrsinaefolia*		*C. cuspidata*	
	R	*p*	*R*	*p*	*R*	*p*	*R*	*p*
N	0.217	0.199	−0.350	0.038 *	−0.394	0.020 *	0.138	0.415
P	0.030	0.859	0.435	0.010 *	0.605	<0.001 *	0.777	<0.001 *

The upper row shows the relationships between total dry mass and ectomycorrhizal (ECM) fungal colonization. The lower row shows the relationship between photosynthetic rate at light saturation (P_{sat}) and area-based concentrations of N and P in leaves at the second sampling. * shows a significant correlation at $p < 0.05$.

3.4. Photosynthetic Capacity

The P_{sat} of *Q. glauca* and *Q. salicina* was significantly higher in AH seedlings than in SC and NE seedlings (Figure 5). In *Q. myrsinaefolia*, the P_{sat} in AH and SC seedlings was significantly higher than that in NE seedlings. The P_{sat} of *C. cuspidata* was significantly higher in SC seedlings than in NE seedlings. Comparing the four Fagaceae species, P_{sat} was highest in AH seedlings of *Q. glauca* and lowest in NE seedlings of *C. cuspidata*.

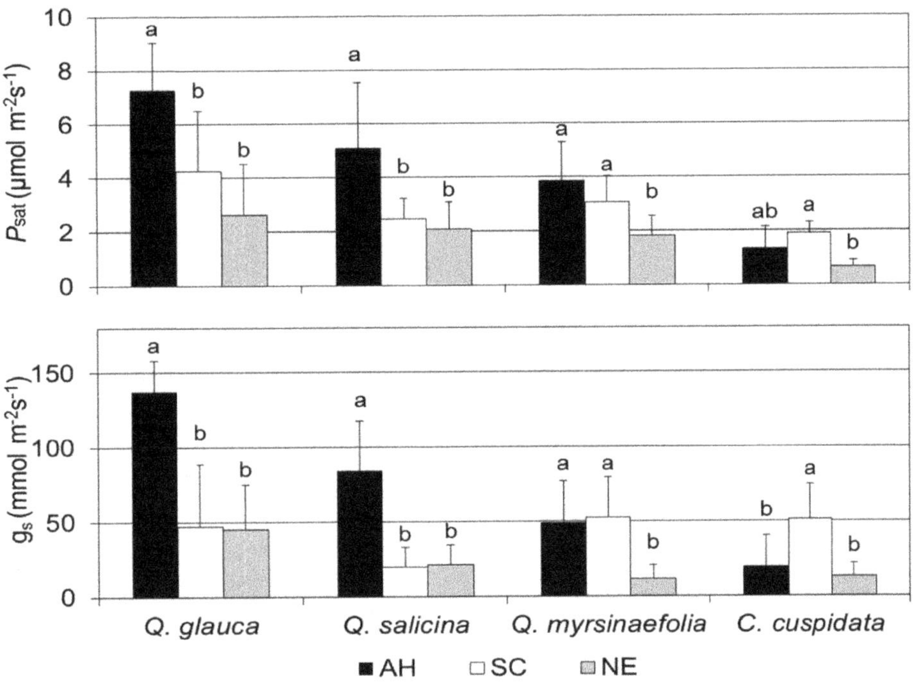

Figure 5. Photosynthetic rate at light saturation (P_{sat}, $\mu mol \cdot m^{-2} \cdot s^{-1}$) and stomatal conductance (g_s, $mol \cdot m^{-2} \cdot s^{-1}$) for seedlings of four Fagaceae species grown in a calcareous soil (mean ± SD, $n = 12$). Different letters indicate significant effects according to a Steel–Dwass test ($p < 0.05$). Mean values followed by the same letter showed no significant difference ($p > 0.05$). P_{sat} and g_s for *Q. glauca* and *Q. salicina* were measured in October 2009, and for *Q. myrsinaefolia* and *C. cuspidata* in October 2010.

A similar trend was observed for g_s as for P_{sat}, and g_s of *Q. glauca* and *Q. salicina* was significantly higher in AH seedlings than in SC and NE seedlings. g_s in AH and SC seedlings of *Q. myrsinaefolia* was significantly higher than that in NE seedlings. In *C. cuspidata*, the g_s in SC seedlings was higher than that in AH and NE seedlings. In particular, the g_s in AH seedlings of *Q. glauca* was higher compared with other seedlings.

3.5. Element Concentrations in Leaves

The concentrations of elements in leaves at the second sampling are shown in Table 3. The concentration of P was significantly higher in AH seedlings of *Q. salicina* and *Q. myrsinaefolia* than in SC seedlings. In particular, AH seedlings of *Q. salicina* showed the highest P concentration between the four Fagaceae species. In *C. cuspidata*, the concentration of P was highest for SC seedlings. In contrast, other elements did not show high concentrations in AH seedlings despite their greater growth. The concentration of N in AH seedlings was significantly lower for *Q. salicina* and *C. cuspidata* than in SC and NE seedlings. Similarly, the concentration of K was significantly lower in AH seedlings of *Q. glauca* and *Q. myrsinaefolia* compared with in SC and NE seedlings. Moreover, NE seedlings had the highest Mg concentration in *Q. salicina* and *C. cuspidata*.

For the relationship between the concentrations of N or P and P_{sat}, we found a significant positive correlation between the concentrations of P and P_{sat} of *Q. salicina*, *Q. myrsinaefolia*, and *C. cuspidata* (Table 2, $p < 0.05$). In contrast, there were significant negative correlations between concentration of N and P_{sat} of *Q. salicina* and *Q. myrsinaefolia* ($p < 0.05$).

Table 3. Concentrations of elements (N, P, K, Ca, Mg, Fe, and Mn) in leaves of seedlings of four Fagaceae species grown in a calcareous soil sampled at the second sampling (mean ± SD, $n = 12$).

Element, Inoculation		*Q. glauca*	*Q. salicina*	*Q. myrsinaefolia*	*C. cuspidata*
N (mg·g^{-1})	AH	10.7 ± 3.4 a	9.4 ± 2.2 b	9.2 ± 2.8 a	9.7 ± 1.8 b
	SC	11.1 ± 1.5 a	13.2 ± 2.3 a	11.2 ± 3.8 a	15.4 ± 3.4 a
	NE	10.5 ± 2.8 a	15.3 ± 2.0 a	11.0 ± 2.41 a	12.2 ± 1.10 a
P (mg·g^{-1})	AH	1.43 ± 0.33 a	1.98 ± 0.30 a	1.81 ± 0.27 a	0.84 ± 0.33 a
	SC	1.40 ± 0.28 a	1.34 ± 0.35 b	1.41 ± 0.12 b	1.11 ± 0.19 a
	NE	1.35 ± 0.52 a	1.41 ± 0.62 ab	0.69 ± 0.29 c	0.30 ± 0.12 b
K (mg·g^{-1})	AH	5.54 ± 0.80 b	4.52 ± 1.12 a	3.91 ± 0.50 b	5.20 ± 2.12 ab
	SC	8.58 ± 1.37 a	5.39 ± 0.83 a	4.78 ± 1.48 b	3.96 ± 0.65 b
	NE	8.65 ± 1.15 a	6.17 ± 2.22 a	7.15 ± 1.18 a	5.81 ± 1.18 a
Ca (mg·g^{-1})	AH	25.3 ± 7.1 a	19.6 ± 6.5 a	26.6 ± 3.3 a	15.9 ± 2.8 a
	SC	19.2 ± 5.8 a	14.9 ± 3.2 a	26.8 ± 5.1 a	16.2 ± 2.7 a
	NE	20.6 ± 4.8 a	14.6 ± 4.2 a	23.8 ± 3.30 a	17.3 ± 5.11 a
Mg (mg·g^{-1})	AH	1.40 ± 0.43 a	1.12 ± 0.35 b	2.30 ± 0.31 a	1.86 ± 0.65 ab
	SC	1.43 ± 0.40 a	1.23 ± 0.20 b	2.47 ± 0.40 a	1.36 ± 0.25 b
	NE	1.88 ± 0.60 a	3.50 ± 2.66 a	2.68 ± 0.41 a	2.18 ± 0.52 a
Fe (μg·g^{-1})	AH	102 ± 20 a	77 ± 12 b	85 ± 19 a	155 ± 139 a
	SC	110 ± 44 a	159 ± 51 a	79 ± 17 a	105 ± 26 a
	NE	182 ± 137 a	215 ± 139 a	94 ± 21 a	180 ± 115 a
Mn (μg·g^{-1})	AH	204 ± 31 b	202 ± 76 ab	273 ± 65 a	454 ± 239 a
	SC	140 ± 67 b	163 ± 73 b	200 ± 115 a	167 ± 58 b
	NE	434 ± 189 a	363 ± 151 a	206 ± 67 a	507 ± 222 a

Different letters indicate significant effects according to a Steel-Dwass test ($p < 0.05$). Mean values followed by the same letter showed no significant difference ($p > 0.05$). AH, seedlings inoculated with *Astraeus hygrometricus*; SC, seedlings inoculated with *Scleroderma citrinum*; NE, non-ectomycorrhizal seedlings.

3.6. Element Concentrations and Contents in Roots

The concentrations of elements in roots at the second sampling are shown in Table 4. The concentrations of P in *Q. glauca*, *Q. salicina*, and *Q. myrsinaefolia* were significantly higher in AH seedlings than in NE seedlings. AH seedlings also exhibited high concentrations of Fe and Mn in *Q. myrsinaefolia* than in SC seedlings. In *C. cuspidata*, the concentrations of N and P were highest in SC seedlings. The concentration of Ca was significantly lower in SC seedlings of *Q. myrsinaefolia* and *C. cuspidata*. In addition, the concentration of Fe in SC seedlings of *Q. salicina* and *Q. myrsinaefolia* was significantly lower. In NE seedlings of *Q. myrsinaefolia*, the concentration of Mg was significantly higher than in AH and SC seedlings. Across the four Fagaceae species, the concentration of P was the highest for AH seedlings of *Q. salicina*.

Table 4. Concentrations of elements (N, P, K, Ca, Mg, Fe, and Mn) in roots of seedlings of four Fagaceae species grown in a calcareous soil sampled at the second sampling (mean ± SD, $n = 12$).

Element	Inoculation	*Q. glauca*		*Q. salicina*		*Q. myrsinaefolia*		*C. cuspidata*	
N ($mg \cdot g^{-1}$)	AH	8.3 ± 1.8	a	9.8 ± 1.7	a	7.5 ± 1.9	a	7.5 ± 1.9	b
	SC	7.3 ± 1.3	a	7.6 ± 0.8	b	6.3 ± 1.9	a	12.2 ± 1.4	a
	NE	6.8 ± 1.6	a	7.7 ± 1.6	ab	5.5 ± 1.8	a	5.8 ± 1.1	b
P ($mg \cdot g^{-1}$)	AH	2.34 ± 0.49	a	3.76 ± 0.70	a	2.63 ± 0.74	a	1.20 ± 0.92	ab
	SC	1.84 ± 0.76	ab	1.21 ± 0.30	b	1.95 ± 1.02	ab	2.45 ± 0.75	a
	NE	1.13 ± 0.29	b	1.17 ± 0.29	b	0.97 ± 0.17	b	0.39 ± 0.17	b
K ($mg \cdot g^{-1}$)	AH	2.74 ± 0.42	a	2.65 ± 0.42	a	2.75 ± 0.36	a	2.17 ± 0.50	a
	SC	2.70 ± 1.40	a	2.28 ± 0.45	a	2.21 ± 0.42	a	2.44 ± 0.34	a
	NE	2.07 ± 0.54	a	2.49 ± 0.60	a	2.52 ± 0.42	a	1.96 ± 0.37	a
Ca ($mg \cdot g^{-1}$)	AH	14.8 ± 2.3	a	15.0 ± 2.2	a	13.4 ± 1.0	a	15.3 ± 2.3	a
	SC	13.8 ± 2.2	a	13.8 ± 1.1	a	11.7 ± 1.0	b	12.5 ± 1.2	b
	NE	17.0 ± 2.2	a	15.7 ± 1.6	a	13.5 ± 1.7	ab	15.1 ± 1.81	a
Mg ($mg \cdot g^{-1}$)	AH	1.85 ± 0.21	ab	1.64 ± 0.23	a	1.54 ± 0.18	b	1.46 ± 0.18	a
	SC	1.70 ± 0.59	b	1.62 ± 0.27	a	1.51 ± 0.28	b	1.25 ± 0.12	bc
	NE	2.10 ± 0.32	a	1.71 ± 0.26	a	1.94 ± 0.19	a	1.36 ± 0.16	ab
Fe ($\mu g \cdot g^{-1}$)	AH	0.99 ± 0.49	a	1.45 ± 0.53	a	1.38 ± 0.42	a	0.76 ± 0.87	ab
	SC	0.82 ± 0.33	a	0.76 ± 0.19	b	0.64 ± 0.20	b	0.69 ± 0.26	a
	NE	1.49 ± 0.77	a	1.32 ± 0.33	a	0.98 ± 0.42	ab	0.38 ± 0.18	b
Mn ($\mu g \cdot g^{-1}$)	AH	63 ± 17	b	70 ± 20	a	39 ± 9	a	31 ± 13	a
	SC	73 ± 11	b	67 ± 18	a	28 ± 6	b	33 ± 8	a
	NE	138 ± 78	a	80 ± 20	a	34 ± 10	ab	24 ± 5	a

Different letters indicate significant effects according to a Steel-Dwass test ($p < 0.05$). Mean values followed by the same letter showed no significant difference ($p > 0.05$).

3.7. Vector Analysis between Relative Dry Mass and Nutrient Concentration

The relationships between relative leaf dry mass and relative concentrations in leaves are shown in Figure 6. For the relationship of Ca, uptake of each Fagaceae species showed enhanced or luxury consumption. In contrast, Mn showed dilution or efficient use by each Fagaceae species. For N, most plots showed dilution or sufficiency, whereas those for SC seedlings of *Q. salicina* and SC and NE seedlings of *C. cuspidate* showed enhanced uptake. In the case of NE seedlings of *Q. salicina*, the plot showed toxicity.

For P and Fe in the leaves, most plots of each Fagaceae species showed dilution or efficient use for leaves and roots. For P in leaves of NE seedlings of *Q. glauca* and *Q. salicina*, and Fe in leaves of NE seedlings of *Q. salicina*, the plots showed nutrient deficiency. The trend of K in leaves showed various patterns. Plots of AH seedling of *Q. glauca* and *Q. salicina* showed sufficiency, but the plots of *C. cuspidata* and AH and SC seedlings of *Q. myrsinaefolia* showed dilution. Other plots showed enhanced or luxury consumption. For the relationship of Mg, the plots showed sufficiency or dilution for each Fagaceae species. In the case of NE seedlings of *Q. salicina*, the plot showed toxicity.

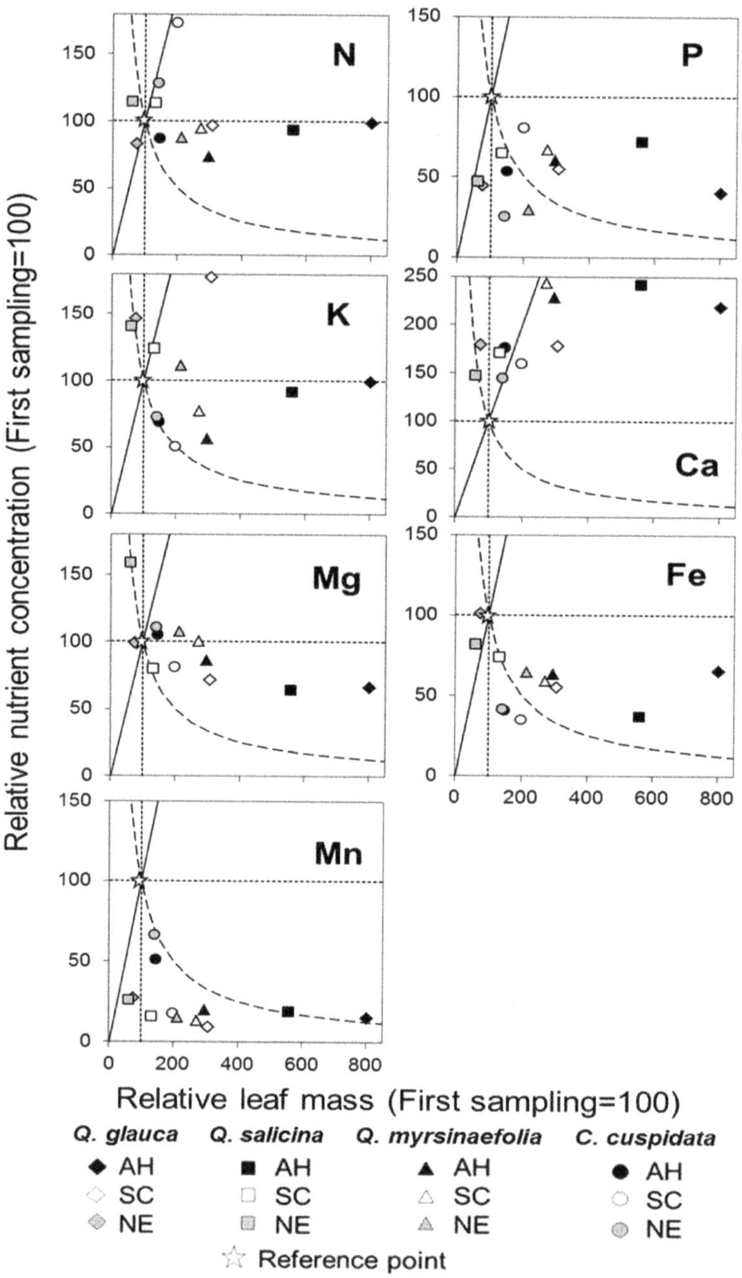

Figure 6. Comparison of relative nutrient concentration in leaves and leaf dry masses for seedlings of four Fagaceae species grown in a calcareous soil. The content isograms represent the combination of dry mass and concentration required for these values at the second sampling relative to the first sampling. A value of 100 represents the dry mass and concentration at the first sampling.

4. Discussion

4.1. Growth and Photosynthesis

After 6 months of cultivation in calcareous soil, we confirmed that symbiosis with ECM fungi was essential for growth acceleration in calcareous soil (Figure 2). Symbiosis with *A. hygrometricus* contributed to growth acceleration in the three *Quercus* species compared with that of *S. citrinum*. In *C. cuspidata*, only *S. citrinum* inoculation showed high values of dry mass of each organ (Figure 3). In general, compatibility with ECM fungi and effects on growth acceleration are different among woody species [57,58]. Unlike *Quercus* species, *C. cuspidata* may have better compatibility with *S. citrinum*. We have already confirmed that *C. cuspidata* inoculated with *S. citrinum* showed accelerated growth [40].

Moreover, AH seedlings of *Q. glauca* and *Q. salicina* showed the highest values of P_{sat} and g_s (Figure 5). Woody species inoculated with ECM fungi exhibit accelerated water uptake from external hyphae [59]. AH seedlings probably had accelerated water uptake, and stomatal closure did not occur. Accordingly, the P_{sat} and g_s were high in AH seedlings. Similar trends were shown for AH and SC seedlings of *Q. myrsinaefolia*, which had high values for the percentage of ECM colonization (Figures 4 and 5).

Comparing AH seedlings, *Q. salicina* showed remarkable growth acceleration (Figure 2). *Q. salicina* have high acclimation capacity in calcareous soil, and symbiosis with *A. hygrometricus* can take place in calcareous soil. *Q. salicina* grows at high altitudes in the karst region in southwestern Japan [17,18,21]. Our results reflected the distribution of *Q. salicina*.

Compared with *Q. salicina* and *Q. glauca*, growth acceleration by inoculation with ECM fungi was not obvious for *Q. myrsinaefolia* (Figure 2). Acclimation capacity in calcareous soil was lower for *Q. myrsinaefolia* than for *Q. salicina* and *Q. glauca*. In contrast, the total dry mass for *C. cuspidata* was lowest even in SC seedlings (Figure 2). The leaves of *C. cuspidata* yellowed after transplantation, indicating the occurrence of lime-induced chlorosis [3]. In addition, lime-induced chlorosis can readily appear in calcifuge plants [7]. Based on growth and lime-induced chlorosis, *C. cuspidata* was considered as a calcifuge plant. Conversely, *C. cuspidata* inoculated with ECM fungi accelerated its growth in acidic soil [40], indicating that this species has a preference for an acidic environment.

4.2. Nutrient Relationships

For the nutrients in calcareous soil, the concentration of exchangeable Ca was 2.6 $g \cdot kg^{-1}$ (Table 1). The values of soil pH and Ca concentration differ among calcareous soils in Japan [12,13,60–63], with ranges of 5.8 to 8.3 for pH and 0.8 to 12.1 $g \cdot kg^{-1}$ for Ca. Compared with other regions, the soil used for our experiment did not have high soil pH, and did not exhibit an extreme excess of Ca. In contrast, the concentration of N in calcareous soil used in our experiment was lower than those in other regions [60–63]. However, the concentration of N in leaves of the four Fagaceae species was not in the deficient range compared with other woody species grown in calcareous soil [13,63]. For other nutrients, the concentrations of K and Mg were higher than typical forest soil in this region [64]. There was no previous information available on P in calcareous soil in Japan from analyses by the Olsen method. Compared with the data from other countries, the concentration of P was higher in calcareous soil used in our experiment [65,66]. Thus, the concentrations of N, P, K, and Mg in calcareous soil in our experiment were not insufficient.

AH and SC seedlings of the four Fagaceae species did not show high N concentrations in leaves (Table 3). Thus, uptake of N in leaves is not related to ECM colonization. Comparing traits among *Quercus* species, the concentration of N in roots was higher for *Q. salicina* and *Q. glauca* than that for *Q. myrsinaefolia* (Table 4). *Q. salicina* and *Q. glauca* had higher adaptability to various poor habitats [19,67]. *Q. glauca* and *Q. salicina* have high capacity to acquire N from calcareous soil. In contrast, the habitats of natural forests of *Q. myrsinaefolia* include deep, well-drained soil [68]. In the karst region, *Q. myrsinaefolia* has been found on a site with deep calcareous soil [19]. Differences in acclimation capacity for *Q. myrsinaefolia* may originate from the inherent nutrient requirements. There was a similar case for *Picea abies*, which has low adaptability to calcareous soil [69], and suppressed uptake of N in roots increased in calcareous soil [70].

On the uptake of P in roots, NE seedlings showed low P concentrations compared with AH seedlings of *Q. glauca*, *Q. salicina* and *Q. myrsinaefolia* (Table 4). Thus, uptake of P is linked with symbiosis with ECM fungi. On calcareous soil, symbiosis with ECM fungi is important for the uptake of P [27,28]. Symbiosis between *Quercus* species and *A. hygrometricus* produced a high capacity for the acquisition of P.

Meanwhile, we confirmed that the uptake of P, Fe and Mn in leaves was suppressed in each Fagaceae species (Table 3, Figure 6). Uptake of micronutrients (e.g., Fe, Mn, Cu, and Zn) and P is suppressed by low availability at high pH in calcareous soil [3–5,70]. Our results suggested that uptake of Mn is suppressed for the Fagaceae species in calcareous soil. For P and Fe in leaves, AH seedlings of three *Quercus* species exhibited increased contents; however, concentration was decreased and showed evidence of dilution (Figure 6). This trend suggests that the uptake of P and Fe in leaves is suppressed even after inoculation with *A. hygrometricus*. In particular, NE seedlings of *Q. glauca* and *Q. salicina* showed a P deficiency in leaves (Figure 6). We considered that NE seedlings of *Q. glauca* and *Q. salicina* suffered from serious nutrient suppression due to lower availability. *C. cuspidata* showed an obvious decrease of the concentration of Fe in leaves, and showed efficient use of little Fe (Figure 6). When calcifuge woody plants are inoculated with ECM fungi, the effects of the fungi are probably quite limited in calcareous soil.

In contrast, Ca in leaves and roots showed a high value (Tables 3 and 4) compared with seedlings of Fagaceae species grown on acidic colluvial soil [40]. Moreover, many plots showed enhancement of uptake or luxury consumption (Figure 6). However, the actual concentration of Ca in plant organs was almost the same among AH, SC, and NE seedlings for each Fagaceae species (Tables 3 and 4). Thus, we considered that the differences in Ca were not the main factor for growth differences among AH, SC, and NE seedlings.

4.3. Relationships among Variables

In examining the factors affecting differences in growth between AH, SC and NE seedlings, ECM colonization was considered one of the important factors. The percentages of ECM colonization were higher in AH seedlings of the three *Quercus* species and in SC seedlings of *C. cuspidata* (Figure 4). Additionally, there were positive relationships between total dry mass and ECM colonization (Table 2). Thus, different percentages of ECM colonization were probably related to differences in growth.

With respect to the relationships between growth and nutrient uptake, N and P in roots showed enhancement of uptake in AH seedlings of *Q. salicina* (Table 4). The storage of N and P in belowground organs has a significant role in plant growth, and plants with high N and P in roots show high growth rates [71–73]. Based on these results, it appeared that the amount of stored N and P resulted in the high growth of AH seedlings of *Q. salicina*.

The photosynthetic rate showed positive correlations between P_{sat} and the concentrations of P in leaves of *Q. salicina*, *Q. myrsinaefolia*, and *C. cuspidata* (Table 2). In general, the photosynthetic rate is closely related to P concentration [55,56]. Thus, the concentration of P in leaves is considered as an important factor that affects the photosynthetic rate. In contrast, *Q. glauca* did not show a positive correlation between P_{sat} and the concentration of P (Table 2). Other factors that affect differences in P_{sat} include high value of g_s (Figure 5). AH seedlings of *Q. glauca* could absorb water through developed hyphae (Figure 4); as a result, the stomata stay open and absorption of CO_2 through the stomata is also increased.

The concentration of N in leaves showed a negative correlation with P_{sat} for *Q. salicina* and *Q. myrsinaefolia* (Table 2), and our results contradicted the general trend [7,54]. The cause of this negative trend was probably related to the high concentration of N in leaves and the low values of P_{sat} in NE seedlings. In the case of NE seedlings, they had no hyphae of ECM fungi and a lower capacity for water uptake. The stomata of NE seedlings may have closed in order to avoid water stress; as a result, NE seedlings showed lower P_{sat} and g_s and decreased absorption of CO_2 through stomata. Similar trends were shown for NE seedlings of Fagaceae species planted on acidic soil [40].

With respect to the relationship among nutrients for *C. cuspidata* and AH seedlings of *Q. myrsinaefolia*, the concentrations of K were decreased in leaves (Table 3) and were diluted (Figure 6). High Ca content in soil causes ion-antagonism with K [70]. As a result, the suppression of was confirmed for several woody species (*Picea abies* and several Lauraceae species), in which growth was suppressed on calcareous soil [13,70]. Compared with AH seedlings of *Q. glauca* and *Q. salicina*, growth levels of *C. cuspidata* and AH seedlings of *Q. myrsinaefolia* were inferior. Woody species that show less adaptability to calcareous soil can probably suppress uptake of K.

5. Conclusions

When seedlings of the four Fagaceae species were planted on calcareous soil, the uptake of various nutrients, such as P, Fe, and Mn in leaves was suppressed. However, Fagaceae seedlings inoculated with ECM fungi showed accelerated growth. In particular, Fagaceae seedlings inoculated with ECM fungi had increased concentration of P in roots, and this trait was important for growth acceleration. Additionally, the concentration of P in leaves was an important factor and affected the differences in photosynthetic rate. The most suitable ECM fungi differed among the four Fagaceae species, and higher growth levels were shown from inoculation with *A. hygrometricus* for the three *Quercus* species, and with *S. citrinum* for *C. cuspidata*. Growth levels of the seedlings inoculated with ECM fungi were different among the four Fagaceae species, and *C. cuspidata*, which showed the lowest growth, was considered as a calcifuge plant. From their growth characteristics, we concluded that *Q. salicina* and *Q. glauca* inoculated with *A. hygrometricus* showed the highest growth levels, and these species are most suitable for growth on calcareous soil.

These findings are expected to contribute to new methods for reforestation of limestone quarries in warm-temperature zones in East Asia. In particular, *Q. salicina* and *Q. glauca* inoculated with *A. hygrometricus* may be useful species for reforestation. The effects on growth acceleration without fertilization are advantageous for reforestation in limestone environments. It should be noted that our experiment used sterilized calcareous soil; therefore, further field cultivation experiments with ECM-fungi inoculated plants will be needed.

Acknowledgments: We thank researchers of the Kyushu Research Center, FFPRI, for their encouragement. We also thank the staff of Todaka Mining Co. for the collection of calcareous soil. We thank Yosuke Matsuda for his cooperation in the collection of ECM fungi. The phytotron experiment was carried out with the cooperation of Nahoko Aoki, to whom we express our thanks. Moreover, we thank Yoshimi Narimatsu and Takeo Kokuryo for analysis of Fagaceae seedlings. The soil analyses were carried out at JIRCAS. Thanks are also due to the staff of JIRCAS and Medical English Service, Kyoto, and Forestry Division, for English proofreading. This research was supported by a grant from the Ministry of Agriculture, Forestry and Fishery (Research and Development Projects for Application in Promoting New Policy of Agriculture, Forestry and Fishery, No. 1904: Development of Regeneration Technology for Leading Artificial Coniferous Forest to Broadleaf Forests). In addition, this study was supported by a grant-in-aid for Scientific Research from the Japan Society of Promotion of Science (No. 20780124: Physiological Characteristics of Trees Grown on a Limestone Region and Application to Revegetation Technology of Limestone Quarry).

Author Contributions: Masazumi Kayama conceived the experiments. Takashi Yamanaka raised seedlings of all Fagaceae species and inoculated them with ectomycorrhizal fungi. Masazumi Kayama performed the experiments, measured photosynthetic rates, and analyzed various nutrients. Masazumi Kayama and Takashi Yamanaka discussed the results, and co-wrote the paper.

References

1. Ford, D.C.; Williams, P.W. *Karst Geomorphology and Hydrology*; Chapman & Hall: London, UK, 1989; p. 620.
2. Haleem, A.A.; Loeppert, R.H.; Anderson, W.B. Role of soil carbonate and iron oxide in iron nutrition of soybean in calcareous soils of Egypt and the United States. In *Iron Nutrition in Soils and Plants*; Abadía, J., Ed.; Kluwer Academic Publishers: Dordrecht, The Netherland, 1995; pp. 307–314.
3. Marschner, H. *Mineral Nutrition of Higher Plants*; Academic Press: New York, NY, USA, 1995; p. 889.
4. Chen, Y.; Barak, P. Iron nutrition of plants in calcareous soils. *Adv. Agron.* **1982**, *35*, 217–240.
5. Tyler, G. Mineral nutrient limitation of calcifuge plants in phosphate sufficient limestone soil. *Ann. Bot.* **1996**, *77*, 649–656. [CrossRef]
6. Jefferies, R.L.; Willis, A.J. Studies on the calcicole-calcifuge habit: II. The influence of calcium on the growth and establishment of four species in soil and sand cultures. *J. Ecol.* **1964**, *52*, 691–707. [CrossRef]
7. Larcher, W. *Physiological Plant Ecology*, 4th ed.; Springer: Berlin, Germany, 2003; p. 513.

8. Zohlen, A.; Tyler, G. Soluble inorganic tissue phosphorus and calcicole-calcifuge behaviour of plants. *Ann. Bot.* **2004**, *94*, 427–432. [CrossRef]
9. Oates, T. Lime and limestone. In *Kirk-Othmer Encyclopedia of Chemical Technology*; Kirk-Othmer, Ed.; John Wiley and Sons: Hoboken, NJ, USA, 2010; pp. 1–53.
10. Ursic, K.A.; Kenkel, N.C.; Larson, D.W. Revegetation dynamics of cliff faces in abandoned limestone quarries. *J. Appl. Ecol.* **1997**, *34*, 289–303. [CrossRef]
11. Duan, W.J.; Ren, H.; Fu, S.L.; Wang, J.; Yang, L.; Zhang, J. Natural recovery of different areas of a deserted quarry in south China. *J. Environ. Sci.* **2008**, *20*, 476–481. [CrossRef]
12. Kayama, M. Growth characteristics of Lauraceae tree species planted on limestone quarry. *Ann. Rep. Kyushu Res. Cen. FFPRI* **2011**, *23*, 16–17. (In Japanese)
13. Kayama, M. Growth characteristics of four Lauraceae tree species planted on calcareous and brown forest soils. *Kyushu J. For. Res.* **2013**, *66*, 63–66. (In Japanese)
14. Kira, T. Forest ecosystems of east and Southeast Asia in a global perspective. *Ecol. Res.* **1991**, *6*, 185–200. [CrossRef]
15. Fang, J.Y.; Song, Y.C.; Liu, H.Y.; Piao, S.L. Vegetation-climate relationship and its application in the division of vegetation zone in China. *Acta Bot. Sin.* **2002**, *44*, 1105–1122.
16. Menitsky, Y.L. *Oaks of Asia*; Science Publishers: Enfield, CT, USA, 2005; p. 549.
17. Yamanaka, T. The limestone vegetation in middle Kyushu. *Res. Rep. Kochi Univ. Nat. Sci.* **1966**, *15*, 1–9. (In Japanese with English summary)
18. Yamanaka, T. The forest and scrub vegetation in limestone areas of Shikoku, Japan. *Vegetatio* **1969**, *19*, 286–307. [CrossRef]
19. Miyawaki, A. *Vegetation of Japan, Chugoku*; Shibundo: Tokyo, Japan, 1983; p. 540. (In Japanese with English summary)
20. Xu, Z.R. A study of the vegetation and floristic affinity of the limestone forest in southern and southwestern China. *Ann. Mo. Bot. Gard.* **1995**, *82*, 570–580.
21. Miyawaki, A. *Vegetation of Japan, Kyushu*; Shibundo: Tokyo, Japan, 1981; p. 484. (In Japanese with English summary)
22. Liu, J.M. The reproductive and regenerative counter measures of the main woody species in Maolan karst forest. *Sci. Silvae Sin.* **2000**, *36*, 115–122. (In Chinese with English summary)
23. Yao, Y.Q.; Zhang, Z.H.; Liang, S.C.; Bi, X.L.; Li, G.R.; Hu, G. Structure of *Cyclobalanopsis glauca* population on karst hills of Guilin. *J. Zhejiang For. Sci. Technol.* **2008**, *28*, 8–11. (In Chinese with English summary)
24. Feroz, S.M.; Hagihara, A. Comparative studies on community ecology of two types of subtropical forest grown in silicate and limestone habitats in the northern part of Okinawa Island, Japan. *Taiwania* **2008**, *53*, 139–149.
25. Newman, E.I.; Reddell, P. The distribution of mycorrhizas among families of vascular plants. *New Phytol.* **1987**, *106*, 745–751. [CrossRef]
26. Wang, B.; Qiu, Y.-L. Phylogenic distribution and evolution of mycorrhizas in land plants. *Mycorrhiza* **2006**, *16*, 299–363. [CrossRef]
27. Wallander, H. Uptake of P from apatite by *Pinus sylvestris* seedlings colonised by different ectomycorrhizal fungi. *Plant Soil* **2000**, *218*, 249–256. [CrossRef]
28. Ström, L.; Owen, A.G.; Godbold, D.L.; Jones, D.L. Organic acid behaviour in a calcareous soil: Sorption reactions and biodegradation rates. *Soil Biol. Biochem.* **2001**, *33*, 2125–2133. [CrossRef]
29. Tyler, G. Some ecophysiological and historical approaches to species richness and calcicole/calcifuge behavior—Contribution to a debate. *Folia Geobot.* **2003**, *38*, 419–428. [CrossRef]
30. Van der Heijden, E.W.; Vosatka, M. Mycorrhizal association of *Salix repens* L. communities in succession of dune ecosystems. II. Mycorrhizal dynamics and interactions of ectomycorrhizal and arbuscular mycorrhizal fungi. *Can. J. Bot.* **1999**, *77*, 1833–1841. [CrossRef]
31. Wiemken, V.; Laczko, E.; Ineichen, K.; Boller, T. Effects of elevated carbon dioxide and nitrogen fertilization on mycorrhizal fine roots and the soil microbial community in beech-spruce ecosystems on siliceous and calcareous soil. *Microb. Ecol.* **2001**, *42*, 126–135.
32. Kennedy, K.H.; Maxwell, J.F.; Lumyong, S. Fire and the production of *Astraeus odoratus* (Basidiomycetes) sporocarps in deciduous dipterocarp-oak forests in northern Thailand. *Maejo Int. J. Sci. Technol.* **2012**, *6*, 483–504.

33. Lapeyrie, F. The role of ectomycorrhizal fungi in calcareous soil tolerance by "symbiocalcicole" woody plants. *Ann. Sci. For.* **1990**, *21*, 579–589. [CrossRef]
34. Taste, F.P.; Schmidt, M.G.; Berch, S.M.; Bulmer, C.; Egger, K.N. Effects of ectomycorrhizal inoculants on survival and growth of interior Douglas-fir seedlings on reforestation sites and partially rehabilitated landings. *Can. J. For. Res.* **2004**, *34*, 2074–2088. [CrossRef]
35. Marx, D.H.; Maul, S.B.; Cordell, C.E. Application of specific ectomycorrhizal fungi in world forestry. In *Frontiers in Industrial Mycology*; Leatham, G.F., Ed.; Chapman and Hall: New York, NY, USA, 1992; pp. 78–98.
36. Endo, M. Higher fungi of the Hirao-karst topography. *Res. Bull. Cult. Asset Kitakyushu* **1973**, *13*, 53–56. (In Japanese)
37. Yoshimi, S. Taxonomic study of the Japanese taxa of *Scleroderma* Pers. *Jpn. J. Mycol.* **2002**, *43*, 3–18. (In Japanese with English summary)
38. Fangfuk, W.; Okada, K.; Petchang, R.; To-anun, C.; Fukuda, M.; Yamada, A. In vitro mycorrhization of edible *Astraeus* mushrooms and their morphological characterization. *Mycoscience* **2010**, *51*, 234–241. [CrossRef]
39. Makita, N.; Hirano, Y.; Yamanaka, T.; Yoshimura, K.; Kosugi, Y. Ectomycorrhizal-fungal colonization induces physio-morphological changes in *Quercus serrata* leaves and roots. *J. Plant Nutr. Soil Sci.* **2012**, *17*, 900–906. [CrossRef]
40. Kayama, M.; Yamanaka, T. Growth characteristics of ectomycorrhizal seedlings of *Quercus glauca*, *Quercus salicina*, and *Castanopsis cuspidata* planted on acidic soil. *Trees* **2014**, *28*, 569–583. [CrossRef]
41. Akama, K.; Okabe, H.; Yamanaka, T. Growth of ectomycorrhizal fungi on various culture media. *Bull. FFPRI* **2008**, *7*, 165–181. (In Japanese with English summary)
42. Ohta, A. A new medium for mycelial growth of mycorrhizal fungi. *Trans. Mycol. Soc. Jpn.* **1990**, *31*, 323–334.
43. Oita Prefecture. *Fundamental land classification survey, Usuki and Hotojima*; Fuji Micro-service Center: Kumamoto, Japan, 1980; p. 51. (In Japanese)
44. Van Reeuwijk, L.P. *Procedures for Soil Analysis*, 6th ed.; International Soil Reference and Information Centre: Wagningen, The Netherlands, 2002; p. 100.
45. Sparks, D.L.; Page, A.L.; Helmke, P.A.; Loeppert, R.H.; Soltanpour, P.N.; Tabatabai, M.A.; Johnson, C.T.; Sumner, M.E. *Methods of Soil Analysis, Part 3. Chemical Methods*; Soil Science Society of America Inc.: Madison, WI, USA, 1996; p. 1390.
46. American Public Health Association; American Water Works Association; Water Environment Federation. *Standard Methods for the Examination of Water and Wastewater*, 20th ed.; American Public Health Association: Washington, DC, USA, 1998; p. 1220.
47. Japan Meteorological Agency. Climate Statistics. Available online: http://www.data.jma.go.jp/obd/stats/etrn/index.php (access on 31 October 2010).
48. Iwasaki, T.; Okuyama, T.; Sakuratani, T. Climatic estimation of photosynthetically active radiation (PAR) in Kanto and Tokai districts. *Bull. Natl. Inst. Agro Environ. Sci.* **1986**, *1*, 1–35. (In Japanese with English summary)
49. Mohan, V.; Natarajan, K.; Ingleby, K. Anatomical studies of ectomycorrhizas. III. The ectomycorrhizas produced by *Rhizopogon luteolus* and *Scleroderma citrinum* on *Pinus patula*. *Mycorrhiza* **1993**, *3*, 51–56. [CrossRef]
50. Quoreshi, A.M.; Timmer, V.R. Exponential fertilization increases nutrient uptake and ectomycorrhizal development of black spruce seedlings. *Can. J. For. Res.* **1998**, *28*, 674–682. [CrossRef]
51. Goto, S. Digestion method. In *Manual of Plant Nutrition*; Editorial Committee of Methods for Experiments in Plant Nutrition. Hakuyusha: Tokyo, Japan, 1990; pp. 125–128. (In Japanese)
52. Valentine, D.W.; Allen, H.L. Foliar responses to fertilization identify nutrient limitation in loblolly pine. *Can. J. For. Res.* **1990**, *20*, 144–151. [CrossRef]
53. Scagel, C.F. Growth and nutrient use of ericaceous plants grown in media amended with sphagnum moss peat or coir dust. *HortScience* **2003**, *38*, 46–54.
54. Evans, J.R. Photosynthesis and nitrogen relationships in leaves of C_3 plants. *Oecologia* **1989**, *78*, 9–19. [CrossRef]
55. Raaimakers, D.; Boot, R.G.A.; Dijkstra, P.; Pot, S.; Pons, T. Photosynthetic rates in relation to leaf phosphorus content in pioneer versus climax tropical rainforest trees. *Oecologia* **1995**, *102*, 120–125. [CrossRef]

56. Bown, H.E.; Watt, M.S.; Clinton, P.W.; Mason, E.G.; Richardson, B. Partitioning concurrent influences of nitrogen and phosphorus supply on photosynthetic model parameters of *Pinus radiata*. *Tree Physiol.* **2007**, *27*, 335–344. [CrossRef]
57. Parladé, J.; Peta, J.; Alvarez, I.F. Inoculation of containerized *Pseudotsuga menziesii* and *Pinus pinaster* seedlings with spores of five species of ectomycorrhizal fungi. *Mycorrhiza* **1996**, *6*, 237–245. [CrossRef]
58. Chen, Y.L.; Dell, B.; Malajczuk, N. Effects of *Scleroderma* spore density and age on mycorrhiza formation and growth of containerized *Eucalyptus globulus* and *E. urophylla* seedlings. *New For.* **2006**, *31*, 453–467. [CrossRef]
59. Lehto, T.; Zwiazek, J.J. Ectomycorrhizas and water relations of trees: A review. *Mycorrhiza* **2011**, *21*, 71–90. [CrossRef]
60. Morita, S.; Nagata, S. The causes and prevention of the leaf dying of the melon in the calcareous soil. *Jpn. J. Soil Sci. Plant Nutr.* **1995**, *66*, 217–222. (In Japanese with English summary)
61. Terai, A. The effects of organic fertilizer application on limestone-derived soils in Miyakojima Islands, Okinawa. *Ryukoku J. Econ. Stud.* **2008**, *47*, 125–138. (In Japanese)
62. Kayama, M. Concentrations of nutrients for trees grown on limestone region. *Ann. Rep. Kyushu Res. Cen. FFPRI* **2009**, *21*, 9–10. (In Japanese)
63. Kayama, M.; Aoki, N.; Hirayama, D.; Tateno, R.; Kawaji, M.; Yoneda, T. Growth characteristics of *Quercus miyagii* and *Q. glauca* var. *amamiana* seedlings planted on two types of soil. *Kyushu J. For. Res.* **2012**, *65*, 64–67. (In Japanese)
64. Maesako, S.; Yonemaru, S.; Sakai, M. Comparative studies of the physical and chemical properties of surface soils in a sugi plantation and it's two adjacent broad-leaved forest stands. *Bull. Kagoshima Prefect. For. Exp. Stn.* **2002**, *7*, 14–22. (In Japanese)
65. Sardans, J.; Rodà, F.; Peñuelas, J. Phosphorus limitation and competitive capacities of *Pinus halepensis* and *Quercus ilex* subsp. *rotundifolia* on different soils. *Plant Ecol.* **2004**, *174*, 305–317.
66. Pascual, S.; Olarieta, J.R.; Rodoríguez-Ochoa, R. Development of *Queucus ilex* is related to soil phosphorus availability on shallow calcareous soils. *New For.* **2012**, *43*, 805–814. [CrossRef]
67. Ito, S.; Ohtsuka, K.; Yamashita, T. Ecological distribution of seven evergreen *Quercus* species. *Veg. Sci.* **2007**, *24*, 53–63.
68. Miyawaki, A. *Vegetation of Japan, Kanto*; Shibundo: Tokyo, Japan, 1986; p. 641. (In Japanese with English summary)
69. Ewald, J. Ecological background of crown condition, growth and nutritional status of *Picea abies* (L.) Karst. In the Bavarian Alps. *Eur. J. For. Res.* **2005**, *124*, 9–18. [CrossRef]
70. Baier, R.; Ettl, R.; Hahn, C.; Göttlein, A. Early development and nutrition of Norway spruce (*Picea abies* (L.) Karst.) seedlings on different seedbeds in the Bavarian limestone alps—A bioassay. *Ann. For. Sci.* **2006**, *63*, 339–348. [CrossRef]
71. Volenec, J.J.; Ourry, A.; Joern, B.C. A role for nitrogen reserves in forage regrowth and stress tolerance. *Physiol. Plant.* **1996**, *97*, 185–193. [CrossRef]
72. Gloser, V. The consequences of lower nitrogen availability in autumn for internal nitrogen reserves and spring growth of *Calamagrostis epigejos*. *Plant Ecol.* **2005**, *179*, 119–126. [CrossRef]
73. Peri, P.L.; Gargaglione, V.; Pastur, G.M. Above- and belowground nutrients storage and biomass accumulation in marginal *Nothofagus antarctica* forests in southern Patagonia. *For. Ecol. Manag.* **2008**, *255*, 2502–2511. [CrossRef]

14

Leaf Nitrogen and Phosphorus Stoichiometry of *Cyclocarya paliurus* across China

Yang Liu [1,2], Qingliang Liu [1], Tongli Wang [2] and Shengzuo Fang [1,3,*]

1 College of Forestry, Nanjing Forestry University, Nanjing 210037, China; lyang_188@sina.com (Y.L.); zhuanshag@163.com (Q.L.)
2 Department of Forest and Conservation Sciences, University of British Columbia, 3041-2424 Main Mall, Vancouver, BC V6T 1Z4, Canada; tongli.wang@ubc.ca
3 Co-Innovation Center for Sustainable Forestry in Southern China, Nanjing Forestry University, Nanjing 210037, China
* Correspondence: fangsz@njfu.edu.cn

Abstract: Leaf stoichiometry (nitrogen (N), phosphorus (P) and N:P ratio) is not only important for studying nutrient composition in forests, but also reflects plant biochemical adaptation to geographic and climate conditions. However, patterns of leaf stoichiometry and controlling factors are still unclear for most species. In this study, we determined leaf N and P stoichiometry and their relationship with soil properties, geographic and climate variables for *Cyclocarya paliurus* based on a nation-wide dataset from 30 natural populations in China. The mean values of N and P concentrations and N:P ratios were 9.57 mg g^{-1}, 0.91 mg g^{-1} and 10.51, respectively, indicating that both leaf N and P concentrations in *C. paliurus* forests were lower than those of China and the global flora, and almost all populations were limited in N concentration. We found significant differences in leaf N and P concentrations and N:P ratios among the sampled *C. paliurus* populations. However, there were no significant correlations between soil properties (including organic C, total N and P concentrations) and leaf stoichiometry. The pattern of variation in leaf N concentration across the populations was positively correlated with latitude (24.46° N–32.42° N), but negatively correlated with mean annual temperature (MAT); meanwhile, leaf N concentration and N:P ratios were negatively correlated with mean temperature in January (MT_{min}) and mean annual frost-free period (MAF). Together, these results suggested that temperature-physiological stoichiometry with a latitudinal trend hold true at both global and regional levels. In addition, the relationships between leaf stoichiometry and climate variables provided information on how leaf stoichiometry of this species may respond to climate change.

Keywords: leaf stoichiometry; *Cyclocarya paliurus*; geographic variations; natural populations; climate variables; nitrogen; phosphorus; N:P ratio

---

1. Introduction

Patterns of leaf stoichiometry play a vital role in studying biological nutrient dynamics, biological symbiosis relationship, microbial nutrition, judgment of restrictive elements, consumer-driven nutrient cycle, and global C, N, P biogeochemical cycles [1–5]. The mechanisms of leaf stoichiometry in forests and their relationship to the environment conditions have attracted the attention of many scholars in recent years [6–10]. It is demonstrated that leaf stoichiometry is correlated with both geographic and climate variables such as latitude, temperature and precipitation, of which several hypotheses have been developed [11,12]. One famous hypothesis is the plant physiology hypothesis, which proposes that the developmental processes of plants are temperature sensitive, and plants will increase their nutrient concentrations (including leaf N and P) to compensate for the decreases in the growth rate that happen in lower-temperature or higher-latitude regions [6,11]. Another hypothesis is the biogeochemical hypothesis. This assumes that soil nutrient conditions, which are influenced by precipitation through leaching effects, drives the variation of plant nutrient (e.g., N, P concentrations) [13,14].

Previous studies of leaf stoichiometry at the global or regional levels have revealed a non-linear relation between leaf N concentration and climate factors [6,7,9,15,16]. Based on data across North America, Yin reported that leaf N in forests increased from boreal to temperate regions, and then decreased towards subtropical area [17]. Reich and Oleksyn's study at the global level also showed a similar pattern, where leaf N concentration increased from cold regions (mean annual temperature (MAT): −10 °C), peaked at temperate regions (MAT: 15 °C), and then tended to decrease in areas of high temperature (MAT: 30 °C) [6].

Recently, studies have been focused on the patterns of leaf stoichiometry in individual families [6], genus [9,18], and also species [14]. However, information about leaf stoichiometry at species level is still limited (e.g., among natural forests of a given species), and whether the mechanisms are consistent across different scales is unknown [19–21]. The relationships between leaf stoichiometry and climate factors have been found to differ among plants due to their dissimilarities caused by the ranges of different habitats. For example, Reich and Oleksyn reported that leaf N and P concentrations decreased with MAT in *Calamagrostis*, increased with MAT in birch (*Betula*), but showed a convex curve with MAT in maple (*Acer*) [6]. Wu et al. also demonstrated that leaf N and P concentrations of *Quercus* species across China decreased with mean temperature in January (MT_{min}) [9]. However, in Scots pine (*Pinus sylvestris* Linn.), leaf N concentration appears to decrease with latitude across Europe regions [22]. Thus, the response of leaf stoichiometry to climate change in plants may need to be differentiated by individual species.

Cyclocarya paliurus Batal. is a multiple function woody plant native to China, with a wide distribution from the warm temperate to the sub-tropical areas [23,24] and from the plain to highlands (e.g., up to about 2000 m in altitude in Guangxi province) [24]. Such a wide-range distribution (a broad gradient of both altitude and temperature) provides an opportunity to validate the hypothesis of temperature–plant physiological stoichiometry [25,26]. The nutritional status of *C. paliurus* in both plantations and natural forests has been investigated due to its importance in providing food and drug ingredient for the treatment of diabetes mellitus and hypertension [27–30].

In recent years, cultivation techniques including optimizing soil (NPK fertilizer used) and light environment have also been carried out to improve plant growth and yield of targeted health-promoting substances [31–34]. However, no information on the variation in leaf stoichiometry is available for *C. paliurus* forests or plantations. Therefore, our major objectives in this study were to determine the variation of leaf stoichiometry in *C. paliurus* sampled from different populations across China and the relationships between leaf stoichiometry of *C. paliurus* and geographic origin and climate factors. Soil organic C, total N and total P concentrations were analyzed at the same time to determine whether the differences of leaf stoichiometry were linked to the soil properties (concentrations). The findings from this study not only provide information for the characterization of the pattern of variation in leaf stoichiometry of this species, but also help to understand how leaf stoichiometry of *C. paliurus* populations may respond to climate change.

2. Materials and Methods

2.1. Study Areas and Materials

We investigated a total of 30 *C. paliurus* populations across the major distribution areas in 10 provinces of China (Figure 1). These populations range from 290 to 1798 m a.s.l. in altitude, 24.46° N–32.42° N in latitude and 103.78° E–121.79° E in longitude. Leaf sampling was carried out in September 2014, because leaf nutrients are relatively stable at this stage [9]. The longitude, latitude and altitude were measured by GPS on the spot. The climate data of the populations was obtained from ClimateAP for the historical period 1991–2014, including mean annual precipitation (MAP), mean annual temperature (MAT), mean temperature in July (MT_{max}), mean temperature in January (MT_{min}) and mean annual frost-free period (MAF) (Table 1, http://climateap.net/) [35,36]. The detailed method of sample collection and pre-treatment was described as Liu et al. (2018) [24]. Briefly, each sample consisted of about 400 g mature leaves collected from six to thirty average-sized trees (>20 years) in each population. All samples were dried to a constant weight at 70 °C and ground into fine powder in the lab. Soil samples (0–20 cm, n = 3) were collected from each location at the same time, sealed in polythene bags and brought back to laboratory.

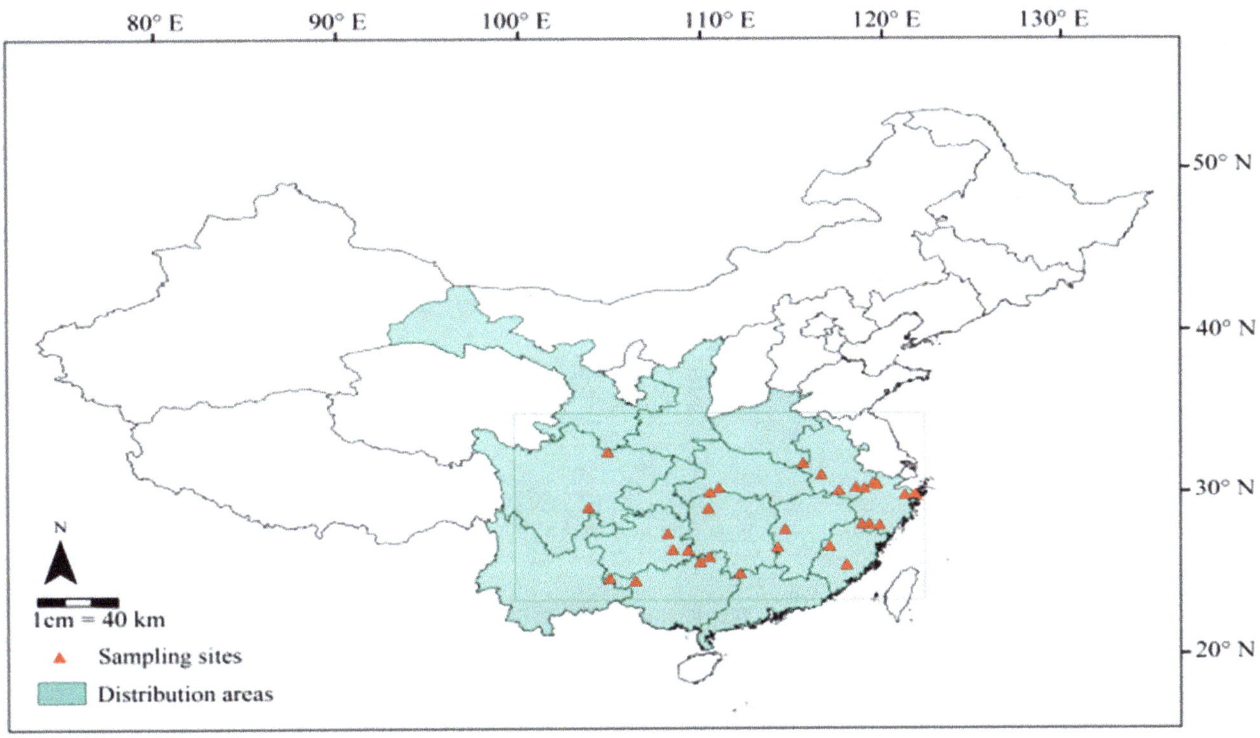

Figure 1. Natural distribution of *C. paliurus* (colored area in line box) and locations of 30 populations sampled (red triangle).

Table 1. List of the 30 sampled populations and their geographic and climate variables, as well as N and P concentrations and N:P ratios in the leaf of *C. paliurus*.

Sample ID	Locations	Longitude (°)	Latitude (°)	Altitude (m)	MAP (mm)	MAT (°C)	MT_{max} (°C)	MT_{min} (°C)	MAF (d)	Leaf N (mg g^{-1})	Leaf P (mg g^{-1})	N:P
HF	Hefeng, Hubei	110.42	29.88	1150	1669	12.7	23.3	1.3	292	8.43 ± 1.38 h~n	0.96 ± 0.03 fg	8.78 ± 1.21 j~o
WF	Wufeng, Hubei	110.90	30.19	959	1434	13.4	24.2	2	298	9.57 ± 0.46 e~i	0.93 ± 0.13 f~h	10.40 ± 1.40 f~k
LS	Longsheng, Guangxi	109.89	25.62	606	1704	14.7	25	3.6	320	8.58 ± 0.49 g~m	0.79 ± 0.01 ij	10.88 ± 0.65 e~h
ZY	Ziyuan, Guangxi	110.38	25.92	820	1915	13.4	23.8	2.3	300	10.47 ± 1.11 c~f	1.07 ± 0.03 cde	9.77 ± 1.04 g~n
BS	Baise, Guangxi	106.34	24.46	1798	1850	15.9	22.6	7.4	352	7.76 ± 0.22 i~o	0.86 ± 0.01 hi	9.04 ± 0.35 h~n
JZS	Jinzhongshan, Guangxi	104.95	24.61	1798	1568	16.2	22.2	7.7	351	9.41 ± 1.18 f~j	1.00 ± 0.01 ef	9.40 ± 1.08 g~n
JGS	Jinggangshan, Jiangxi	114.10	26.51	970	2001	13.8	23.1	3.3	312	7.75 ± 0.82 i~o	1.10 ± 0.04 a~d	7.04 ± 0.65 o
FY	Fenyi, Jiangxi	114.53	27.63	450	1834	17.4	27.6	6.4	343	8.97 ± 0.67 f~l	1.13 ± 0.01 abc	7.92 ± 0.56 no
GNJ	Guniujiang, Anhui	117.53	30.02	1100	2371	14.4	25.7	1.7	306	11.98 ± 1.04 c	0.76 ± 0.01 j	15.81 ± 1.44 b
JD	Jingde, Anhui	118.45	30.23	610	1777	15.2	27	2.4	312	14.48 ± 0.75 ab	1.18 ± 0.01 a	12.32 ± 0.56 cde
SC	Sucheng, Anhui	116.54	31.02	769	1679	13.6	25.4	1.1	275	13.58 ± 1.91 b	1.01 ± 0.06 ef	13.53 ± 1.16 c
QLF	Qingliangfeng, Anhui	118.89	30.15	680	1691	13.2	24.4	0.9	289	14.40 ± 1.68 ab	0.86 ± 0.03 hi	16.77 ± 1.54 b
QC	Qingchuan, Sichuan	104.86	32.42	1383	1093	11.7	21.6	0.7	270	15.18 ± 1.16 a	1.16 ± 0.03 ab	13.09 ± 0.95 cd
MC	Muchuan, Sichuan	103.78	28.97	1100	1448	15.7	24.5	5.7	348	11.15 ± 0.59 cde	0.80 ± 0.09 ij	13.86 ± 1.32 c
LP	Liping, Guizhou	109.24	26.34	727	1322	15.9	25.8	5.1	339	9.21 ± 0.75 f~k	1.10 ± 0.02 bcd	8.38 ± 0.58 l~o
LGS	Leigongshan, Guizhou	108.38	26.37	1178	1574	15.3	24.1	4.8	341	9.89 ± 0.17 e~h	1.16 ± 0.03 ab	8.53 ± 0.07 k~o
BWS	Bawanshan, Guizhou	108.38	26.37	1156	1564	15.3	24.1	4.8	341	10.22 ± 0.39 d~g	0.95 ± 0.02 fg	10.77 ± 0.32 e~i
SQ	Shiqian, Guizhou	108.11	27.35	1239	1268	14.6	23.9	4	332	7.64 ± 0.16 k~o	0.94 ± 0.04 fg	8.11 ± 0.28 mno
YS	Yongshun, Hunan	110.33	28.88	680	1556	15.3	25.9	4	328	9.60 ± 0.28 e~i	0.90 ± 0.02 gh	10.62 ± 0.44 e~j
JH	Jianhe, Hunan	112.03	24.92	845	1740	16.2	26.8	4.8	333	7.31 ± 0.88 l~p	0.89 ± 0.02 gh	8.18 ± 0.95 mno
PC	Pucheng, Fujian	118.76	27.93	650	1921	16.8	26.3	6.8	342	6.67 ± 0.37 op	0.76 ± 0.02 j	8.75 ± 0.48 j~o
NML	Niumulin, Fujian	117.93	25.43	500	1705	18.5	26.4	9.5	354	7.11 ± 1.04 m~p	0.66 ± 0.03 k	10.75 ± 1.19 e~i
MX	Mingxi, Fujian	117.01	26.59	574	1812	17.6	26.5	7.4	340	7.91 ± 1.00 i~o	0.92 ± 0.02 gh	8.63 ± 0.98 k~o
SCH	Shangcheng, Henan	115.55	31.72	910	1906	13.4	25.7	0.7	259	8.35 ± 0.86 h~o	0.76 ± 0.03 j	10.92 ± 0.77 e~h
WC	Wencheng, Zhejiang	119.79	27.88	959	1911	15.3	24.5	6	337	5.89 ± 0.17 p	0.50 ± 0.07 l	11.74 ± 0.83 def
FH	Fenghua, Zhejiang	121.22	29.76	513	1641	15.7	27.3	4.1	329	6.82 ± 0.88 n~p	0.67 ± 0.08 k	10.21 ± 0.71 f~l
LQ	Longquan, Zhejiang	119.19	27.91	1216	2224	14.1	23.3	4.3	323	8.79 ± 0.34 f~m	0.79 ± 0.03 ij	11.13 ± 0.25 efg
NB	Ningbo, Zhejiang	121.79	29.80	290	1452	16.4	27.2	6	344	10.33 ± 0.77 def	1.04 ± 0.01 de	9.92 ± 0.80 f~m
MW	Meiwu, Zhejiang	119.64	30.41	774	1643	13.8	25	2.3	289	8.02 ± 0.11 i~o	0.90 ± 0.08 gh	8.96 ± 0.81 i~n
LWS	Laowangshan, Zhejiang	119.41	30.50	540	1555	14.6	26.4	2.6	297	11.55 ± 0.26 b~f	0.63 ± 0.02 k	18.39 ± 0.39 a

MAP: mean annual precipitation (mm), MAT: mean annual temperature (°C), MT_{max}: mean temperature in July (°C), MT_{min}: mean temperature in January (°C), MAF: mean annual frost-free period (d). Sample ID was defined from the abbreviation of the location name. Different letters indicate significant differences (p <0.05 by Tukey's test) between populations.

2.2. Measurements

All samples were sieved through a mesh screen (1 mm) before measurements of leaf N and P concentrations. Leaf N concentration was detected following the method of Wu et al., using an auto analyzer (Kjeltec 2300 Analyzer Unit, Foss Tecator, Hoganas, Sweden) (n = 3) [9]. Leaf P concentration was measured according to the ammonium molybdate method described by the General Administration of Quality Supervisionin of China (reference code: GBW08513) (n = 3). Soil samples were air dried at 70 °C, grounded, and then sieved through a 2-mm mesh before analysis. Soil organic C, total N and total P concentrations in each sample were calculated following the method of Jiao et al. [37].

2.3. Data Analysis

One-way analysis of variance (ANOVA) was conducted to detect the quantitative differences in leaf stoichiometry and soil properties among different populations followed by Tukey's multiple range tests. All data were expressed as means ± standard deviation (SD). Scatter plots were used to show the relationships between leaf N and P concentrations and N:P ratios and factors studied (including longitude, latitude, altitude, MAT, MT_{max}, MT_{min}, MAP, and MAF), with appropriate regression equations developed. Relationships between leaf stoichiometry and soil properties were evaluated using the Pearson's correlation analysis. All statistical analyses were performed by using SPSS 19.0 software (SPSS, Chicago, IL, USA).

3. Results

3.1. Variation in Leaf Stoichiometry among C. paliurus Populations

Significant differences were found in leaf N and P concentrations and N:P ratios among different populations of *C. paliurus* (Table 1). For the 30 populations studied, the mean values of N and P concentrations and N:P ratios were 9.57 mg g^{-1}, 0.91 mg g^{-1} and 10.51, respectively, with a range of 5.89 (Wencheng, Zhejiang)–15.18 (Qingchuan, Sichuan) mg g^{-1}, 0.50 (Wencheng, Zhejiang)–1.18 (Jingde, Anhui) mg g^{-1}, and 7.04 (Jinggangshan, Jiangxi)–18.39 (Laowangshan, Zhejiang) (Table 1), respectively. Moreover, leaf N concentrations were significantly correlated with leaf P concentrations (R^2 = 0.1815, p = 0.0189) (Figure 2).

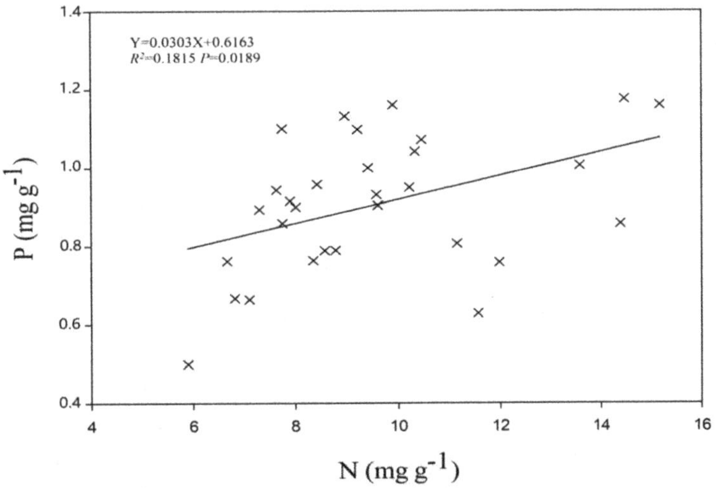

Figure 2. The relationship between leaf N and P concentrations for the 30 *C. paliurus* populations.

3.2. Soil Properties and Their Relationships with Leaf Stoichiometry

Soil organic C, total N and total P concentrations varied significantly among the natural populations, ranging from 24.04 (Jianhe, Hunan) to 75.82 (Leigongshan, Guizhou) mg g^{-1}, 1.94 (Yongshun, Hunan) to 5.84 (Leigongshan, Guizhou) mg g^{-1}, and 0.76 (Baise, Guangxi) to 3.52

(Shangcheng, Henan) mg g^{-1} (Table 2), respectively. Table 3 shows that there were no significant correlations between leaf stoichiometry and the soil properties studied (including soil C, N and P concentrations).

Table 2. Soil organic C, total N and total P concentrations for the 30 *C. paliurus* populations.

Sample ID	Locations	Organic C (mg g^{-1})	Total N (mg g^{-1})	Total P (mg g^{-1})
HF	Hefeng, Hubei	28.41 ± 1.01 ijk	2.07 ± 0.06 gh	1.45 ± 0.23 c~f
WF	Wufeng, Hubei	47.46 ± 6.79 d~h	2.72 ± 0.10 e~h	1.30 ± 0.24 d~g
LS	Longsheng, Guangxi	50.53 ± 1.74 c~f	4.21 ± 0.62 bcd	1.29 ± 0.13 d~g
ZY	Ziyuan, Guangxi	31.05 ± 4.62 ijk	3.80 ± 0.84 b~e	0.88 ± 0.08 efg
BS	Baise, Guangxi	51.33 ± 6.22 c~f	3.48 ± 0.21 c~g	0.76 ± 0.13 g
JZS	Jinzhongshan, Guangxi	52.54 ± 1.97 b~f	4.44 ± 0.82 abc	3.51 ± 0.97 a
JGS	Jinggangshan, Jiangxi	45.64 ± 3.57 e~h	4.94 ± 0.78 ab	1.67 ± 0.16 cde
FY	Fenyi, Jiangxi	45.57 ± 5.57 e~h	3.93 ± 0.75 b~e	1.13 ± 0.02 d~g
GNJ	Guniujiang, Anhui	36.03 ± 2.10 h~k	2.14 ± 0.16 gh	0.80 ± 0.16 fg
JD	Jingde, Anhui	37.49 ± 4.33 g~j	2.17 ± 0.26 fgh	1.83 ± 0.22 bcd
SC	Sucheng, Anhui	49.76 ± 1.43 c~g	4.13 ± 0.27 b~e	2.18 ± 0.16 bc
QLF	Qingliangfeng, Anhui	sm	sm	sm
QC	Qingchuan, Sichuan	31.55 ± 4.91 ijk	2.79 ± 0.24 d~h	0.95 ± 0.05 efg
MC	Muchuan, Sichuan	35.73 ± 2.11 h~k	2.67 ± 0.17 e~h	0.80 ± 0.19 fg
LP	Liping, Guizhou	58.22 ± 3.91 bcd	3.90 ± 0.58 b~e	1.57 ± 0.10 c~f
LGS	Leigongshan, Guizhou	75.82 ± 5.73 a	5.84 ± 0.66a	1.37 ± 0.24 d~g
BWS	Bawanshan, Guizhou	56.21 ± 2.39 b~e	3.98 ± 0.42 b~e	1.23 ± 0.06 d~g
SQ	Shiqian, Guizhou	54.58 ± 2.27 b~e	3.80 ± 0.38 b~e	1.86 ± 0.25 bcd
YS	Yongshun, Hunan	25.99 ± 2.52 jk	1.94 ± 0.19 h	0.96 ± 0.10 efg
JH	Jianhe, Hunan	24.04 ± 0.52 k	1.97 ± 0.23 h	1.49 ± 0.29 c~f
PC	Pucheng, Fujian	47.28 ± 1.40 d~h	3.92 ± 0.25 b~e	1.17 ± 0.15 d~g
NML	Niumulin, Fujian	40.55 ± 5.23 f~i	3.34 ± 0.13 c~h	0.92 ± 0.22 efg
MX	Mingxi, Fujian	55.25 ± 3.32 b~e	4.12 ± 0.24 b~e	1.12 ± 0.05 d~g
SCH	Shangcheng, Henan	61.30 ± 3.38 bc	4.64 ± 0.16 abc	3.52 ± 0.20 a
WC	Wencheng, Zhejiang	64.82 ± 4.20 ab	3.45 ± 0.23 c~g	0.85 ± 0.03 fg
FH	Fenghua, Zhejiang	47.97 ± 2.28 d~h	4.05 ± 0.75 b~e	2.57 ± 0.25 b
LQ	Longquan, Zhejiang	36.21 ± 1.20 h~k	2.71 ± 0.07 e~h	0.79 ± 0.06 fg
NB	Ningbo, Zhejiang	58.83 ± 5.47 bcd	4.23 ± 0.50 bcd	1.18 ± 0.19 d~g
MW	Meiwu, Zhejiang	64.10 ± 0.86 ab	4.79 ± 0.57 abc	2.17 ± 0.17 bc
LWS	Laowangshan, Zhejiang	61.36 ± 2.41 bc	3.60 ± 0.46 b~f	1.39 ± 0.02 def

Note: QLF sample missing (sm). Sample ID was defined from the abbreviation of the location name. Different letters indicate significant differences (p <0.05 by Tukey's test) between populations.

Table 3. Pearson correlation coefficients between leaf stoichiometry and soil properties (n = 90).

Leaf Stoichiometry	Soil Properties		
	Organic C (mg g^{-1})	Total N (mg g^{-1})	Total P (mg g^{-1})
Leaf N (mg g^{-1})	−0.222	−0.245	−0.063
Leaf P (mg g^{-1})	−0.102	0.152	0.065
Leaf N:P	−0.073	−0.337	−0.140

3.3. *Leaf Stoichiometry in Relation to Geographic and Climate Variables*

Leaf stoichiometry of *C. paliurus populations* was significantly related to both geographic and climate variables of their origins. Leaf N concentration was positively correlated with latitude (R^2 = 0.2747 and p = 0.0030 for the linear fit) (Figure 3A), but negatively correlated with mean annual temperature (MAT) (R^2 = 0.2130 and p = 0.0103 for the linear fit) (Figure 4A), mean temperature in January (MT_{min}) (R^2 = 0.3077 and p = 0.0015 for the linear fit) (Figure 4C), and mean annual frost-free period (MAF) (R^2 = 0.2319 and p = 0.0071 for the linear fit) (Figure 5B). Leaf P concentration was negatively related to longitude (R^2 = 0.1660 and p = 0.0255 for the linear fit) (Figure 3E). Leaf N:P ratios were positively correlated with latitude (R^2 = 0.2921 and p = 0.0020 for linear fit) (Figure 3G), mean temperature in January (MT_{min}) (R^2 = 0.1741 and p = 0.0218 for the linear fit) (Figure 4I), and mean annual frost-free period (MAF) (R^2 = 0.1655 and p = 0.0257 for the linear fit) (Figure 5F).

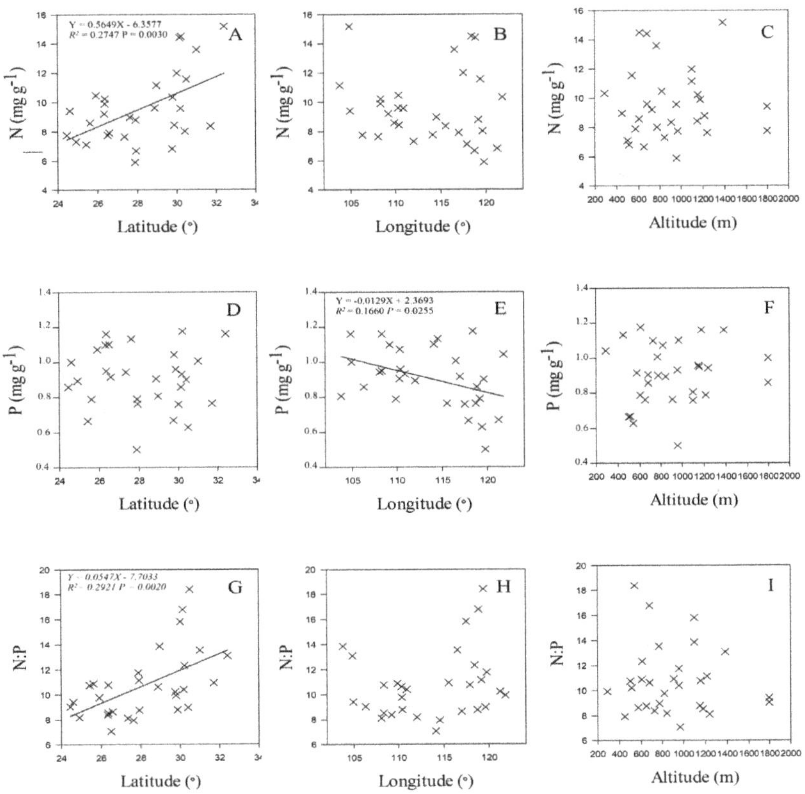

Figure 3. Relationships between leaf N (**A–C**), P (**D–F**), N:P ratio (**G–I**) and geographic origin (latitude, longitude and altitude) of *C. paliurus* populations.

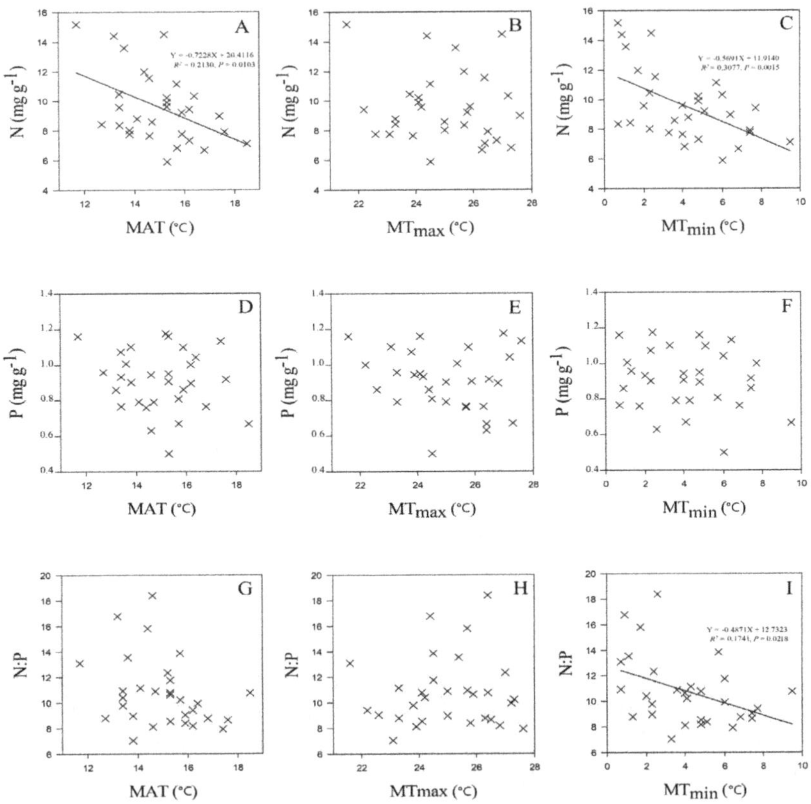

Figure 4. Relationships between leaf N (**A–C**), P (**D–F**), N:P ratio (**G–I**) and temperature (MAT: mean annual temperature, MT_{max}: mean temperature in July, and MT_{min}: mean temperature in January) for *C. paliurus* populations.

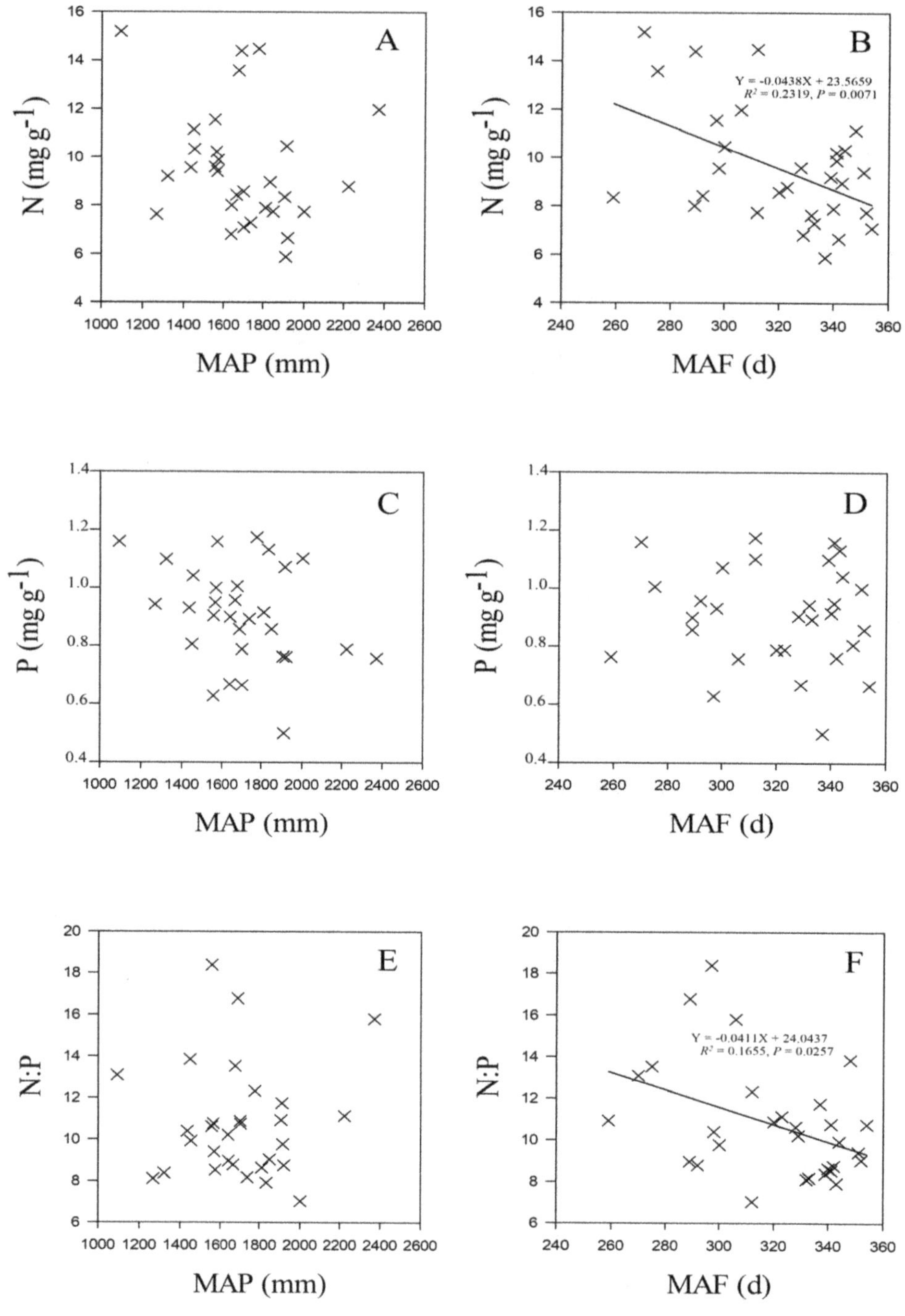

Figure 5. Relationships between leaf N (**A**,**B**), P (**C**,**D**), N:P ratio (**E**,**F**) and mean annual precipitation (MAP), and mean annual frost-free period (MAF) of *C. paliurus* populations.

4. Discussion

4.1. Patterns of Leaf Stoichiometry of C. paliurus across China

The ranges of both leaf N and P concentrations and N:P ratio of *C. paliurus* populations were smaller than those species of Juglandaceae family that have been studied previously (*Carya cathayensis* Sarg. in China and *Juglans nigra* Linn. in the USA) [38,39], and are also smaller than those of China and the global flora (Table 4) [6,7]. Leaf mean N concentration was also smaller than that of China and the global flora, which might be due to the relatively lower nitrogen resorption capacity in woody plants than that of herbs [7,40]. Previous studies have reported that leaf P concentration in China was much lower than that at the global level, which was confirmed in this study [6,7]. Lower leaf P concentrations in China have been demonstrated to be related to the lower soil available P concentration (3.83 mg kg^{-1} for China vs 7.65 mg kg^{-1} for the globe) [7,41,42]. Thus, soil conditions in *C. paliurus* populations, especially soil available P, needs to be further studied to confirm this internal relationship. Leaf N concentrations were significantly and positively correlated with leaf P concentrations in *C. paliurus* populations, which was inconsistent with previous studies [43]. This is mainly because N and P elements have similar biochemical pathways in plants, of which both are necessary in the basic biological processes such as protein synthesis, cell division and photosynthesis [44].

Table 4. Leaf N and P concentrations and N:P ratio for *C. paliurus* in the present study and other species of Juglandaceae family studied for China's and the global flora.

Data Source	Leaf N (mg g^{-1})		Leaf P (mg g^{-1})		N:P		References
	Range	Mean	Range	Mean	Range	Mean	
Present study	5.89–15.18	9.57	0.50–1.18	0.91	7.04–18.39	10.51	
Carya cathayensis Sarg. in China	13.10–24.65	19.92	1.24–2.70	1.75	8.83–15.31	11.92	ref. [38]
Juglans nigra Linn. in the USA	-	15.90	-	1.70	-	-	ref. [39]
China's flora	6.25–52.61	20.24	0.05–10.27	1.45	3.28–78.89	16.35	ref. [7]
The global flora	4.10–59.9	20.10	0.10–6.99	1.77	2.60–111.80	13.80	ref. [6]

Leaf N:P is an important index that reflects the limitation of nutrition in plants and ecosystem. For different plants and vegetation in different regions, the threshold of element restriction shown by N:P value is always distinct [9,45]. Based on the effects of N:P ratios on chlorophyll yield, Koerselman and Meuleman have proposed that a leaf N:P <14 indicates N limitation, and leaf N:P >16 indicates P limitation in plants [46]. The mean value of leaf N:P (10.51) over all *C. paliurus* populations suggested that in general *C. paliurus* forests were limited by N supply. However the LWS (Laowangshan) site had an leaf N:P ration of 18.39, which may indicate that it was limited by the supply of P although the concentration of total P in soil was not low compared to other sites (Table 2). The unusually high N:P ration at the LWS site might be due to the differences of plant growth rate caused by microclimate in *C. paliurus* forests.

4.2. *Leaf N and P Stoichiometry in Relation to Soil Properties, Geographic Origins and Climate Factors*

The results from our study showed that leaf stoichiometry of *C. paliurus* populations displayed significant relationships with geographic and climate factors. Meanwhile, results indicated that there were no significant correlations between leaf stoichiometry and the soil properties studied (including soil C, N and P concentrations). Our findings were quite different from those of Aerts and Chapin; they demonstrated that there was a significant correlation between leaf P and soil P concentrations [47]. This might be due to the differences in the species studied and their geographical distributions. Leaf N concentration increased with increasing latitude and decreasing MAT (Figures 3 and 4), which followed similar trends found in China's and the global flora [6,7,43]. Our study further demonstrated that leaf N concentration and N:P ratios increased with decreasing mean temperature in January (MT_{min}) (Figure 4C) and mean annual frost-free period (MAF) (Figure 5B). Together, these results confirmed that temperature-physiological stoichiometry with a latitudinal trend hold true at both global and regional levels [6]. Plants from colder habitats usually have higher values of leaf nutrient (N and P concentrations included) [11], which has been considered an adaptation mechanism to enhance their growth rates under lower temperatures [48,49].

A longitude-P relationship was also observed in *C. paliurus* populations across China (Figure 3E). Our study indicated that leaf P concentration increased with decreasing longitude. This result confirmed the biogeochemical hypothesis, which assumes that soil nutrient conditions, which are affected by leaching effects, drive the variation of plant nutrients [14]. Soil P concentration in China has been reported to decrease from northwest to southeast due to the variation of water distribution caused by precipitation [50,51]. Leaf N and P concentrations have been found strongly correlated with MAP at the global level [12]. However, a weak relationship between leaf P concentration and MAP was found in our study, which might be due to the small range of MAP (1093–2371 mm) in *C. paliurus* populations. Wu et al. also demonstrated that leaf N concentration and N:P ratios of *Quercus acutissima* Caruth. increased with increasing longitude in China [52]; however, this was not observed in our study. This difference further reveals that the responses of leaf stoichiometry to climate factors in plants are species specific. Overall, these results indicated that variation in leaf stoichiometry for a given species was more complicated, and these differences were mainly driven by mean annual temperature and water availability associated with geographic variation.

5. Conclusions

Our study demonstrated significant differences in leaf stoichiometry among *C. paliurus* populations, with the pattern of variation being related to geographic and climate variables. The leaf N concentration showed an increasing trend with increasing latitude, decreasing MAT, MT_{min} and MAF, which supported the hypothesis of temperature-plant physiological stoichiometry. These findings help to understand how the leaf stoichiometry of this species responds to climate variables, which might be useful for predicting the impact of climate change on the leaf stoichiometry of this species in the future. Meanwhile, we found that leaf P concentration increased with decreasing longitude, which might be due to the soil P concentration affected by leaching effects. The relatively low value of mean leaf N:P for *C. paliurus* populations, compared to other species, also suggested *C. paliurus* forests were generally limited by N supply. However, more tests about the relationship between leaf stoichiometry, soil conditions and leaf morphology need to be carried out in the future.

Author Contributions: S.F. conceived and designed the experiments; Q.L. collected the leaf samples, and performed the experiments; Y.L. analyzed the data, and wrote the manuscript; S.F. and T.W. participated in writing the manuscript.

References

1. Allen, A.P.; Gillooly, J.F. Towards an integration of ecological stoichiometry and the metabolic theory of ecology to better understand nutrient cycling. *Ecol. Lett.* **2009**, *12*, 369–384. [CrossRef] [PubMed]
2. Elser, J.J.; Sterner, R.W.; Gorokhova, E.A.; Fagan, W.F.; Markow, T.A.; Cotner, J.B.; Weider, L.W. Biological stoichiometry from genes to ecosystems. *Ecol. Lett.* **2000**, *3*, 540–550. [CrossRef]
3. Venterink, H.O.; Güsewell, S. Competitive interactions between two meadow grasses under nitrogen and phosphorus limitation. *Funct. Ecol.* **2010**, *24*, 877–886. [CrossRef]
4. Pang, D.; Wang, G.; Li, G.; Sun, Y.; Liu, Y.; Zhou, J. Ecological stoichiometric characteristics of two typical plantations in the karst ecosystem of southwestern China. *Forests* **2018**, *9*, 56. [CrossRef]
5. Zhang, P.; Wang, H.; Wu, Q.; Yu, M.; Wu, T. Effect of wind on the relation of leaf N, P stoichiometry with leaf morphology in *Quercus* species. *Forests* **2018**, *9*, 110. [CrossRef]
6. Reich, P.B.; Oleksyn, J. Global patterns of plant leaf N and P in relation to temperature and latitude. *Proc. Natl. Acad. Sci.* **2004**, *101*, 11001–11006. [CrossRef] [PubMed]
7. Han, W.; Fang, J.; Guo, D.; Zhang, Y. Leaf nitrogen and phosphorus stoichiometry across 753 terrestrial plant species in China. *New Phytol.* **2005**, *168*, 377–385. [CrossRef] [PubMed]
8. Rentería, L.Y.; Jaramillo, V.J. Rainfall drives leaf traits and leaf nutrient resorption in a tropical dry forest in Mexico. *Oecologia* **2011**, *165*, 201–211. [CrossRef] [PubMed]
9. Wu, T.; Dong, Y.; Yu, M.; Wang, G.G.; Zeng, D.H. Leaf nitrogen and phosphorus stoichiometry of *Quercus* species across China. *For. Ecol. Manag.* **2012**, *284*, 116–123. [CrossRef]
10. Jiang, P.; Chen, Y.; Cao, Y. C: N: P stoichiometry and carbon storage in a naturally-regenerated secondary *Quercus variabilis* forest age sequence in the Qinling Mountains, China. *Forests* **2017**, *8*, 281. [CrossRef]
11. Chen, Y.; Han, W.; Tang, L.; Tang, Z.; Fang, J. Leaf nitrogen and phosphorus concentrations of woody plants differ in responses to climate, soil and plant growth form. *Ecography* **2013**, *36*, 178–184. [CrossRef]
12. Ordoñez, J.C.; Van Bodegom, P.M.; Witte, J.P.M.; Wright, I.J.; Reich, P.B.; Aerts, R. A global study of relationships between leaf traits, climate and soil measures of nutrient fertility. *Glob. Ecol. Biogeogr.* **2009**, *18*, 137–149. [CrossRef]

13. McGroddy, M.E.; Daufresne, T.; Hedin, L.O. Scaling of C:N:P stoichiometry in forests worldwide: Implications of terrestrial redfield-type ratios. *Ecology* **2004**, *85*, 2390–2401. [CrossRef]
14. Kang, H.; Zhuang, H.; Wu, L.; Liu, Q.; Shen, G.; Berg, B.; Liu, C. Variation in leaf nitrogen and phosphorus stoichiometry in *Picea abies* across Europe: An analysis based on local observations. *For. Ecol. Manag.* **2011**, *261*, 195–202. [CrossRef]
15. Sardans, J.; Rivas-Ubach, A.; Peñuelas, J. The C:N:P stoichiometry of organisms and ecosystems in a changing world: A review and perspectives. *Perspect. Plant Ecol.* **2012**, *14*, 33–47. [CrossRef]
16. Zhang, S.B.; Zhang, J.L.; Slik, J.F.; Cao, K.F. Leaf element concentrations of terrestrial plants across China are influenced by taxonomy and the environment. *Glob. Ecol. Biogeogr.* **2012**, *21*, 809–818. [CrossRef]
17. Yin, X. Variation in foliar nitrogen concentration by forest type and climatic gradients in North America. *Can. J. For. Res.* **1993**, *23*, 1587–1602. [CrossRef]
18. Gotelli, N.J.; Mouser, P.J.; Hudman, S.P.; Morales, S.E.; Ross, D.S.; Ellison, A.M. Geographic variation in nutrient availability, stoichiometry, and metal concentrations of plants and pore-water in ombrotrophic bogs in New England, USA. *Wetlands* **2008**, *28*, 827–840. [CrossRef]
19. Townsend, A.R.; Cleveland, C.C.; Asner, G.P.; Bustamante, M.M. Controls over foliar N:P ratios in tropical rain forests. *Ecology* **2007**, *88*, 107–118. [CrossRef]
20. Elser, J.J.; Fagan, W.F.; Kerkhoff, A.J.; Swenson, N.G.; Enquist, B.J. Biological stoichiometry of plant production: Metabolism, scaling and ecological response to global change. *New Phytol.* **2010**, *186*, 593–608. [CrossRef] [PubMed]
21. Hall, E.K.; Maixner, F.; Franklin, O.; Daims, H.; Richter, A.; Battin, T. Linking microbial and ecosystem ecology using ecological stoichiometry: a synthesis of conceptual and empirical approaches. *Ecosystems* **2011**, *14*, 261–273. [CrossRef]
22. Oleksyn, J.; Reich, P.B.; Zytkowiak, R.; Karolewski, P.; Tjoelker, M.G. Nutrient conservation increases with latitude of origin in European *Pinus sylvestris* populations. *Oecologia* **2003**, *136*, 220–235. [CrossRef] [PubMed]
23. Fang, S.; Wang, J.; Wei, Z.; Zhu, Z. Methods to break seed dormancy in *Cyclocarya paliurus* (Batal) Iljinskaja. *Sci. Hortic.* **2006**, *110*, 305–309. [CrossRef]
24. Liu, Y.; Fang, S.; Zhou, M.; Shang, X.; Yang, W.; Fu, X. Geographic variation in water-soluble polysaccharide content and antioxidant activities of *Cyclocarya paliurus* leaves. *Ind. Crop. Prod.* **2018**, *121*, 180–186. [CrossRef]
25. Ladanai, S.; Ågren, G.I. Temperature sensitivity of nitrogen productivity for Scots pine and Norway spruce. *Trees* **2004**, *18*, 312–319. [CrossRef]
26. Briceno-Elizondo, E.; Garcia-Gonzalo, J.; Peltola, H.; Matala, J.; Kellomäki, S. Sensitivity of growth of Scots pine, Norway spruce and silver birch to climate change and forest management in boreal conditions. *For. Ecol. Manag.* **2006**, *232*, 152–167. [CrossRef]
27. Liu, Y.; Cao, Y.; Fang, S.; Wang, T.; Yin, Z.; Shang, X.; Fu, X. Antidiabetic effect of *Cyclocarya paliurus* leaves depends on the contents of antihyperglycemic flavonoids and antihyperlipidemic triterpenoids. *Molecules* **2018**, *23*, 1042. [CrossRef]
28. Liu, Y.; Chen, P.; Zhou, M.; Wang, T.; Fang, S.; Shang, X.; Fu, X. Geographic variation in the chemical composition and antioxidant properties of phenolic compounds from *Cyclocarya paliurus* (Batal) Iljinskaja leaves. *Molecules* **2018**, *23*, 2440. [CrossRef]
29. Xie, J.H.; Dong, C.J.; Nie, S.P.; Li, F.; Wang, Z.J.; Shen, M.Y.; Xie, M.Y. Extraction, chemical composition and antioxidant activity of flavonoids from *Cyclocarya paliurus* (Batal.) Iljinskaja leaves. *Food Chem.* **2015**, *186*, 97–105. [CrossRef]
30. Fang, S.; Yang, W.; Chu, X.; Shang, X.; She, C.; Fu, X. Provenance and temporal variations in selected flavonoids in leaves of *Cyclocarya paliurus*. *Food Chem.* **2011**, *124*, 1382–1386. [CrossRef]
31. Deng, B.O.; Shang, X.; Fang, S.; Li, Q.; Fu, X.; Su, J. Integrated effects of light intensity and fertilization on growth and flavonoid accumulation in *Cyclocarya paliurus*. *J. Agric. Food Chem.* **2012**, *60*, 6286–6292. [CrossRef] [PubMed]
32. Liu, Y.; Fang, S.; Yang, W.; Shang, X.; Fu, X. Light quality affects flavonoid production and related gene expression in *Cyclocarya paliurus*. *J. Photochem. Photobiol. B Biol.* **2018**, *179*, 66–73. [CrossRef] [PubMed]
33. Liu, Y.; Qian, C.; Ding, S.; Shang, X.; Yang, W.; Fang, S. Effect of light regime and provenance on leaf characteristics, growth and flavonoid accumulation in *Cyclocarya paliurus* (Batal) Iljinskaja coppices. *Bot. Stud.* **2016**, *57*, 28. [CrossRef] [PubMed]
34. Liu, Y.; Wang, T.; Fang, S.; Zhou, M.; Qin, J. Responses of morphology, gas exchange, photochemical activity of photosystem II, and antioxidant balance in *Cyclocarya paliurus* to light spectra. *Front. Plant Sci.* **2018**, *9*, 1704. [CrossRef] [PubMed]

35. Wang, T.; Wang, G.; Innes, J.L.; Seely, B.; Chen, B. ClimateAP: An application for dynamic local downscaling of historical and future climate data in Asia Pacific. *Front. Agric. Sci. Eng.* **2017**, *4*, 448–458. [CrossRef]
36. Wang, T.; Campbell, E.M.; O'Neill, G.A.; Aitken, S.N. Projecting future distributions of ecosystem climate niches: Uncertainties and management applications. *For. Ecol. Manag.* **2012**, *279*, 128–140. [CrossRef]
37. Jiao, F.; Shi, X.R.; Han, F.P.; Yuan, Z.Y. Increasing aridity, temperature and soil pH induce soil CNP imbalance in grasslands. *Sci. Rep.* **2016**, *6*, 19601. [CrossRef] [PubMed]
38. Yan, D.L.; Huang, Y.J.; Jin, S.H.; Huang, J.Q. Temporal variation of C, N, P stoichiometric in functional organs rootlets, leaves of *Carya cathayensis* and forest soil (in Chinese). *J. Soil Water Conserv.* **2013**, *5*, 256–259.
39. Niklas, K.J.; Cobb, E.D. Biomass partitioning and leaf N, P-stoichiometry: comparisons between tree and herbaceous current-year shoots. *Plant Cell Environ.* **2006**, *29*, 2030–2042. [CrossRef] [PubMed]
40. Yuan, Z.Y.; Li, L.H.; Han, X.G.; Huang, J.H.; Jiang, G.M.; Wan, S.Q.; Chen, Q.S. Nitrogen resorption from senescing leaves in 28 plant species in a semi-arid region of northern China. *J. Arid Environ.* **2005**, *63*, 191–202. [CrossRef]
41. Batjes, N.H. *A Homogenized Soil Data File for Global Environmental Research: A Subset of FAO, ISRIC and NRCS Profiles (Version 1.0) (No. 95/10b)*; ISRIC: Wageningen, The Netherlands, 1995.
42. Han, W.X.; Fang, J.Y.; Reich, P.B.; Ian Woodward, F.; Wang, Z.H. Biogeography and variability of eleven mineral elements in plant leaves across gradients of climate, soil and plant functional type in China. *Ecol. Lett.* **2011**, *14*, 788–796. [CrossRef] [PubMed]
43. He, J.S.; Wang, L.; Flynn, D.F.; Wang, X.; Ma, W.; Fang, J. Leaf nitrogen: Phosphorus stoichiometry across Chinese grassland biomes. *Oecologia* **2008**, *155*, 301–310. [CrossRef] [PubMed]
44. Zhang, L.X.; Bai, Y.F.; Han, X.G. Application of N:P stoichiometry to ecology studies. *Acta Bot. Sin.* **2003**, *45*, 1009–1018.
45. Li, H.; Crabbe, M.J.C.; Xu, F.; Wang, W.; Niu, R.; Gao, X.; Chen, H. Seasonal variations in carbon, nitrogen and phosphorus concentrations and C:N:P stoichiometry in the leaves of differently aged *Larix principis-rupprechtii* Mayr. plantations. *Forests* **2017**, *8*, 373. [CrossRef]
46. Koerselman, W.; Meuleman, A.F. The vegetation N:P ratio: A new tool to detect the nature of nutrient limitation. *J. Appl. Ecol.* **1996**, *33*, 1441–1450. [CrossRef]
47. Aerts, R.; Chapin, F.S. The mineral nutrition of wild plants revisited: A reevaluation of processes and patterns. *Adv. Ecol. Res.* **2000**, *30*, 1–67.
48. Woods, H.A.; Makino, W.; Cotner, J.B.; Hobbie, S.E.; Harrison, J.F.; Acharya, K.; Elser, J.J. Temperature and the chemical composition of poikilothermic organisms. *Funct. Ecol.* **2003**, *17*, 237–245. [CrossRef]
49. Van de Waal, D.B.; Verschoor, A.M.; Verspagen, J.M.; van Donk, E.; Huisman, J. Climate-driven changes in the ecological stoichiometry of aquatic ecosystems. *Front. Ecol. Environ.* **2010**, *8*, 145–152. [CrossRef]
50. Zhang, C.; Tian, H.; Liu, J.; Wang, S.; Liu, M.; Pan, S.; Shi, X. Pools and distributions of soil phosphorus in China. *Glob. Biogeochem. Cycles* **2005**. [CrossRef]
51. Wendroth, O.; Vasquez, V.; Matocha, C.J. Field experimental approach to bromide leaching as affected by scale-specific rainfall characteristics. *Water Resour. Res.* **2011**. [CrossRef]
52. Wu, T.; Wang, G.G.; Wu, Q.; Cheng, X.; Yu, M.; Wang, W.; Yu, X. Patterns of leaf nitrogen and phosphorus stoichiometry among *Quercus acutissima* provenances across China. *Ecol. Complex.* **2014**, *17*, 32–39. [CrossRef]

15

The Effects of Fertilization on the Growth and Physiological Characteristics of *Ginkgo biloba* L.

Jing Guo, Yaqiong Wu, Bo Wang, Yan Lu, Fuliang Cao and Guibin Wang *

Nanjing Forestry University, Co-Innovation Center for Sustainable Forestry in Southern China, 159 Longpan Road, Nanjing 210037, China; guojingg0921@163.com (J.G.); ya_qiong@126.com (Y.W.); wangboo984@163.com (B.W.); 18260093866@163.com (Y.L.); fuliangcaonjfu@163.com (F.C.)

* Correspondence: guibinwang99@163.com

Academic Editors: Scott X. Chang and Xiangyang Sun

Abstract: *Ginkgo biloba* L. is one of the most extensively planted and productive commercial species in temperate areas around the world, but slow-growth is the most limiting factor for its utilization. Fertilization is one of the key technologies for high quality and high forest yield. To better understand the impacts of fertilization on *Ginkgo* productivity, the effects of fertilization treatments (single fertilizer and combined fertilizer) on growth, nutrient content in *Ginkgo* leaves, and photosynthesis characteristics were studied in a 10-year-old *Ginkgo* plantation over two years. The single factor experiments suggested that DBH (diameter at breast height), H (height), NSL (length of new shoots), and V (trunk volume) showed significant differences between the different levels of single nitrogen (N) or phosphate (P) fertilizer application. Orthogonal test results showed that the nine treatments all promoted the growth of *Ginkgo*, and the formula (N: 400 g·tree^{-1}, P: 200 g·tree^{-1}, potassium (K): 90 g·tree^{-1}) was the most effective. G_s (stomatal conductance) and P_n (net photosynthesis rate) showed significant differences between the different amounts of single N or P fertilizer application, while single K fertilizer only affected P_n. Combined N, P, and K fertilizer had significant promoting effects on C_i (intercellular CO_2 concentration), G_s and P_n. N and P contents in *Ginkgo* leaves showed significant differences between the different amounts of a single N fertilizer application. A single P fertilizer only improved foliar P contents in *Ginkgo* leaves. A single K fertilizer application improved N and K content in *Ginkgo* leaves. The effects of different N, P, and K fertilizer treatments on the nutrient content of *Ginkgo* leaves were different.

Keywords: *Ginkgo*; fertilization; growth; photosynthesis; nutrient content

---

1. Introduction

Nutrient limitations develop when a stand's potential nutrient use cannot be met by the soil nutrient supply [1]. Improving stand nutrient supply through fertilization is a viable silvicultural option [2]. As an intervention strategy, large-scale application of fertilizers to forests has been implemented to accelerate growth of existing stands and shorten rotation times to overcome future projected timber shortfalls [3]. It is well established that fertilization, in particular nitrogen (N) and phosphate (P) applications, on nutrient limited sites increases tree productivity by increasing photosynthesis in the short term [4,5] and leaf area over the long term [6]. N is a vital constituent in proteins, nucleic acids, chlorophylls, and many secondary metabolites of plants [7]; therefore, it plays an essential role in the enzymatic activities of photosynthetic processes [8]. P plays a central role in almost all aspects of plant metabolism and is one of the nutrients that most commonly limits growth [9]. Potassium (K) is one of important nutrient elements and plays a major role in growth, modifying abundant enzyme activations and controlling cell osmoregulation [10]. Gross primary production (GPP) was also greatly increased by K fertilization as a result of lower stomatal and mesophyll resistance to CO_2 diffusion and higher photosynthetic capacity in the leaves [11,12]. In addition, potassium can also alleviate the harmful effects of abiotic stress [13–15].

The advantage of using chemical fertilizers is that nutrients are soluble and immediately available to the plants; therefore, the effect is usually direct and fast. Several studies have reported an increase in growing efficiency and higher enzymatic activities following fertilization [16,17]. Many studies have found that growing efficiency is largely unaffected by nutrient additions [18,19]. Different plants do not have the same nutrient demand and specific fertility targets [20]. The fast-growing broad-leaved tree *Populus alba* × *P. glandulosa* is more sensitive to increasing N availability [21], while the slow-growing *P. popularis* is more sensitive to decreasing N availability [22]. Among conifers, the fast-growing spruce families show more plasticity in biomass allocation than do the slowly growing ones under different nitrogen supplies [23], and the fast growing species *Pinus radiata* allocates more to the aboveground biomass under N and P fertilization [24].

Ginkgo is commonly referred to as Gongsun Tree or Duck Foot Tree, is one of the most ancient gymnosperms in the world, and is native to the temperate forests of China [25]. *Ginkgo* is an eco-economic tree species which is valuable for food, health care, medicine, timber, landscape, ecological protection, and scientific research. The great environmental adaptability and beauty of *Ginkgo* have made the tree a favorite species for planting as an ornamental tree throughout the world [26]. Its widespread use has been facilitated by its tolerance of air pollution and soil compaction. *Ginkgo* has a long history in leaf and nut production. However, it is still in the initial stage for timber cultivation in China. *Ginkgo* wood has also been highly valued for making furniture and handicraft articles, and *Ginkgo* wood is also an ideal material for making musical instruments [27]. Nowadays, *Ginkgo* wood is only used for carving and chopping blocks. The compression of *Ginkgo* wood was found to be stiffer than that of the other species [28]. Burgert et al. (2004) consequently concluded that there has been an evolutionary trend towards much more flexible compression wood [29]. Gong et al. (2009) reported higher lignin content in *Ginkgo* than in conifers [30]. *Ginkgo* wood was once used for pillars and rafters in the palaces and temples in ancient China. *Ginkgo* also has an anti-dust function, purifying the environment. One important role of *Ginkgo* is planting in rows in corridors, squares, or on both sides of a street as shade trees, as well as planting in gardens and vestibule entrances as landscape trees. *Ginkgo* wood has value for many purposes from the help of science and technology. In recent years, with the development of the *Ginkgo* industry, *Ginkgo* cultivation area continues to expand; many researchers have developed a variety of optimum fertilization schemes for *Ginkgo* plantations (especially leaf-producing plantations, nuts plantations, and seedlings) in China. However, studies concentrated on timber plantations are comparatively insufficient.

This study was conducted to examine the relationship between fertilization and growth, photosynthesis characteristics, and foliage nutrient content of pure *Ginkgo* timber plantations. The main objectives were to assess the effect of fertilization on growth, nutrient content in *Ginkgo* foliage, and photosynthesis characteristics of new *Ginkgo* shoots in a 10-year-old *Ginkgo* plantation over two years. Information generated from this study is expected to be of great value for providing optimal fertilization measures for improving the yield of *Ginkgo* timber forests and realizing balanced nutrient management for its fast growth. In addition, the application of a suitable fertilizer ratio after foliage nutrient content analysis, photosynthesis, and tree growth characteristics are some of the most effective ways to reduce cost and fertilizer waste.

2. Materials and Methods

2.1. The Study Site

The experimental field was located in Yellow Sea Forest Park, the east-central region of the Jiangsu Province, Eastern China (32°51′ N, 120°51′ E, 5 m above sea level). The study area belongs to a transition zone between a subtropical zone and warm temperate zone, with seasonal pluvial heat and significant monsoon activity. The study area is characterized by a mean annual temperature 15.0 °C, an annual rainfall of 1061.2 mm, an annual sunshine duration of 2209 h, and an annual frost-free period of 220 days. For determining the site fertility, soil samples from the experimental field were collected for physical and chemical characteristics before actual experiments. The soil was characterized as coastal sandy saline-alkali soil with a pH of 8.42, a bulk density of 1.28 $g \cdot cm^{-3}$, a total nitrogen of 0.75 $mg \cdot g^{-1}$, a total phosphorus of 0.26 $mg \cdot g^{-1}$, and a total potassium of 5.25 $mg \cdot g^{-1}$.

2.2. Materials and Experimental Design

This study was carried out in 2014 and 2015. The test forest was pure *Ginkgo* forest planted in 2005, with a vine space and row space of 2 m × 8 m. The trees came from the same variety. The experiment with different fertilizer treatments was designed based on a single factor design (EXP.1) and an orthogonal design (EXP.2). Fertilizer was applied to a depth of approximately 10–15 cm (to prevent losses to rain or wind) in several spots in a 70 cm radius circle around each tree in early April 2014 and early April 2015.

Single factor designs (unit: $g \cdot tree^{-1}$): EXP.1 was designed as single-factor experiment. The dosage of four treatments of single N fertilizer was 600, 400, 200, and 100, respectively. The dosage of four treatments of single P fertilizer was 800, 600, 400, and 200, respectively. The dosage of four treatments of single K fertilizer was 200, 90, 40, and 15, respectively. The amounts of fertilizers were in descending order. The fertilizers used in the experiments were urea (N), superphosphate (P), and potassium sulfate (K). There were a total of 13 treatments, and one of them was CK_1 (control, without fertilizer application). Each treatment had 10 *Ginkgo* trees with three replications in the trial. There were a total of 39 plots and 390 *Ginkgo* trees (average diameter at breast height was 13.37 cm, average height was 8.04 m, and their growth was uniform.) in the single factor tests. Additionally, there were two guard rows around the test plots.

Orthogonal designs (unit: $g \cdot tree^{-1}$): Second, the experiment with different fertilizer treatments (EXP.2) was designed based on orthogonal designs with three levels of urea, superphosphate, or potassium sulfate, for a total of 10 treatments one of them was CK_2 (control), without fertilizer application, which differed from the CK_1 above. Treatments were replicated three times (Table 1). There were a total of 30 plots and 300 *Ginkgo* trees (average diameter at breast height was 11.62 cm, average height was 5.78 m, and their growth was uniform) in the orthogonal tests. Orthogonal tests plots were in close proximity to the single-factor test spots.

Table 1. Orthogonal trial fertilizer treatments.

Treatment No.	Fertilization Level ($g \cdot tree^{-1}$)		
	Urea	Superphosphate	Potassium Sulfate
1	100	200	15
2	100	400	40
3	100	600	90
4	200	200	40
5	200	400	90
6	200	600	15
7	400	200	90
8	400	400	15
9	400	600	40
CK_2	0	0	0

CK_2: control in EXP.2.

2.3. Tree Growth Indictors' Measurement

Tree diameter at breast height (DBH), height (H), and length of new shoots (NSL) were measured in November 2015. Tree height was measured using a hypsometer (SGQ-1, Harbin, China). DBH and NSL were measured using diameter tape. We selected 21 *Ginkgo* trees of different diameter classes (2, 4, 6, 8, 10, 14, and 20) around the experiment plots, cut them down, and measured their DBH and H. Then, according to the standard volume equation and formula (Equation (1)):

$$V = a_0 D^{a1} H^{a2} \quad (1)$$

we used the R programming language (version 3.0, The University of Auckland, Auckland, New Zealand) software to obtain the trunk volume determination formula (Equation (2)):

$$LnV = 2.0135 \times lnD + 0.5685 \times lnH - 9.2452 \ (R^2 = 0.9819) \quad (2)$$

V: trunk volume (m^3), D: DBH (cm), H: height (m). We used this formula to calculate the *Ginkgo* trunk volume (V) on the test plots.

2.4. Determination of Photosynthetic Indexes

The photosynthetic characteristics were measured with a CIRAS-2 photosynthetic instrument (Hansatech, Norfolk, VA, USA). We selected sunny days in early August 2015 (08:30–11:30), and set the parameters: leaf temperature was not controlled, relative humidity was 85%, cylinder supply CO_2 concentration was 380 $\mu mol \cdot mol^{-1}$, and the light intensity of an artificial light source was 1200 $\mu mol \cdot m^{-2} \cdot s^{-1}$ (exceeding the *Ginkgo* light saturation point). Five trees were randomly selected from each replicate of each treatment on the test plots. Then, we selected mature healthy leaves of the fully-extended branches in the high position of the canopy. The net photosynthetic rate (P_n), transpiration rate (T_r), stomatal conductance (G_s), and intercellular CO_2 concentration (C_i) were measured.

2.5. Determination of Nutrient Concentration in Leaves

The sampling time was in August 2015. We selected five trees randomly from each replicate of each treatment in the test plots. We mined the new shoots in the middle part of the canopy with pole tree pruners. We collected the healthy 8–12 pieces of functional leaves without diseases and pest attack, and immediately put the leaves into an ice box, respectively. After the leaves were brought to the laboratory, the leaves were washed with deionized water, rapid fixed for 15 min at 105 °C, then dried to counterweigh at 70 °C, and, finally, crushed and passed through a 100-mesh sieve, severally (a total of 69 samples). Then, we weighed each sample to 0.1000 g and used concentrated H_2SO_4-$HClO_4$ to digest them. The nitrogen and phosphorus were analyzed with an AA3 continuous flow analytical system (Bran + Luebbe, Hamburg, Germany); potassium was analyzed via flame atomic absorption spectrometry (WFX-210, Beijing, China).

2.6. Statistical Analysis

Data are reported as the mean of three replicates ± standard deviation (SD), and all tests were performed using the Statistical Product and Service Solution statistical software program (IBM Inc., Chicago, IL, USA). One-way analysis of variance (ANOVA) was conducted to compare the effect of fertilization treatments on growth, photosynthetic indexes, and the element contents in leaves. ANOVA was performed at the $p < 0.05$ level of significance. Duncan's multiple-range test was performed for each variable (note: different lowercases in the table showed significant differences at the 0.05 level. Treatment means with different lowercases among treatments in the figure are significantly different at the 0.05 level).

3. Results

3.1. Effects of Fertilizer on Growth Indicators

In Yellow Sea Forest Park, DBH, H, NSL, and V in EXP.1 showed significant differences between levels of single N or P fertilizer applications, suggesting that N fertilizer application in excess of 100 $g \cdot tree^{-1}$ or P fertilizer application in excess of 400 $g \cdot tree^{-1}$ may be suitable for tree growth (Table 2; $p < 0.05$). However, growth responses were small and none of the variables showed a significant growth response to K fertilizer (Table 2; $p > 0.05$). Maximum dosage tended to have the greatest influences on the growth indicators for single N, P, and K fertilizers (N = 600 g·tree^{-1}, P = 800 g·tree^{-1}, or K = 200 g·tree^{-1}). After two years of continuous fertilization, a single N fertilizer application rate at 600 g·tree^{-1} resulted in an increase in V greater than 17.7% compared with unfertilized plots (CK_1). Meanwhile, a single P fertilizer application rate at 800 g·tree^{-1} resulted in an increase in V greater than 9.7% compared with unfertilized plots (CK_1). Therefore, single N or P fertilizer treatments promoted *Ginkgo* growth indicators significantly, while the promotion effect of single K fertilizer treatment was not significant. Additionally, orthogonal tests in EXP.2 showed that combined N, P, and K fertilizer had significant promoting effects on growth indictors (Table 3; $p < 0.05$); the maximum effect on the growth of *Ginkgo* was achieved at a high dosage of N. After two years of continuous fertilization, Treatment 7 resulted in an increase in V greater than 27.7% compared with unfertilized plots (CK_2).

Table 2. Effects of two years in EXP.1 of N, P, and K fertilizer (F) on the diameter at breast height (DBH, cm), height (H, m), length of new shoots (NSL, cm), and trunk volume (V, m^3) of *G. biloba* in 2014 and 2015 in Yellow Sea Forest Park in China.

F	DBH	H	NSL	V
N1 600	15.34 ± 0.396 a	9.85 ± 0.383 a	26.59 ± 1.204 a	0.0146 ± 0.000929 a
N2 400	15.11 ± 0.123 a	9.27 ± 0.492 b	25.56 ± 1.284 ab	0.0135 ± 0.000755 ab
N3 200	14.96 ± 0.301 a	9.17 ± 0.131 b	24.64 ± 2.840 ab	0.0133 ± 0.000448 b
N4 100	14.92 ± 0.407 a	9.14 ± 0.068 b	21.93 ± 2.635 bc	0.0132 ± 0.000369 b
CK_1 0	14.35 ± 0.095 b	8.92 ± 0.158 b	20.23 ± 2.594 c	0.0124 ± 0.000256 b
P1 800	14.84 ± 0.063 a	9.45 ± 0.257 a	25.98 ± 1.33 a	0.0136 ± 0.000398 a
P2 600	14.42 ± 0.255 ab	9.13 ± 0.221 ab	24.02 ± 1.86 ab	0.0127 ± 0.000408 b
P3 400	14.23 ± 0.333 b	9.05 ± 0.182 ab	24.37 ± 0.688 ab	0.0124 ± 0.000514 b
P4 200	14.21 ± 0.292 b	8.83 ± 0.270 b	22.69 ± 0.419 bc	0.0121 ± 0.000464 b
CK_1 0	14.35 ± 0.095 b	8.92 ± 0.158 b	20.23 ± 2.594 c	0.0124 ± 0.000256 b
K1 200	14.91 ± 0.499 a	9.01 ± 0.288 a	24.01 ± 0.844 a	0.0124 ± 0.000965 a
K2 90	14.65 ± 0.428 a	8.93 ± 0.195 a	23.71 ± 1.29 a	0.0117 ± 0.000445 a
K3 40	14.66 ± 0.553 a	8.88 ± 0.326 a	23.35 ± 2.14 a	0.0113 ± 0.000699 a
K4 15	14.38 ± 0.422 a	8.63 ± 0.459 a	24.49 ± 0.967 a	0.0114 ± 0.000993 a
CK_1 0	14.35 ± 0.095 a	8.92 ± 0.158 a	20.23 ± 2.59 a	0.0124 ± 0.000256 a

Note: different lowercase letters in the table reflect significant differences between treatments. CK_1: control in EXP.1.

Table 3. Effects of two years in EXP.2 of combined N, P, and K fertilizer (F) on the diameter at breast height (DBH, cm), height (H, m), length of new shoots (NSL, cm), and trunk volume (V, m^3) of *G. biloba* in 2014 and 2015 in Yellow Sea Forest Park in China.

F	DBH	H	NSL	V
1	13.66 ± 0.908 bcd	7.45 ± 0.022 bc	24.15 ± 0.832 bcd	0.00983 ± 0.000675 cd
2	13.78 ± 0.825 abcd	7.49 ± 0.065 bc	26.606 ± 3.250 abc	0.00997 ± 0.000679 cd
3	13.2 ± 0.172 cd	7.42 ± 0.168 bc	26.00 ± 2.32 abc	0.00947 ± 0.000327 de
4	13.71 ± 0.233 bcd	7.65 ± 0.123 ab	25.90 ± 1.921 abc	0.01013 ± 0.000207 cd
5	13.88 ± 0.173 abc	7.73 ± 0.293 ab	25.26 ± 1.705 abc	0.01036 ± 0.000261 bc
6	13.94 ± 0.370 abc	7.55 ± 0.046 b	22.83 ± 0.656 cd	0.01017 ± 0.000304 cd
7	14.61 ± 0.347 a	8.01 ± 0.239 a	26.77 ± 4.070 abc	0.01130 ± 0.000302 a
8	14.50 ± 0.423 ab	7.81 ± 0.173 ab	28.89 ± 2.014 a	0.01094 ± 0.000240 ab
9	14.47 ± 0.208 ab	7.81 ± 0.205 ab	27.49 ± 1.980 ab	0.01092 ± 0.000304 ab
CK_2	12.92 ± 0.231 d	7.09 ± 0.0346 c	20.23 ± 2.594 d	0.00885 ± 0.000161 e

Note: different lowercase letters in the table reflect significant differences between treatments.

3.2. Effects of Fertilizer on Photosynthesis Indicators

In Yellow Sea Forest Park, P_n and G_s showed significant differences between levels of single N or P fertilizer application ($p < 0.05$), while T_r and C_i were not affected by fertilizer application (Table 4; $p > 0.05$). However, photosynthesis indicators showed no significant response to single K fertilizer treatment except P_n (Table 4; $p > 0.05$). After two years of continuous fertilization, a single N fertilizer application rate at 600 g·tree^{-1} resulted in an increase in P_n greater than 49.5% compared with unfertilized plots (CK_1). Meanwhile, single P fertilizer application rate at 800 g·tree^{-1} resulted in an increase in P_n greater than 27.6% compared with unfertilized plots (CK_1). Therefore, single N or P fertilizer treatments promoted *Ginkgo* P_n significantly, while the promotion effect of single K fertilizer was also significant at 90 g·tree^{-1} (K2). Additionally, orthogonal tests showed that combined N, P, and K fertilizer had a significant promoting effect on P_n and G_s ($p < 0.05$); while fertilization was not so important on T_r and C_i (Table 5; $p > 0.05$) in EXP.2, after two years of continuous fertilization, Treatment 9 resulted in an increase in P_n greater than 36.9% compared with unfertilized plots (CK_2).

Table 4. Average effects of two years N, P, and K fertilizer (F) in EXP.1 on the photosynthesis characteristics, net photosynthetic rate (P_n, μmol·m^{-2}·s^{-1}), transpiration rate (T_r, μmol·m^{-2}·s^{-1}), stomatal conductance (G_s, mol·m^{-2}·s^{-1}), and intercellular CO_2 concentration (C_i, μmol·mol^{-1}) in 2014 and 2015 in Yellow Sea Forest Park in China.

F	C_i	T_r	G_s	P_n
N1 600	278.56 ± 8.57 a	2.71 ± 0.252 a	167.64 ± 3.32 a	12.30 ± 0.472 a
N2 400	277.33 ± 4.98 a	2.68 ± 0.150 a	165.78 ± 2.55 a	11.74 ± 0.597 ab
N3 200	269.78 ± 5.10 a	2.43 ± 0.173 a	157.78 ± 4.43 a	10.25 ± 1.42 b
N4 100	267.79 ± 16.65 a	2.51 ± 0.284 a	156.67 ± 7.86 a	10.57 ± 0.635 b
CK_1 0	267.33 ± 21.73 a	2.53 ± 0.208 a	127.33 ± 15.5 b	8.23 ± 0.833 c
P1 800	289.89 ± 3.29 a	3.03 ± 0.153 a	156.56 ± 5.34 a	10.50 ± 1.19 a
P2 600	275.11 ± 8.26 a	2.80 ± 0.152 a	139.78 ± 4.17 b	10.22 ± 1.06 ab
P3 400	266.89 ± 5.23 a	2.62 ± 0. 135 a	129.89 ± 8.06 b	8.77 ± 1.72 abc
P4 200	275.33 ± 19.50 a	2.67 ± 0.333 a	133.88 ± 6.02 b	7.14 ± 0.635 c
CK_1 0	267.33 ± 21.73 a	2.53 ± 0.208 a	127.33 ± 15.50 b	8.23 ± 0.833 bc
K1 200	256.89 ± 1.92 a	2.60 ± 0.101 a	148.22 ± 11.5 a	10.39 ± 0.27 a
K2 90	251.00 ± 4.58 a	2.54 ± 0.081 a	133.11 ± 10.98 a	10.44 ± 0.44 a
K3 40	265.78 ± 7.93 a	2.46 ± 0.102 a	123.67 ± 15.94 a	8.37 ± 0.62 b
K4 15	266.79 ± 9.25 a	2.27 ± 0.095 a	121.56 ± 16.60 a	7.61 ± 0.37 b
CK_1 0	267.33 ± 21.73 a	2.53 ± 0.208 a	127.33 ± 15.50 a	8.23 ± 0.83 b

Note: different lowercase letters in the table reflect significant differences between treatments.

Table 5. Average effects of two years of N, P, and K fertilizer (F) in EXP.2 on the photosynthesis characteristics, net photosynthetic rate (P_n, μmol·m^{-2}·s^{-1}), transpiration rate (T_r, μmol·m^{-2}·s^{-1}), stomatal conductance (G_s, mol·m^{-2}·s^{-1}), and intercellular CO_2 concentration (C_i, μmol·mol^{-1}) in 2014 and 2015 in Yellow Sea Forest Park in China.

F	C_i	T_r	G_s	P_n
1	239.00 ± 10.40 b	2.56 ± 0.34 a	122.33 ± 13.39 c	9.00 ± 0.91 cd
2	246.78 ± 27.03 b	2.41 ± 0.20 a	127.33 ± 13.00 bc	9.22 ± 0.74 bcd
3	243.00 ± 25.32 b	2.41 ± 0.47 a	136.11 ± 7.31 abc	9.36 ± 1.02 bcd
4	254.00 ± 8.09 b	2.61 ± 0.24 a	133.56 ± 9.52 bc	9.48 ± 0.17 bcd
5	247.33 ± 5.20 b	2.59 ± 0.34 a	141.34 ± 3.51 abc	9.71 ± 1.51 abcd
6	272.11 ± 17.88 ab	2.74 ± 0.15 a	140.11 ± 4.40 abc	10.11 ± 0.23 abc
7	271.78 ± 16.68 ab	2.69 ± 0.25 a	142.67 ± 3.51 abc	10.68 ± 1.22 abc
8	272.55 ± 20.51 ab	2.69 ± 0.08 a	146.89 ± 12.98 ab	10.71 ± 0.71 ab
9	290.33 ± 16.18 a	2.79 ± 0.16 a	154.33 ± 12.20 a	11.27 ± 0.38 a
CK_2	265.33 ± 18.73 ab	2.50 ± 0.11 a	129.13 ± 12.50 bc	8.53 ± 0.58 d

Note: different lowercase letters in the table reflect significant differences between treatments.

3.3. *Effects of Fertilizer on Leaf N, P, and K Contents*

In Yellow Sea Forest Park, N and P contents in leaves showed significant differences between levels of single N fertilizer application ($p < 0.05$), while K content was not affected by N fertilizer application (Figure 1; $p > 0.05$). N and K contents showed no significant response to single P fertilizer ($p > 0.05$), while P content showed a significant difference (Figure 2; $p < 0.05$). N and K contents showed significant differences between the different amounts of single K fertilizer application ($p < 0.05$), while P content was not affected by K fertilizer application (Figure 3; $p > 0.05$). After two years of continuous fertilization, a single N fertilizer application of 600 g·tree^{-1} resulted in an increase in N content greater than 16.4% and an increase in P content greater than 19.9% compared with unfertilized plots (CK_1). Meanwhile, a single P fertilizer application of 800 g·tree^{-1} resulted in an increase in P content greater than 23.8% compared with unfertilized plots (CK_1). A single K fertilizer application of 200 g·tree^{-1} resulted in an increase in N content greater than 12.2% and an increase in K content greater than 21.8% compared with unfertilized plots (CK_1). Therefore, single N fertilizer treatment promoted *Ginkgo* N and P content significantly, single P fertilizer treatment only promoted *Ginkgo* P content significantly, and the promotion effect of single K fertilizer treatment on N and K content were significant. Additionally, orthogonal tests in EXP.2 showed that combined N, P, and K fertilizer treatment had a significant promoting effect on N, P, and K content (Figure 4; $p < 0.05$). After two years of continuous fertilization, Treatment 9 resulted in an increase in N content greater than 13.4% compared with unfertilized plots (CK_2). Treatment 6 resulted in an increase in P content greater than 21.5% compared with unfertilized plots (CK_2). Treatment 7 resulted in an increase in K content greater than 16.7% compared with unfertilized plots (CK_2).

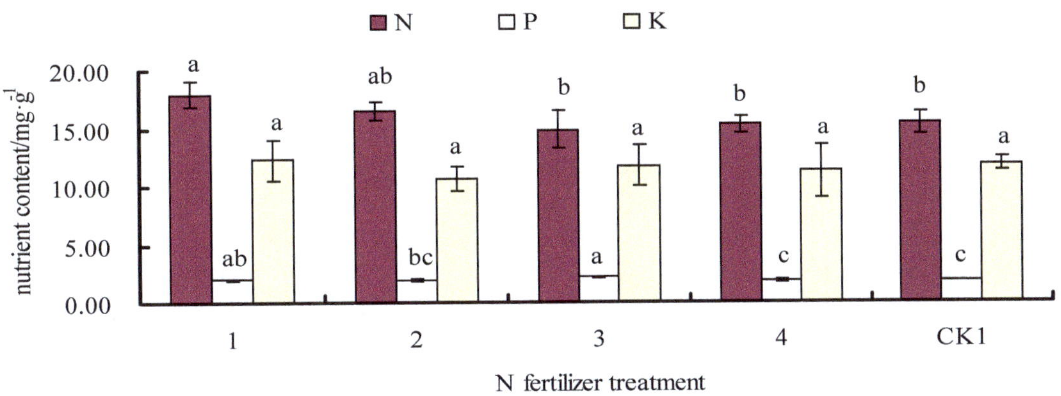

Figure 1. Effects of single N fertilizer application on *G. biloba* leaf N, P, and K contents in EXP.1.

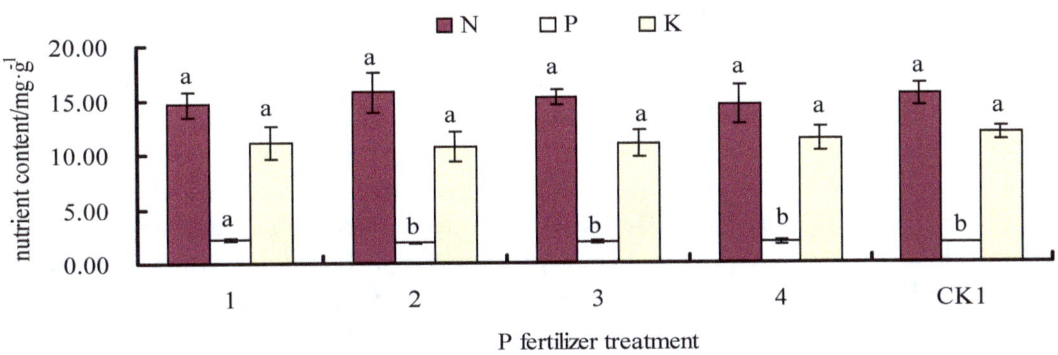

Figure 2. Effects of single P fertilizer application on *G. biloba* leaf N, P, and K contents in EXP.1.

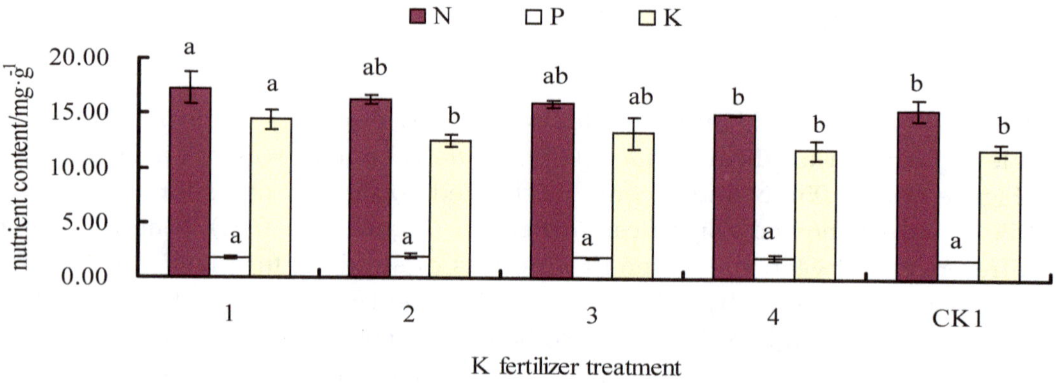

Figure 3. Effects of single K fertilizer application on *G. biloba* leaf N, P, and K contents in EXP.1.

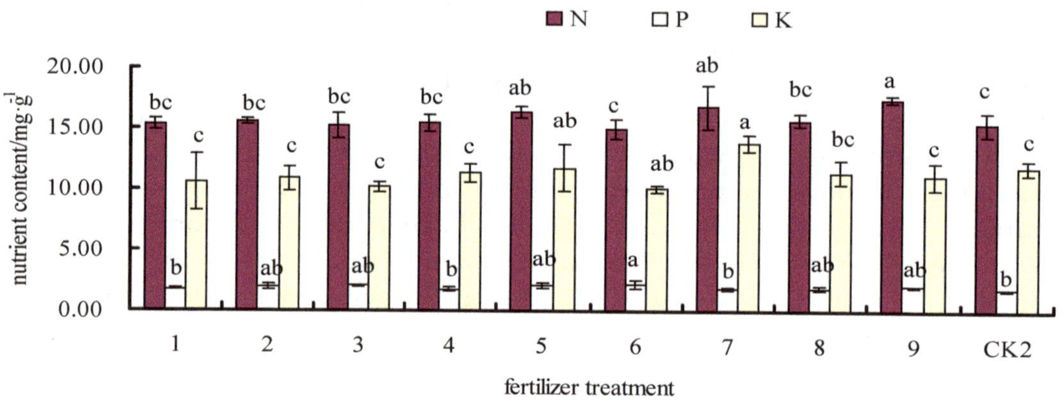

Figure 4. Effects of combined N, P, and K fertilizer application on *G. biloba* leaf N, P, and K contents in EXP.2.

4. Discussion

Nutrient deficiency is generally considered the major factor limiting growth, and fertilization is the main way to supplement the lack of nutrition. The fertilization effects have shown different results under different quantities of fertilizer, the number of years since fertilization treatment, and the nutrition status of the forest site [31]. Our study was carried out in 2014 and 2015. The single factor experiments suggested that DBH, H, NSL, and V showed significant differences between the different levels of single N or P fertilizer application. The orthogonal test results showed that the nine treatments all promoted the growth of *Ginkgo*, and Treatment 7 (N: 400 g·tree^{-1}, P: 200 g·tree^{-1}, K: 90 g·tree^{-1}) was the most effective one. The response to fertilization in *Ginkgo* plantations depended on the degree of mismatch between nutrient supply from the soil and the plant nutrient demand [32]. With an increment in the N fertilizer, there was also an increment in the growth, foliar nutrient status, and photosynthesis rate of the *Ginkgo* plantation. Phosphorus supply affected photosynthesis, leaf metabolites, and allocation to roots versus leaves and growth. In our study, the growth indicators improved with the P fertilizer dosage. However, excess P supply in the soil is a major environmental concern. The accumulation of P in the soil from applications of chemical fertilizer exceed the requirement of plant and can increase the risk of P movement to surface and groundwater [33]. Thus, we should be very circumspect in the application of P fertilizer. Many investigations have shown that improved N and P nutrition increased growth and yield in many tree species [34]. N and P deficiency cause negative effects on plants [35,36]. However, Gotore et al. (2014) proposed that single N application had negative effects on tree growth, while single P had positive effects on *Pinus patula* growth at Charter Forests. This may be because N is not a limiting factor to tree growth in the first few years after planting and may be readily available in the soil [19]. Additionally, Moilanen et al.

(2004) reported clearly that N fertilization is unnecessary on the nitrogen-rich sites, and additional N increased growth only at the most barren sites [37]. In our study, the promotion effect of a single K fertilizer on *Ginkgo* growth indicators was not significant, though it was better than the unfertilized plots. A previous study showed that fertilization with K was causing a decreased allocation to roots, which enables increased growth in height and leaf number [38]. Moreover, the viewpoint that an optimal potassium nutrition status can reduce the effects of abiotic stresses, such as drought, heat, high light intensity, or salinity has been well established [14,39,40]. The characteristic of our study site is coastal sandy saline–alkali soil; thus, K fertilizer is essential in Yellow Sea Forest Park. In general, N played a key role in the growth of *Ginkgo*, and P and K were important nutrient elements playing a major role in *Ginkgo* growth and development. According to the classification standard of soil nutrients in China, the analysis of soil elements in our experimental plots reveals that contents of N, P, and K are basically at low levels. Therefore, supplementary fertilization may be required on the experimental plots, particularly as the soil nutrient availability cannot meet the demand of the trees.

Photosynthesis is the basis of growth and yield formation, and the effects of nutrition on growth may be driven primarily by photosynthesis [41]. Fertilization can increase the photosynthetic efficiency, prolong the duration of photosynthesis, and enlarge the leaf area index (LAI). In the present study, G_s and P_n showed significant differences between the different levels of single N or P fertilizer applications, while single K fertilizer only affected P_n. Combined N, P, and K fertilizers had significant promoting effects on C_i, G_s, and P_n. N played a vital role in the enzymatic activities of photosynthetic processes, and N addition improved the *Ginkgo* photosynthesis rate and growth response. Among species there is often a strong positive correlation between maximum rates of photosynthesis and N [42]. Greater application of N fertilizer and greater allocation of N to metabolically active tissues, such as leaves and new stems, in the previous study promoted P_n by increasing LAI and chlorophyll concentration of leaves, which, in turn, supported better plant growth [43]. There was also a strong connection between P and the maximum rate of photosynthesis [44]; a partial explanation may lie in a positive relationship between the concentration of P and the amount of Rubisco. Inorganic phosphate (Pi) is a prerequisite for RuBP and also a competitive inhibitor of RUBP [45,46]. Additionally, there was a strong correlation between K concentrations and Rubisco [21]. Zorb et al. (2014) also reported that rates of photosynthesis were positively correlated with application rates of K [47]. K deficiency reduces photosynthesis by decreasing stomatal conductance [48]. However, the differences of stomatal conductance were not significant between different amounts of K fertilizer in our research. Possibly owing to the high mobility of K in both soils and plants, growth is expected to improve in response to added K, as shown in sandy soils [49]. Under this circumstance, deficiency of K may be so severe that fertilization is necessary to sustain tree growth until the end of rotation [50,51]. Transpiration and leaf water status are coordinated to minimize plant water loss and avoid hydraulic failure [52]. T_r of different levels of fertilization had no significant difference in our study. Probably because the climate characteristic of the site belongs to seasonal pluvial heat and significant monsoon, the rainfalls mainly occur in summer. Many studies have shown that growth of *Populus* was related to increasing water availability and fertilization on nutrient-limited soils [53,54]. Thus, sufficient rainfall and appropriate fertilization contributed to better growth in *Ginkgo* plantations in Yellow Sea Forest Park.

Nutrient concentrations of foliage have been accepted as adequate indicators of growth and soil fertility in forest stands [55,56]. Maintaining near-optimal foliar nutrient status of trees is especially critical to the maintenance of growth, the establishment of cold tolerance, and the evaluation of soil fertility at sites [56,57]. Our results suggested that N and P contents showed a significant influence of single N fertilizer application. Single P fertilizer only improved foliar P content, and high P fertilizer dosage had an inhibiting effect on K content. Single K fertilizer application improved N and K content in *Ginkgo* leaves. Effects of different combined N, P, and K fertilizer treatments on the nutrient content of *Ginkgo* leaves were different. In this study, we found that foliar N concentration was significantly correlated with N fertilizer application, which might result primarily from the supply status of soil nutrient elements. Moreover, foliar N concentration was significantly correlated with K fertilizer application, which might result from K-increased N use efficiency [58]. A previous study has showed that the maximum photosynthetic carbon assimilation rate positively correlated with foliar N concentrations [59]. A greater allocation of N to metabolically-active tissues, such as leaves and new stems in the fertilization treatment, increases the LAI and chlorophyll concentration of leaves, further supporting a better growth of *Ginkgo* plantations. In N-poor sites, particularly during cold growing seasons, the availability of N for trees may also be low and N deficiencies may be quite common [60,61]. Thus, it is an adaptation to store N in excess of current requirements for photosynthesis that can later be recycled and utilized during periods of lower N supply from the soil. P fertilization improved foliar P contents, and a high P fertilizer dosage had an inhibiting effect on K content. Our findings were consistent with the trend of a previous study of *Phoebe bournei* seedlings, where foliar N and P concentrations increased, but K concentration decreased as the phosphorus addition increased [62]. The results from Crous et al. (2008) showed a trend of increasing foliar K with an increase in the application of K when no phosphate fertilizer was applied. When P was applied with K on the plots with residual fertilizer, no such trend was observed [63]. Nakashgir (1992) reported nitrogen utilization of maize was accentuated when K application was supplemented [64]. We found that foliar K concentration was significantly correlated with K fertilizer application, which might owe to the high mobility of K in plants and soil. Additionally, light affects the internal redistribution of potassium, leading to higher concentrations in sun-exposed branches than in shaded ones [65,66]. The previous studies reported that potassium plays an important role in stomatal function by maintaining turgor pressure [67]. Growth is expected to improve in response to added K as shown in sandy soils [49]. According to the nutrient content contained in the *Ginkgo* leaves, the largest demand is N, followed by K, then P in this study. These results suggest that a new compound ratio of fertilizer is needed to optimize the growth of *Ginkgo* timber forest planted in coastal sandy saline–alkali soil because of different foliar N, P, and K concentrations under different fertilization treatments. According to the actual situation, we can suggest an optimal fertilization formula for improving the yield and quality of *Ginkgo* timber forests in Yellow Sea Forest Park.

5. Conclusions

The beneficial influences of single N, P, and K fertilization and combined N, P, and K fertilization on *Ginkgo* nutrition, photosynthesis, and growth were very obvious. This study confirmed the positive foliar nutrient content and the increased photosynthesis and growth responses with sufficient data, which indicated significantly better growth in fertilized stands compared to unfertilized stands. Single N, P, and K fertilizers and the combined N, P, and K fertilizer had positive effects on the growth of *Ginkgo*. Combined N, P, and K fertilizer and optional formulation are the better choices for achieving balanced nutrition of *Ginkgo* on our experimental plots. Sufficient nutrition supplements in barren places are significant to guarantee the growth of the forests. Efficient fertilizer regimes require an accurate knowledge of the nutrient status during the growth period to avoid any unforeseen deficiencies.

Acknowledgments: This study was supported by a Grant from the research program "Research on key technology of the cultivation of *Ginkgo* timber and medicinal forest" (2012BAD21B04) provided by National Science and technology support program, "Research and demonstration of sustainable management technology of *Ginkgo* timber forest in coastal beach" (BE2013443) provided by Science and technology support program of Jiangsu Province, the Priority Academy Program Development of Jiangsu Higher Education Institution (PAPD) and the Doctorate Fellowship Foundation of Nanjing Forestry University.

Author Contributions: Jing Guo and Yan Lu designed the study and conducted the field trial. Jing Guo, Yaqiong Wu, and Bo Wang were responsible for the samples collection and laboratory analysis. Jing Guo was responsible for the statistical analyses, with contributions from Fuliang Cao and Guibin Wang. Jing Guo wrote the paper.

References

1. Pathak, H.; Aggarwal, P.K.; Roetter, R.; Kalra, N.; Bandyopadhaya, S.K.; Prasad, S.; Keulch, H.V. Modelling the quantitative evaluation of soil nutrient supply, nutrient use efficiency, and fertilizer requirements of wheat in India. *Nutr. Cycl. Agroecosyst.* **2003**, *65*, 105–113. [CrossRef]
2. Fox, T.R.; Allen, H.L.; Albaugh, T.J.; Rubilar, R.; Carlson, C.A. Tree Nutrition and Forest Fertilization of Pine Plantations in the Southern United States. *South. J. Appl. For.* **2006**, *31*, 5–11.
3. Sophie, W.; Leigh, A.K.K.; Grayston, S.J. Effects of long-term fertilization of forest soils on potential nitrification and on the abundance and community structure of ammonia oxidizers and nitrite oxidizers. *FEMS Microbiol. Ecol.* **2011**, *79*, 142–154.
4. Gough, C.M.; Seiler, J.R.; Maier, C.A. Short-term effects of fertilization on loblolly pine (*Pinus taeda* L.) physiology. *Plant Cell Environ.* **2004**, *27*, 876–886. [CrossRef]
5. King, N.T.; Seiler, J.R.; Fox, T.R.; Johnsen, K.H. Post-fertilization physiology and growth performance of loblolly pine clones. *Tree Physiol.* **2008**, *28*, 703–711. [CrossRef] [PubMed]
6. Samuelson, L.J.; Butnor, J.; Maier, C.; Stokes, T.A.; Johnsen, K.; Kane, M. Growth and physiology of loblolly pine in response to long-term resource management: Defining growth potential in the southern United States. *Can. J. For. Res.* **2008**, *38*, 721–732. [CrossRef]
7. Luo, J.; Qin, J.J.; He, F.F.; Li, H.; Liu, T.X.; Polle, A.; Peng, C.H.; Luo, Z.B. Net fluxes of ammonium and nitrate in association with H^+ fluxes in fine roots of *Populus popularis*. *Planta* **2013**, *237*, 19–31. [CrossRef] [PubMed]
8. Güsewell, S. N:P ratios in terrestrial plants: Variation and functional significance. *New Phytol.* **2004**, *164*, 243–266. [CrossRef]
9. Warren, C.R. How does P affect photosynthesis and metabolite profiles of *Eucalyptus globulus*? *Tree Physiol.* **2011**, *31*, 727–739. [CrossRef] [PubMed]
10. Coskun, D.; Britto, D.T.; Kronzucker, H.J. The physiology of channel-mediated K^+, acquisition in roots of higher plants. *Physiol. Plant.* **2014**, *151*, 305–312. [CrossRef] [PubMed]
11. Battie-Laclau, P.; Laclau, J.P.; Beri, C.; Mietton, L.; Muniz, M.R.A.; Arenque, B.C.; Piccolo, M.C.; Jordan-Meille, L.; Bouillet, J.P.; Nouvellon, Y. Photosynthetic and anatomical responses of *Eucalyptus grandis* leaves to potassium and sodium supply in a field experiment. *Plant Cell Environ.* **2014**, *37*, 70–81. [CrossRef] [PubMed]
12. Christina, M.; Maire, G.L.; Battie-Laclau, P.; Nouvellon, Y.; Bouillet, J.P.; Jourdan, C.; Goncalves, J.L.M.; Laclau, J.P. Measured and modeled interactive effects of potassium deficiency and water deficit on gross primary productivity and light-use efficiency in *Eucalyptus grandis* plantations. *Glob. Chang. Biol.* **2015**, *21*, 2022–2039. [CrossRef] [PubMed]
13. Mäser, P.; Gierth, M.; Schroeder, J.I. Molecular mechanisms of potassium and sodium uptake in plants. *Plant Soil* **2002**, *247*, 43–54. [CrossRef]
14. Cakmak, I. The role of potassium in alleviating detrimental effects of abiotic stresses in plants. *J. Plant Nutr. Soil Sci.* **2005**, *168*, 521–530. [CrossRef]
15. Li, X.Y.; Mu, C.S.; Lin, J.X.; Wang, Y.; Li, X.J. Effect of alkaline potassium and sodium salts on growth, photosynthesis, ions absorption and solutes synthesis of wheat seedlings. *Exp. Agric.* **2014**, *50*, 144–157. [CrossRef]
16. Albaugh, T.J.; Allen, H.L.; Dougherty, P.M.; Kress, L.W.; King, J.S. Leaf area and above- and belowground growth responses of loblolly pine to nutrient and water additions. *For. Sci.* **1998**, *44*, 317–328.
17. Balster, N.J.; Marshall, J.D. Eight-year responses of light interception, effective leaf area index, and stemwood production in fertilized stands of interior Douglas-fir (*Pseudotsuga menziesii* var. *glauca*). *Can. J. For. Res.* **2000**, *30*, 733–743. [CrossRef]
18. Samuelson, L.; Stokes, T.; Cooksey, T. Production efficiency of loblolly pine and sweetgum in response to four years of intensive management. *Tree Physiol.* **2001**, *21*, 369–376. [CrossRef] [PubMed]
19. Gotore, T.; Murepa, R.; Gapare, W.J. Effects of Nitrogen, Phosphorus and Potassium on the Early Growth of *Pinus patula* and *Eucalyptus grandis*. *J. Trop. For. Sci.* **2014**, *26*, 22–31.
20. Salifu, K.F.; Jacobs, D.F.; Birge, Z.K.D. Nursery nitrogen loading improves field performance of bareroot oak seedlings planted on abandoned mine lands. *Restor. Ecol.* **2009**, *17*, 339–349. [CrossRef]

21. Li, H.; Li, M.; Luo, J.; Cao, X.; Qu, L.; Gai, Y.; Jiang, X.; Liu, T.; Bai, H.; Janz, D. N fertilization has different effects on the growth, carbon and nitrogen physiology, and wood properties of slow-and fast-growing *Populus* species. *J. Exp. Bot.* **2012**, *63*, 6173–6185. [CrossRef] [PubMed]
22. Luo, J.; Li, H.; Liu, T.; Polle, A.; Peng, C.; Luo, Z. Nitrogen metabolism of two contrasting poplar species during acclimation to limiting nitrogen availability. *J. Exp. Bot.* **2013**, *64*, 4207–4224. [CrossRef] [PubMed]
23. Miller, B.D.; Hawkins, B.J. Ammonium and nitrate uptake, nitrogen productivity and biomass allocation in interior spruce families with contrasting growth rates and mineral nutrient preconditioning. *Tree Physiol.* **2007**, *27*, 901–909. [CrossRef] [PubMed]
24. Bown, H.E.; Watt, M.S.; Clinton, P.W.; Mason, E.G.; Whitehead, D. The influence of N and P supply and genotype on carbon flux and partitioning in potted *Pinus radiata* plants. *Tree Physiol.* **2009**, *29*, 857–868. [CrossRef] [PubMed]
25. Son, Y. Effects of Nitrogen Fertilization on Foliar Nutrient Dynamics in *Ginkgo* Seedlings. *J. Plant Nutr.* **2002**, *25*, 93–102. [CrossRef]
26. Son, Y.; Kim, H.W. Above-ground biomass and nutrient distribution in a 15-year-old ginkgo (*Ginkgo* biloba) plantation in Central Korea. *Bioresour. Technol.* **1998**, *63*, 173–177. [CrossRef]
27. Yuan, J.; Li, Q.; Xiao, G.L.; Zhu, S.H. Discussion on the utilization and development of ginkgo wood. *China For. Sci. Technol.* **2008**, *16*, 6–8.
28. Andersson, S.; Wang, Y.; Ponni, R.; Hanninen, T.; Mononen, M.; Ren, H.; Serimaa, R.; Saranpää, P. Cellulose structure and lignin distribution in normal and compression wood of the Maidenhair tree (*Ginkgo biloba* L.). *J. Integr. Plant Biol.* **2015**, *57*, 388–395. [CrossRef] [PubMed]
29. Burgert, I.; Fruhmann, K.; Keckes, J.; Fratzl, P.; Stanzl-Tschegg, S. Structure-function relationships of four compression wood types: Micromechanical properties at the tissue and fibre level. *Trees* **2004**, *18*, 480–485. [CrossRef]
30. Gong, Q.L.; Hu, A.H.; Xing, S.Y.; Wang, F. Research on systematic evolution of *Ginkgo biloba* based on chemical composition of wood. *Spectrosc. Spectr. Anal.* **2009**, *29*, 1512–1516.
31. Moilanen, M.; Hökkä, H. The growth response of scots pine to PK fertilization depends on the nutrient status of the stand on drained peatlands. *Suo* **2009**, *60*, 111–119.
32. Smethurst, P.; Baillie, C.; Cherry, M.; Holz, G. Fertilizer effects on LAI and growth of four *Eucalyptus nitens* plantations. *For. Ecol. Manag.* **2003**, *176*, 531–542. [CrossRef]
33. Grant, C.; Bittman, S.; Montreal, M.; Plenchette, C.; Morel, C. Soil and fertilizer phosphorus: Effects on plant P supply and mycorrhizal development. *Can. J. Plant Sci.* **2005**, *85*, 3–14. [CrossRef]
34. Bussi, C.; Smith, M.A.L. Effects of nitrogen and potassium fertilization on the growth, yield and pitburn of apricot (cv. Bergeron). *J. Hortic. Sci. Biotechnol.* **1998**, *73*, 387–392. [CrossRef]
35. Boyce, R.L.; Larson, J.R.; Sanford, R.L. Phosphorus and nitrogen limitations to photosynthesis in Rocky Mountain bristlecone pine (*Pinus aristata*) in Colorado. *Tree Physiol.* **2006**, *26*, 1477–1486. [CrossRef] [PubMed]
36. Turnbull, T.L.; Warren, C.R.; Adams, M.A. Novel mannose-sequestration technique reveals variation in subcellular orthophosphate pools do not explain the effects of phosphorus nutrition on photosynthesis in *Eucalyptus globulus* seedlings. *New Phytol.* **2007**, *176*, 849–861. [CrossRef] [PubMed]
37. Moilanen, M.; Silfverberg, K.; Hökkä, H.; Issakainen, J. Comparing effects of wood ash and commercial PK fertiliser on the nutrient status and stand growth of Scots pine on drained mires. *Balt. For.* **2004**, *10*, 2–10.
38. Santiago, L.S.; Wright, S.J.; Harms, K.E.; Yavitt, J.B.; Korine, C.; Garcia, M.N.; Turner, B.L. Tropical tree seedling growth responses to nitrogen, phosphorus and potassium addition. *J. Ecol.* **2012**, *100*, 309–316. [CrossRef]
39. Oosterhuis, D.M.; Loka, D.A.; Raper, T.B. Potassium and stress alleviation: Physiological functions and management of cotton. *J. Plant Nutr. Soil Sci.* **2013**, *76*, 331–343. [CrossRef]
40. Römheld, V.; Kirkby, E.A. Research on potassium in agriculture: Needs and prospects. *Plant Soil* **2010**, *335*, 155–180. [CrossRef]
41. Wang, X.G.; Zhao, X.H.; Jiang, C.J.; Li, C.H.; Cong, S.; Wu, D.; Chen, Y.Q.; Yu, H.Q.; Wang, C.Y. Effects of potassium deficiency on photosynthesis and photo-protection mechanisms in soybean (*Glycine max* (L.) Merr.). *J. Integr. Agric.* **2015**, *14*, 856–863. [CrossRef]
42. Warren, C.R.; Adams, M.A.; Chen, Z.L. Is photosynthesis related to concentrations of nitrogen and Rubisco in leaves of Australian native plants? *Funct. Plant Biol.* **2000**, *27*, 407–416. [CrossRef]

43. Pokharel, P.; Chang, S.X. Exponential fertilization promotes seedling growth by increasing nitrogen retranslocation in trembling aspen planted for oil sands reclamation. *For. Ecol. Manag.* **2016**, *372*, 35–43. [CrossRef]
44. Warren, C.R.; Adams, M.A. Phosphorus affects growth and partitioning of nitrogen to Rubisco in *Pinus pinaster*. *Tree Physiol.* **2002**, *22*, 11–19. [CrossRef] [PubMed]
45. Kumagai, E.; Araki, T.; Kubota, F. Effects of nitrogen supply restriction on gas exchange and photosystem 2 function in flag leaves of a traditional low-yield cultivar and a recently improved high-yield cultivar of rice (*Oryza sativa* L.). *Photosynthetica* **2007**, *45*, 489–495. [CrossRef]
46. Zafar, M.; Abbasi, M.K.; Khaliq, A. Effect of combining organic materials with inorganic phosphorus sources on growth, yield, energy content and phosphorus uptake in maize at Rawalakot Azad Jammu and Kashmir, Pakistan. *Arch. Appl. Sci. Res.* **2011**, *3*, 199–212.
47. Zorb, C.; Senbayram, M.; Peiter, E. Potassium in agriculture: Status and perspectives. *J. Plant Physiol.* **2014**, *171*, 656–669. [CrossRef] [PubMed]
48. Terry, N.; Ulrich, A. Effects of phosphorus deficiency on the photosynthesis and respiration of leaves of sugar beet. *Plant Physiol.* **1973**, *51*, 43–47. [CrossRef] [PubMed]
49. Warren, C.R.; McGrath, J.F.; Adams, M.A. Differential effects of N, P and K on photosynthesis and partitioning of N in *Pinus pinaster* needles. *Ann. For. Sci.* **2005**, *62*, 1–8. [CrossRef]
50. Moilanen, M.; Silfverberg, K.; Hokkanen, T.J. Effects of wood-ash on the tree growth, vegetation and substrate quality of a drained mire: A case study. *For. Ecol. Manag.* **2002**, *171*, 321–338. [CrossRef]
51. Silfverberg, K.; Issakainen, J.; Moilanen, M. Growth and nutrition of scots pine on drained and fertilized purple moor grass fens in central Finland. *Balt. For.* **2011**, *17*, 91–101.
52. Whitehead, D.; Beadle, C.L. Physiological regulation of productivity and water use in *Eucalyptus*: A review. *For. Ecol. Manag.* **2004**, *193*, 113–140. [CrossRef]
53. Dickmann, D.I.; Nguyen, P.V.; Pregitzer, K.S. Effects of irrigation and coppicing on above-ground growth, physiology, and fine-root dynamics of two field-grown hybrid poplar clones. *For. Ecol. Manag.* **1996**, *80*, 163–174. [CrossRef]
54. Thornton, F.C.; Bock, B.R.; Behel, A.D.; Houston, A.; Tyler, D.D. Utilization of waste materials to promote hardwood tree growth. *South. J. Appl. For.* **2000**, *24*, 230–237.
55. Barrongafford, G.A.; Will, R.E.; Burkes, E.C.; Shiver, B.; Teskey, R.O. Nutrient concentrations and contents, and their relation to stem growth, of intensively managed *Pinus taeda* and *Pinus elliottii* stands of different planting densities. *For. Sci.* **2003**, *49*, 291–300.
56. Tausz, M.; Trummer, W.; Wonisch, A.; Goessler, W.; Grill, D.; Jiménez, M.S.; Morales, D. A survey of foliar mineral nutrient concentrations of *Pinus canariensis* at field plots in Tenerife. *For. Ecol. Manag.* **2004**, *189*, 49–55. [CrossRef]
57. Jennifer, A.B.; Boerner, R.E.J. Nitrogen fertilization effects on foliar nutrient dynamics and autumnal resorption in maidenhair tree (*Gingko biloba* L.). *J. Plant Nutr.* **1994**, *17*, 433–443.
58. Pettigrew, W.T.; Meredith, W.R. Dry matter production, nutrient uptake, and growth of cotton as affected by potassium fertilization. *J. Plant Nutr.* **1997**, *20*, 531–548. [CrossRef]
59. Wright, I.J.; Reich, P.B.; Westoby, M.; Ackerly, D.D.; Baruch, Z.; Bongers, F.; Cavender-Bares, J.; Chapin, T.; Cornelissen, J.H.C.; Diemer, M. The worldwide leaf economics spectrum. *Nature* **2004**, *428*, 821–827. [CrossRef] [PubMed]
60. Sundström, E.; Magnusson, T.; Hånell, B. Nutrient conditions in drained peatlands along a north-south climatic gradient in Sweden. *For. Ecol. Manag.* **2000**, *126*, 149–161. [CrossRef]
61. Pietiläinen, P.; Kaunisto, S. The effect of peat nitrogen concentration and fertilization on the foliar nitrogen concentration of Scots pine (*Pinus sylvestris* L.) in three temperature sum regions. *Suo* **2003**, *54*, 1–13.
62. Wang, D.G.; Yin, G.T.; Yang, J.C.; Li, R.S.; Zou, W.T.; Jia, R.F. Effects of phosphorus fertilization on growth and foliar nutrient (N, P, K) of *Phoebe bournei* seedlings. *J. Nanjing For. Univ. (Nat. Sci. Ed.)* **2014**, *38*, 40–44.
63. Crous, J.W.; Morris, A.R.; Scholes, M.C. Growth and foliar nutrient response to recent applications of phosphorus (P) and potassium (K) and to residual P and K fertiliser applied to the previous rotation of *Pinus patula*, at Usutu, Swaziland. *For. Ecol. Manag.* **2008**, *256*, 712–721. [CrossRef]
64. Nakashgir, G.H. Influence of potassium on nitrogen utilization by maize under dryland conditions as affected by water storage. *Adv. Plant Sci.* **1992**, *5*, 134–142.

65. Nardini, A.; Grego, F.; Trifilò, P.; Salleo, S. Changes of xylem sap ionic content and stem hydraulics in response to irradiance in *Laurus nobilis*. *Tree Physiol.* **2010**, *30*, 136–140. [CrossRef] [PubMed]
66. Sellin, A. Experimental evidence supporting the concept of light-mediated modulation of stem hydraulic conductance. *Tree Physiol.* **2010**, *30*, 1528–1535. [CrossRef] [PubMed]
67. Pervez, H.; Ashraf, M.; Makhdum, M.I. Influence of potassium nutrition on gas exchange characteristics and water relations in cotton (*Gossypium hirsutum* L.). *Photosynthetica* **2004**, *42*, 251–255. [CrossRef]

16

Nitrogen and Phosphorus Resorption in Planted Forests Worldwide

Dalong Jiang [1,2], Qinghong Geng [1], Qian Li [3], Yiqi Luo [4], Jason Vogel [2], Zheng Shi [1], Honghua Ruan [1] and Xia Xu [1,*]

1 College of Biology and the Environment, Co-Innovation Center for Sustainable Forestry in Southern China, Nanjing Forestry University, Nanjing 210037, China; dalong.jiang1988@outlook.com (D.J.); gengqh0127@163.com (Q.G.); zheng.shi@ou.edu (Z.S.); hhruan@njfu.edu.cn (H.R.)
2 School of Forest Resources and Conservation, University of Florida, Gainesville, FL 32608, USA; jvogel@ufl.edu
3 Advanced Analysis and Testing Center, Nanjing Forestry University, Nanjing 210037, China; liqian.1987@outlook.com
4 Center for Ecosystem Science and Society, Northern Arizona University, Flagstaff, AZ 86011, USA; yiqi.luo@nau.edu
* Correspondence: xuxia.1982@njfu.edu.cn

Abstract: Nutrient resorption from senescing leaves is one of the plants' essential nutrient conservation strategies. Parameters associated with resorption are important nutrient-cycling constraints for accurate predictions of long-term primary productivity in forest ecosystems. However, we know little about the spatial patterns and drivers of leaf nutrient resorption in planted forests worldwide. By synthesizing results of 146 studies, we explored nitrogen (N) and phosphorus (P) resorption efficiency (NRE and PRE) among climate zones and tree functional types, as well as the factors that play dominant roles in nutrient resorption in plantations globally. Our results showed that the mean NRE and PRE were 58.98% ± 0.53% and 60.21% ± 0.77%, respectively. NRE significantly increased from tropical to boreal zones, while PRE did not significantly differ among climate zones, suggesting differential impacts of climates on NRE and PRE. Plant functional types exert a strong influence on nutrient resorption. Conifer trees had higher PRE than broadleaf trees, reflecting the adaptation of the coniferous trees to oligotrophic habitats. Deciduous trees had lower PRE than evergreen trees that are commonly planted in P-limited low latitudes and have long leaf longevity with high nutrient use efficiency. While non-N-fixing trees had higher NRE than N-fixing trees, the PRE of non-N-fixing trees was lower than that of N-fixing trees, indicating significant impact of the N-fixing ability on the resorption of N and P. Our multivariate regression analyses showed that variations in NRE were mainly regulated by climates (mean annual precipitation and latitude), while variations in PRE were dominantly controlled by green leaf nutrient concentrations (N and P). Our results, in general, suggest that the predicted global warming and changed precipitation regimes may profoundly affect N cycling in planted forests. In addition, green leaf nutrient concentrations may be good indicators for PRE in planted forests.

Keywords: nutrient resorption; nitrogen and phosphorous; planted forests; climate zones; plant functional types; precipitation; green leaf nutrient

---

1. Introduction

Nutrient availability is a critical constraint in plant productivity and carbon (C) stocks in terrestrial ecosystems [1,2]. Nutrient resorption, through which nutrients are reabsorbed before leaf senescence and reused for plant growth directly, is an essential component of nutrient conservation strategies [3,4].

It affects key ecosystem processes such as nutrient uptake [5,6], plant competition [7,8], C cycling and resource-use efficiency [9,10], and hence productivity [1,11]. From a physiological perspective, resorption efficiency is an important issue in nutrient conservation [6,12]. In ecosystem modeling, reliable estimates of resorption efficiency are of key importance for modeling nutrient cycling and for quantifying ecosystem productivity, particularly in the new generation of coupled global models [13,14]. While previous synthesis studies mainly focus on the resorption efficiency of natural forests [1,4,5,9], we know little about nutrient resorption and its associated drivers in planted forests worldwide.

Increasing demands for timber products have promoted more research on plantations, which require more nutrients for their rapid growth [15,16]. Nutrient resorption in planted forests is less studied, since many synthesis studies excluded data of nutrient resorption from plantations [5,7,17]. Additionally, few nutrient-resorption studies have separated planted forests from other forest types [18–21]. Planted forests may have different nutrient resorption efficiencies in comparison to natural forests, since they are commonly pioneer tree species and grow on nutrient-poor soils [4,22]. Recent studies show that plantations had significantly higher nitrogen resorption efficiency (NRE) than the adjacent natural forests in northeast China and inland Hokkaido, Japan [16,23]. Phosphorous resorption efficiency (PRE) was found to be lower in plantations than in natural forests [24]. Plant productivity is largely constrained by nutrient availability [25,26], making exploration of nutrient resorption efficiency of high necessity in planted forests worldwide.

Empirical and regional studies have undoubtedly advanced our understanding of nutrient resorption in planted forests [4,27–30]. A synthesis study showed that while phosphorous resorption efficiency (PRE) was higher in subtropical than in temperate zones, NRE did not differ between the two climate zones in China's planted forests [24]. In contrary, Zhang et al. found that NRE increased and PRE first increased and then decreased along latitude in *Metasequoia glyptostroboides* forests in the east coast of China [27]. In terms of plant functional types, while Lal et al. found that deciduous species had higher NRE and PRE than evergreen species in a dry tropical environment [31], Machado et al. showed that evergreen species presented higher PRE, but not NRE, than deciduous species in plantations in the Brazilian Amazon [29]. Few studies have explored NRE in N-fixing trees vs. non-N-fixing trees in planted forests, though Yuan et al. found lower NRE in N-fixing trees than in the non-N-fixing trees (39% vs. 50%) in a semi-arid region of northern China [32]. A site study in the Karst ecosystem of southwestern China showed higher NRE and lower PRE for the conifer *Pinus yunnanensis* Franch. than for the broadleaf *Eucalyptus maideni* F. Muell. [33]. Much uncertainty still surrounds the global patterns of nutrient resorption efficiency of plantations across different climate zones and functional groups. This lack of clear understanding limits the accuracy in predicting long-term primary productivity in forest ecosystems. Therefore, it is urgent to examine nutrient resorption of planted forests regarding spatial patterns and drivers.

Planted forests, accounting for 7% of forest areas globally in 2015, are one of the essential components of terrestrial forest ecosystems providing us economic and social benefits [34,35]. Moreover, plantations have been advocated as important C sinks to mitigate future climate change [36,37]. Nutrient resorption enables plants to store reabsorbed nutrients for rapid and sustained growth at the beginning of the next growing season [38,39]. Understanding the nutrient resorption patterns in plantations is thus beneficial for improving nutrient conservation and management of planted forests. In this systematic review, we aimed to: (1) explore the global patterns of nutrient resorption (N and P) in planted forests and (2) identify various factors that play important roles in nutrient resorption worldwide.

2. Materials and Methods

2.1. Data Description

In this study, we compiled a dataset of 149 independent sampling sites based on 146 peer-reviewed papers about NRE and PRE in planted forests worldwide (Figure 1, Supplementary materials 1 and

2). In total, 643 observations for NRE and 539 observations for PRE were included in the dataset. We utilized the searching tools of ISI Web of Science and Google Scholar for retrieving articles on nutrient resorption, and the China National Knowledge Infrastructure (CNKI) to search for papers published in Chinese. Both methods included combinations of the terms 'nutrient resorption' or 'nutrient retranslocation' or 'nutrient reabsorption' and 'concentration', 'forest', and 'tree' as the searching keywords. We selected articles and extracted data (using Graph Digitizer 2.24, http://getdata-graph-digitizer.com/) with the following criteria: (1) N and P concentrations based on dry mass in green and senesced leaves were directly available or could be calculated based on presented tables or graphs; (2) we only selected data of trees from plots identified as planted forests; (3) any data from greenhouse, nursery, fertilized, and polluted sites were excluded; and (4) any data from cases of possible 'premature senescence' (e.g., drought, pests, …) were also eliminated.

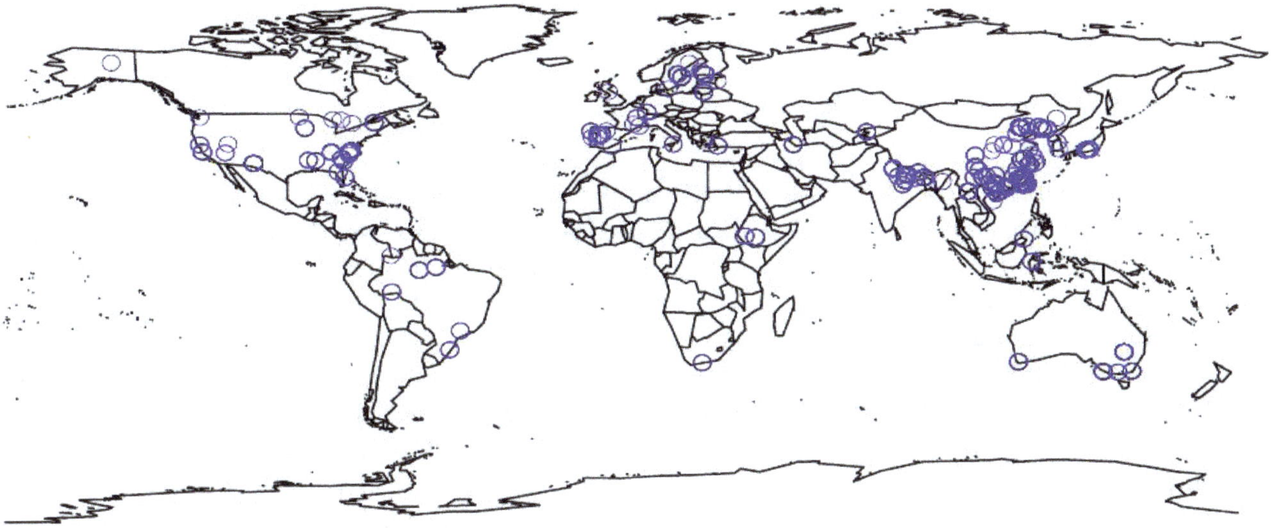

Figure 1. Global distribution of 149 independent sampling sites included in this global data synthesis.

NRE and PRE were estimated by percent reduction between green and senesced leaves [5,7,40,41]. However, to eliminate the underestimations of nutrient resorption, we used a mass loss correction factor (MLCF) with the calculation as follows [1,42]:

$$\text{NuRE }(\%) = \left(1 - \frac{\text{Nu}_{\text{senesced}}}{\text{Nu}_{\text{green}}}\text{MLCF}\right) \times 100\% \quad (1)$$

where NuRE is nutrient resorption efficiency; Nu_{green} and $\text{Nu}_{\text{senesced}}$ are nutrient concentrations in green and senesced leaves, respectively. The MLCF was calculated from the mass of green and senesced leaves or from the percentage of leaf mass loss during senescence. If green and senesced leaf mass and percentage of leaf mass loss were not available from a specific study, we used 0.745 for conifers, 0.780 for evergreen broad-leaved species, and 0.784 for deciduous broad-leaved species [1].

For studies that reported leaf nutrient concentrations during the growing season, we used maximum values for green leaf nutrient concentrations [5]. We also collected climate variables including latitude (°, LAT), longitude (°), mean annual temperature (°C, MAT), and mean annual precipitation (mm, MAP) and forest properties (e.g., tree species, functional groups, and stand ages) from those papers. Climate zones were subdivided into boreal, temperate, subtropical, and tropical regions [13]. Functional groups were separated into coniferous vs. broadleaf trees, deciduous vs. evergreen trees, and non-N-fixing vs. N-fixing trees.

2.2. *Statistical Analysis*

Before statistical analysis, the normality and homoscedasticity of the NRE and PRE were verified by the Kolmogorov–Smirnov test and Levene's test, respectively. One-way analysis of variance (ANOVA) was performed to analyze the differences between NRE and PRE. Linear mixed-effects models (LMEs) were applied to quantify the effects of climate zones and different functional groups on NRE and PRE worldwide. Climate zones and functional groups were treated as fixed factors, and studies were treated as a random factor. The significance of fixed effects terms was assessed via ANOVA with F tests. If the differences were significant, post hoc multiple comparisons were subsequently conducted using Duncan's test ($p = 0.05$). Stepwise regression analyses were used to explore the controls on the variation of NRE and PRE associated with climate variables, leaf nutrient status, and stand ages. All statistical analyses were performed using SPSS 22.0 for Windows (SPSS Inc., Chicago, IL, USA).

3. Results

3.1. *Nutrient Resorption Patterns in Planted Forests*

Globally, mean values of NRE and PRE in the planted forests were 58.98% ± 0.53% and 60.21% ± 0.77%, respectively, with no significant differences between them ($p > 0.05$, Figure 2). Generally, variations in both NRE and PRE showed remarkable differences among different climate zones and tree functional groups. While NRE significantly increased with increasing latitude ($p < 0.05$, Table 1, Figure 3a), PRE did not differ among climate zones ($p > 0.05$, Table 1, Figure 3b). Additionally, NRE did not differ between conifer and broadleaf trees ($p > 0.05$, Table 1, Figure 4a) and the PRE of conifer trees was higher than that of broadleaf trees ($p < 0.05$, Table 1, Figure 4b). Though NRE did not differ between deciduous and evergreen trees ($p > 0.05$, Table 1, Figure 4a), PRE was significantly higher in evergreen than deciduous trees ($p < 0.001$, Table 1, Figure 4b). The NRE of N-fixing trees was much lower than that of non-N-fixing trees ($p < 0.001$, Table 1, Figure 4a) and the PRE of N-fixing trees was much higher than that of non-N-fixing trees ($p < 0.05$, Table 1, Figure 4b).

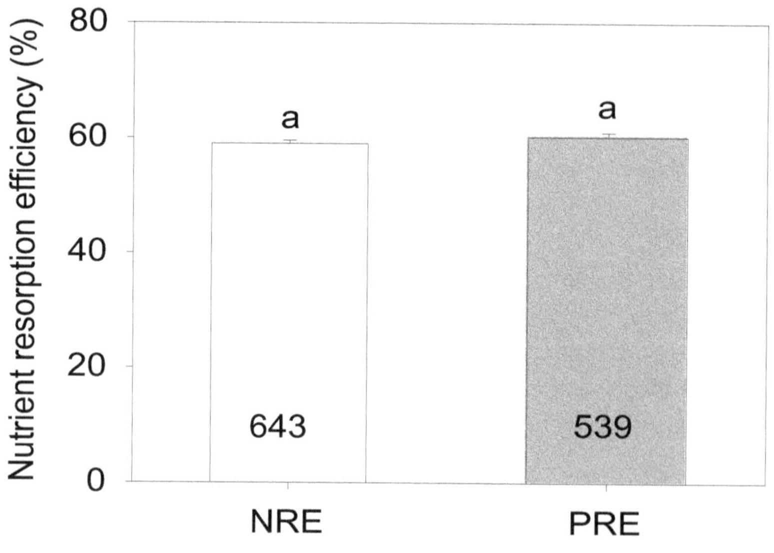

Figure 2. N and P resorption efficiency (%, NRE and PRE) in planted forests (mean ± SE). Values in the bars represent the number of observations. Different lowercase letters represent statistical differences at $p = 0.05$ between NRE and PRE.

Table 1. ANOVA table of the linear mixed-effects models for N and P resorption efficiency (%, NRE and PRE). *: <0.05, ***: <0.001.

Resorption	Fixed Term	*df*	*F, p*
NRE	Climate zones	3	3.56 *
	Broadleaf vs. Conifer	1	0.56
	Deciduous vs. Evergreen	1	2.48
	Non-N-fixing tree vs. N-fixing tree	1	15.85 ***
PRE	Climate zones	3	0.79
	Broadleaf vs. Conifer	1	4.74 *
	Deciduous vs. Evergreen	1	12.93 ***
	Non-N-fixing tree vs. N-fixing tree	1	5.06 *

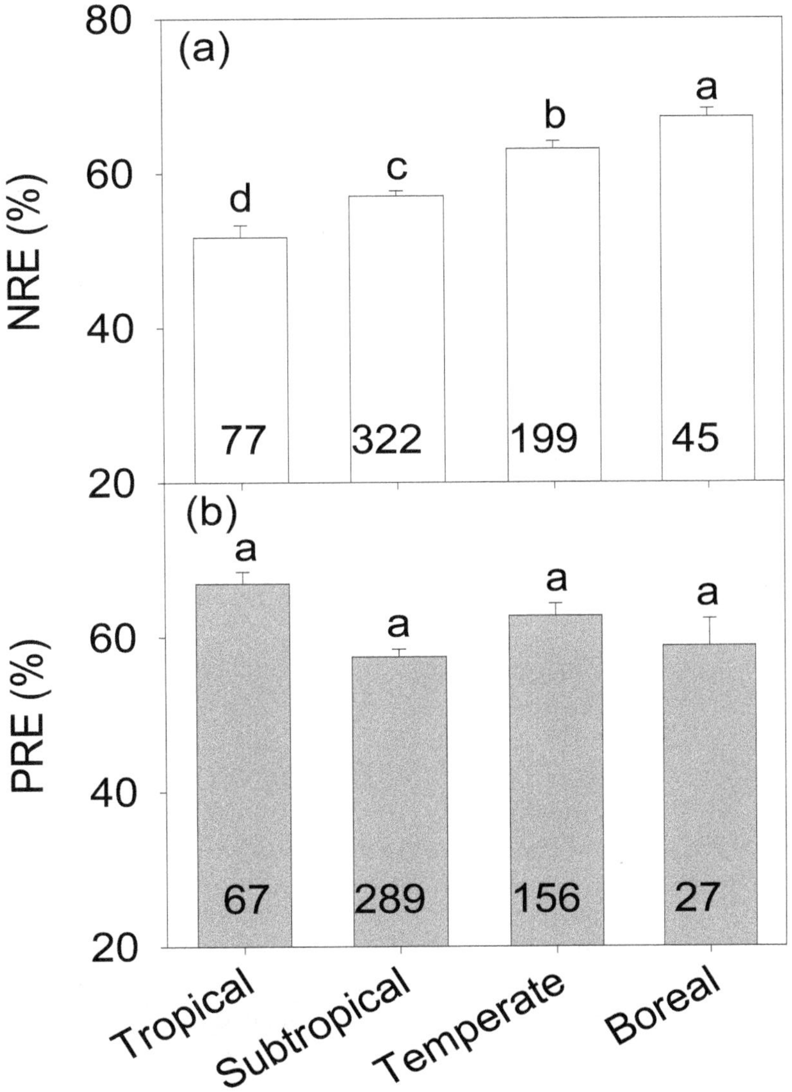

Figure 3. N and P resorption efficiency (%, NRE (**a**) and PRE (**b**)) among different climate zones in planted forests (mean ± SE). Values in the bars represent the number of observations. Different lowercase letters represent statistical differences at $p = 0.05$ among the climate zones.

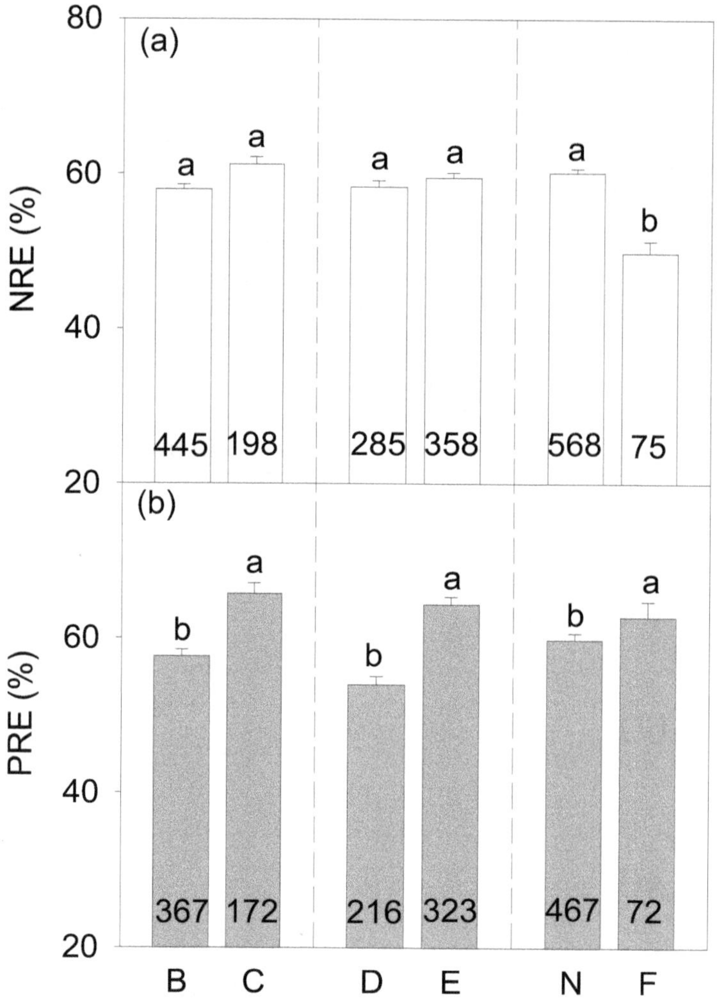

Figure 4. N and P resorption efficiency (%, NRE (**a**) and PRE (**b**)) between broadleaf (B) vs. conifer (C), deciduous (D) vs. evergreen (E), and non-N-fixing (N) vs. N-fixing (F) tree groups in planted forests (mean ± SE). Values in the bars represent the number of observations. Different lowercase letters represent statistical differences at $p = 0.05$ between the different functional types.

3.2. Controls of Nutrient Resorption in Planted Forests

Variations in NRE and PRE at the global scale were not controlled by the same explanatory variables. Our multivariate regression analyses showed that NRE was regulated negatively by MAP and positively by green leaf P concentration (P_{green}) and latitude (LAT) ($p < 0.001$, Table 2). For PRE, green leaf N concentration (N_{green}) and P_{green} had positively and negatively impacted on it, respectively ($p < 0.001$, Table 2). The stand ages of forests had little impact on both NRE and PRE.

Table 2. Stepwise regressions of N and P resorption efficiency (%, NRE and PRE) with stand age (year, Age), absolute latitude (°, LAT), mean annual precipitation (mm, MAP), and green leaf N and P concentrations (mg g^{-1}, N_{green} and P_{green}, respectively) in planted forests. ***: <0.001.

	Variables	Regression	n	r^2	Excluded Variables
NRE	MAP	$y = 67.48 - 0.01 \times MAP$	476	0.09 ***	Age, N_{green}, P_{green}, LAT
	MAP, P_{green}	$y = 62.65 - 0.01 \times MAP + 2.70 \times P_{green}$	476	0.12 ***	Age, N_{green}, LAT
	MAP, P_{green}, LAT	$y = 54.68 - 0.01 \times MAP + 2.36 \times P_{green} + 0.21 \times LAT$	476	0.14 ***	Age, N_{green}
PRE	P_{green}	$y = 66.11 - 4.12 \times P_{green}$	476	0.05 ***	Age, N_{green}, LAT, MAP
	P_{green}, N_{green}	$y = 60.31 - 5.33 \times P_{green} + 0.41 \times N_{green}$	476	0.07 ***	Age, LAT, MAP

4. Discussion

4.1. Nutrient Resorption Patterns in Planted Forests

Our results showed that the mean values of NRE and RPE were 58.98% ± 0.53% and 60.21% ± 0.77%, respectively, for the planted forests worldwide—much higher than the assumed 50% for both NRE and PRE in most models [1]. Consistent with Zhang et al. that the NRE of *Metasequoia glyptostroboides* increased along latitude at the regional scale [27], we found that NRE significantly increased from tropical to boreal zones with increasing latitude. This pattern may result from three main reasons associated with low temperatures at high latitudes. First, decomposition and nutrient mineralization are high at low latitudes, which may subsequently enhance the availability of soil N [43,44]. Second, warmer temperatures and higher water availability could stimulate nutrient movement as well as root N uptake at low latitudes [45]. In high-latitude regions, however, the delivery of mass-flow nutrients from the soil to plant tissues is inhibited, which in turn, may reduce the process of nutrient resorption [20,46]. Third, the metabolic activity of plants may be inhibited at high latitudes [27]. Nutrient resorption is recognized as one of the efficient nutrient conservation mechanisms for plants to adapt to environmental conditions [12]. Plants at high latitudes usually grow quickly in order to finish development in relatively short growing seasons via high nutrient retrieval strategies that can enhance growth rate and lower dependence on the supply of soil nutrients [1,5,47]. Interestingly, PRE did not linearly increase or decrease from tropical to boreal zones, which is in line with the findings of Zhang et al., who observed a curved correlation of PRE with latitude [27]. PRE may largely depend on P availability in soils [13,48]. For example, a negative correlation between PRE and soil P was found in planted forests [27,33,49]. Those results suggest that NRE and PRE are differentially regulated by climate and soil nutrient availability.

The resorption of N and P varied among functional groups. First, the PREs of coniferous trees were significantly higher than those of broadleaf trees. Similar observations were also found in natural forests [18,20]. Coniferous trees are usually planted in nutrient-poor environments. The higher nutrient resorption could thus help them survive via reducing their dependence on soil nutrient supplies [16,50]. High nutrient resorption reflects the adaptation of plants to oligotrophic habitats [10,51]. Interestingly, NRE did not differ between coniferous and broadleaf trees. This may result from the limited number of NRE observations in coniferous planted forests, as also reported by Yuan and Chen [5]. Second, we also found that evergreen tree species presented higher PRE, whereas the NRE of deciduous species did not differ from that of the evergreen species, which is completely in line with the findings in the forest plantations in the Brazilian Amazon [29]. Possible reasons may include that evergreen species (1) are mostly planted in P-limited low latitudes [3,20] and (2) have high nutrient use efficiency and thus low nutrient loss rates associated with prolonged nutrient retention time in leaf biomass and leaf longevity [12,52]. Third, our results showed that N-fixing trees had significantly lower NRE but higher PRE than non-N-fixing trees. N-fixing trees can acquire N from the atmosphere, and they are thus less dependent on the internal N cycling process [53,54]. However, N-fixing trees may contain higher leaf N and may thus enhance PRE in order to maintain N to P stoichiometric homeostasis. These findings indicate that plant functional types exert a strong influence on nutrient resorption.

4.2. Controls of Nutrient Resorption in Planted Forests

In this study, the global-scale NRE and PRE patterns were not mainly regulated by the same ecological factors. Our multivariate regression analyses showed that NRE was primarily controlled by climate variables, supporting the observations of some previous studies [1,20,27,48]. Specially, we found that NRE was regulated negatively by MAP and positively by LAT. Climates (e.g., temperature and precipitation) most commonly affect soil N availability via microbial activities [44]. That is, soil N availability is lower under hostile climates (e.g., dry and cold at high latitude) than under climates that are favored by microbes (e.g., wet and warm at low latitude). Soil nutrient availability usually has a negative impact on nutrient resorption efficiency [41,55], leading to the phenomena that NRE correlated positively with LAT and negatively with MAP. Contrary to what we found here, Achat et al. showed that climates had only minor effects on the NRE of forests in 102 permanent forest sites across France [9]. The gap may result from the rather narrow range of climatic conditions in the study by Achat et al. [9] compared with the global climate scale considered in our study. In addition, NRE was also regulated positively by green leaf P (P_{green}). N and P in plants are usually closely related, which means changes of one nutrient could alter the other [56,57]. For example, plants may elevate NRE to maintain their N and P stoichiometric homeostasis by increasing P_{green}, and vice versa. Generally, our results suggest that the predicted global warming and changed precipitation regimes may profoundly affect N cycling in planted forests.

For the resorption of P, N_{green} had positive impacts on it, which is in accordance with the results of Wang et al. [58] and Yan et al. [4], suggesting that coupled relationships between N and P may exist in planted forests. Furthermore, we found that P_{green} had negative effects on PRE, which is supported by the findings by Kobe et al. [59]. Green leaf nutrient status is supposed to represent soil nutrient availability, whose negative impacts on nutrient resorption efficiency is widely accepted [41,59,60]. Therefore, green leaf nutrient status may negatively relate to nutrient resorption efficiency. We also found that climate variables were not the main explanatory factors for PRE, supporting the idea that soil properties and soil parent materials other than climate had more impact on soil P availability [48]. Green leaf nutrient concentration may be a good indicator for PRE in planted forests.

5. Conclusions

To our best knowledge, this synthesis provides the first comprehensive analysis of NRE and PRE in planted forests worldwide. Generally, NRE significantly increased along climate zones, while no significant differences were observed for PRE among climate zones. These results suggest differential impacts of climates on NRE and PRE. In terms of plant functional groups, the PRE of conifer trees was higher than that of broadleaf trees; evergreen trees had higher PRE than deciduous trees; non-N-fixing trees had higher NRE but lower PRE than N-fixing trees. These findings indicate that plant functional types exert strong impacts on nutrient resorption. Additionally, multivariate regression analyses showed that variations in NRE were mainly regulated by mean annual precipitation (MAP) and latitude, indicating that the predicted global warming and changed precipitation regimes may profoundly affect N cycling in planted forests. Variations in PRE were dominantly controlled by green leaf nutrient concentrations (N and P), suggesting green leaf nutrient concentrations may be good indicators for PRE in planted forests.

Author Contributions: Conceptualization: X.X.; Data curation: D.J.; Formal analysis: D.J.; Funding acquisition: D.J., H.R., and X.X.; Project administration: X.X.; Supervision: X.X.; Writing—original draft: D.J.; Writing—review and editing: Q.G., Q.L., Y.L., J.V., Z.S., H.R., and X.X.

Acknowledgments: We thank many lab members for their help with the manuscript preparation.

References

1. Vergutz, L.; Manzoni, S.; Porporato, A.; Novais, R.F.; Jackson, R.B. Global resorption efficiencies and concentrations of carbon and nutrients in leaves of terrestrial plants. *Ecol. Monogr.* **2012**, *82*, 205–220. [CrossRef]
2. Sokolov, A.P.; Kicklighter, D.W.; Melillo, J.M.; Felzer, B.S.; Schlosser, C.A.; Cronin, T.W. Consequences of considering carbon-nitrogen interactions on the feedbacks between climate and the terrestrial carbon cycle. *J. Clim.* **2008**, *21*, 3776–3796. [CrossRef]
3. Aerts, R. Nutrient Resorption from Senescing Leaves of Perennials: Are there General Patterns? *J. Ecol.* **1996**, *84*, 597–608. [CrossRef]
4. Yan, T.; Lu, X.T.; Zhu, J.J.; Yang, K.; Yu, L.Z.; Gao, T. Changes in nitrogen and phosphorus cycling suggest a transition to phosphorus limitation with the stand development of larch plantations. *Plant Soil* **2018**, *422*, 385–396. [CrossRef]

5. Yuan, Z.Y.; Chen, H.Y.H. Global-scale patterns of nutrient resorption associated with latitude, temperature and precipitation. *Glob. Ecol. Biogeogr.* **2009**, *18*, 11–18. [CrossRef]
6. Wang, Z.Q.; Fan, Z.X.; Zhao, Q.; Wang, M.C.; Ran, J.Z.; Huang, H.; Niklas, K.J. Global data analysis shows that soil nutrient levels dominate foliar nutrient resorption efficiency in herbaceous species. *Front. Plant Sci.* **2018**, *9*, 1–8. [CrossRef] [PubMed]
7. Killingbeck, K.T. Nutrients in senesced leaves: Keys to the search for potential resorption and resorption proficiency. *Ecology* **1996**, *77*, 1716–1727. [CrossRef]
8. Schmidt, M.; Veldkamp, E.; Corre, M.D. Tree-microbial biomass competition for nutrients in a temperate deciduous forest, central Germany. *Plant Soil* **2016**, *408*, 227–242. [CrossRef]
9. Achat, D.L.; Pousse, N.; Nicolas, M.; Augusto, L. Nutrient remobilization in tree foliage as affected by soil nutrients and leaf life span. *Ecol. Monogr.* **2018**, *88*, 408–428. [CrossRef]
10. Aerts, R.; Chapin, F.S. The mineral nutrition of wild plants revisited: A re-evaluation of processes and patterns. *Adv. Ecol. Res.* **2000**, *30*, 1–67. [CrossRef]
11. Deng, M.F.; Liu, L.L.; Jiang, L.; Liu, W.X.; Wang, X.; Li, S.P.; Yang, S.; Wang, B. Ecosystem scale trade-off in nitrogen acquisition pathways. *Nat. Ecol. Evol.* **2018**, *2*, 1724–1734. [CrossRef] [PubMed]
12. Brant, A.N.; Chen, H.Y.H. Patterns and mechanisms of nutrient resorption in plants. *Crit. Rev. Plant Sci.* **2015**, *34*, 471–486. [CrossRef]
13. Yuan, Z.Y.; Chen, H.Y.H. Negative effects of fertilization on plant nutrient resorption. *Ecology* **2015**, *96*, 373–380. [CrossRef] [PubMed]
14. Sun, Z.Z.; Liu, L.L.; Peng, S.S.; Penuelas, J.; Zeng, H.; Piao, S.L. Age-related modulation of the nitrogen resorption efficiency response to growth requirements and soil nitrogen availability in a temperate Pine plantation. *Ecosystems* **2016**, *19*, 698–709. [CrossRef]
15. Sedjo, R.A. The potential of High-Yield Plantation Forestry for Meeting Timber Needs. In *Planted Forests: Contributions to the Quest for Sustainable Societies*; Boyle, J.R., Winjum, J.K., Kavanagh, K., Jensen, E.C., Eds.; Springer: Dordrecht, The Netherlands, 1999; pp. 339–359.
16. Yan, T.; Lu, X.T.; Yang, K.; Zhu, J.J. Leaf nutrient dynamics and nutrient resorption: A comparison between larch plantations and adjacent secondary forests in Northeast China. *J. Plant Ecol.* **2016**, *9*, 165–173. [CrossRef]
17. Sun, X.; Kang, H.; Chen, H.Y.H.; Bjorn, B.; Samuel, B.F.; Liu, C. Biogeographic patterns of nutrient resorption from *Quercus variabilis* Blume leaves across China. *Plant Biol.* **2016**, *18*, 505–513. [CrossRef] [PubMed]
18. Tang, L.Y.; Han, W.X.; Chen, Y.H.; Fang, J.Y. Resorption proficiency and efficiency of leaf nutrients in woody plants in eastern China. *J. Plant Ecol.* **2013**, *6*, 408–417. [CrossRef]
19. Zhang, H.Y.; Lu, X.T.; Hartmann, H.; Keller, A.; Han, X.G.; Trumbore, S.; Phillips, R.P. Foliar nutrient resorption differs between arbuscular mycorrhizal and ectomycorrhizal trees at local and global scales. *Glob. Ecol. Biogeogr.* **2018**, *27*, 875–885. [CrossRef]
20. Yan, T.; Zhu, J.J.; Yang, K. Leaf nitrogen and phosphorus resorption of woody species in response to climatic conditions and soil nutrients: A meta-analysis. *J. For. Res.* **2018**, *29*, 905–913. [CrossRef]
21. Han, W.X.; Tang, L.Y.; Chen, Y.H.; Fang, J.Y. Relationship between the relative limitation and resorption efficiency of nitrogen vs phosphorus in woody plants. *PLoS ONE* **2013**, *8*, e83366. [CrossRef] [PubMed]
22. Zhang, H.; Wang, J.N.; Wang, J.Y.; Guo, Z.W.; Wang, G.G.; Zeng, D.H.; Wu, T.G. Tree stoichiometry and nutrient resorption along a chronosequence of *Metasequoia glyptostroboides* forests in coastal China. *For. Ecol. Manag.* **2018**, *430*, 445–450. [CrossRef]
23. Maeda, Y.; Tashiro, N.; Enoki, T.; Urakawa, R.; Hishi, T. Effects of species replacement on the relationship between net primary production and soil nitrogen availability along a topographical gradient: Comparison of belowground allocation and nitrogen use efficiency between natural forests and plantations. *For. Ecol. Manag.* **2018**, *422*, 214–222. [CrossRef]
24. Xu, S.; Zhou, G.; Tang, X.; Wang, W.; Wang, G.; Ma, K.; Han, S.; Du, S.; Li, S.; Yan, J.; et al. Different spatial patterns of nitrogen and phosphorus resorption efficiencies in China's forests. *Sci. Rep.* **2017**, *7*, 10584. [CrossRef] [PubMed]
25. Yuan, Z.Y.; Li, L.H. Soil water status influences plant nitrogen use: A case study. *Plant Soil* **2007**, *301*, 303–313. [CrossRef]
26. Austin, A.T.; Yahdjian, L.; Stark, J.M.; Belnap, J.; Porporato, A.; Norton, U.; Ravetta, D.A.; Schaeffer, S.M. Water pulses and biogeochemical cycles in arid and semiarid ecosystems. *Oecologia* **2004**, *141*, 221–235. [CrossRef] [PubMed]

27. Zhang, H.; Guo, W.H.; Yu, M.K.; Wang, G.G.; Wu, T.G. Latitudinal patterns of leaf N, P stoichiometry and nutrient resorption of *Metasequoia glyptostroboides* along the eastern coastline of China. *Sci. Total Environ.* **2018**, *618*, 1–6. [CrossRef] [PubMed]
28. Zhou, L.L.; Addo-Danso, S.D.; Wu, P.F.; Li, S.B.; Zou, X.H.; Zhang, Y.; Ma, X.Q. Leaf resorption efficiency in relation to foliar and soil nutrient concentrations and stoichiometry of *Cunninghamia lanceolata* with stand development in southern China. *J. Soils Sediments* **2016**, *16*, 1448–1459. [CrossRef]
29. Machado, M.R.; Sampaio, P.D.; Ferraz, J.; Camara, R.; Pereira, M.G. Nutrient retranslocation in forest species in the Brazilian Amazon. *Acta Sci.-Agron.* **2016**, *38*, 93–101. [CrossRef]
30. Ye, G.F.; Zhang, S.J.; Zhang, L.H.; Lin, Y.M.; Wei, S.D.; Liao, M.M.; Lin, G.H. Age-related changes in nutrient resorption patterns and tannin concentration of *Casuarina equisetifolia* plantations. *J. Trop. For. Sci.* **2012**, *24*, 546–556. [CrossRef]
31. Lal, C.B.; Annapurna, C.; Raghubanshi, A.S.; Singh, J.S. Foliar demand and resource economy of nutrients in dry tropical forest species. *J. Veg. Sci.* **2001**, *12*, 5–14. [CrossRef]
32. Yuan, Z.Y.; Li, L.H.; Han, X.G.; Huang, J.H.; Jiang, G.M.; Wan, S.Q.; Zhang, W.H.; Chen, Q.S. Nitrogen resorption from senescing leaves in 28 plant species in a semi-arid region of northern China. *J. Arid Environ.* **2005**, *63*, 191–202. [CrossRef]
33. Pang, D.B.; Wang, G.Z.; Li, G.J.; Sun, Y.L.; Liu, Y.G.; Zhou, J.X. Ecological stoichiometric characteristics of two typical plantations in the Karst ecosystem of southwestern China. *Forests* **2018**, *9*, 56. [CrossRef]
34. MacDicken, K.; Jonsson, Ö.; Piña, L.; Marklund, L.; Maulo, S.; Contessa, V.; Adikari, Y.; Garzuglia, M.; Lindquist, E.; Reams, G.; et al. *Global Forest Resources Assessment 2015: How Are the World'S Forests Changing?* Food and Agriculture Organization of the United Nations (FAO): Rome, Italy, 2015; pp. 1–54.
35. Van Dijk, A.I.J.M.; Keenan, R.J. Planted forests and water in perspective. *For. Ecol. Manag.* **2007**, *251*, 1–9. [CrossRef]
36. Fang, J.Y.; Chen, A.P.; Peng, C.H.; Zhao, S.Q.; Ci, L.J. Changes in forest biomass carbon storage in China between 1949 and 1998. *Science* **2001**, *292*, 2320–2322. [CrossRef] [PubMed]
37. Goodale, C.L.; Apps, M.J.; Birdsey, R.A.; Field, C.B.; Heath, L.S.; Houghton, R.A.; Jenkins, J.C.; Kohlmaier, G.H.; Kurz, W.; Liu, S.; et al. Forest carbon sinks in the northern Hemisphere. *Ecol. Appl.* **2002**, *12*, 891–899. [CrossRef]
38. Netzer, F.; Schmid, C.; Herschbach, C.; Rennenberg, H. Phosphorus-nutrition of European beech (*Fagus sylvatica* L.) during annual growth depends on tree age and P-availability in the soil. *Environ. Exp. Bot.* **2017**, *137*, 194–207. [CrossRef]
39. Weatherall, A.; Proe, M.F.; Craig, J.; Cameron, A.D.; Midwood, A.J. Internal cycling of nitrogen, potassium and magnesium in young Sitka spruce. *Tree Physiol.* **2006**, *26*, 673–680. [CrossRef] [PubMed]
40. Chen, F.S.; Niklas, K.J.; Liu, Y.; Fang, X.M.; Wan, S.Z.; Wang, H. Nitrogen and phosphorus additions alter nutrient dynamics but not resorption efficiencies of Chinese fir leaves and twigs differing in age. *Tree Physiol.* **2015**, *35*, 1106–1117. [CrossRef] [PubMed]
41. Reed, S.C.; Townsend, A.R.; Davidson, E.A.; Cleveland, C.C. Stoichiometric patterns in foliar nutrient resorption across multiple scales. *New Phytol.* **2012**, *196*, 173–180. [CrossRef] [PubMed]
42. Van Heerwaarden, L.M.; Toet, S.; Aerts, R. Current measures of nutrient resorption efficiency lead to a substantial underestimation of real resorption efficiency: Facts and solutions. *Oikos* **2003**, *101*, 664–669. [CrossRef]
43. Oleksyn, J.; Reich, P.B.; Zytkowiak, R.; Karolewski, P.; Tjoelker, M.G. Nutrient conservation increases with latitude of origin in European *Pinus sylvestris* populations. *Oecologia* **2003**, *136*, 220–235. [CrossRef] [PubMed]
44. Bengtson, P.; Falkengren-Grerup, U.; Bengtsson, G. Relieving substrate limitation-soil moisture and temperature determine gross N transformation rates. *Oikos* **2005**, *111*, 81–90. [CrossRef]
45. Reich, P.B.; Oleksyn, J. Global patterns of plant leaf N and P in relation to temperature and latitude. *Proc. Natl. Acad. Sci. USA* **2004**, *101*, 11001–11006. [CrossRef] [PubMed]
46. Cramer, M.D.; Hawkins, H.J.; Verboom, G.A. The importance of nutritional regulation of plant water flux. *Oecologia* **2009**, *161*, 15–24. [CrossRef] [PubMed]
47. Wright, I.J.; Westoby, M. Nutrient concentration, resorption and lifespan leaf traits of Australian sclerophyll species. *Funct. Ecol.* **2003**, *17*, 10–19. [CrossRef]
48. Augusto, L.; Achat, D.L.; Jonard, M.; Vidal, D.; Ringeval, B. Soil parent material-A major driver of plant nutrient limitations in terrestrial ecosystems. *Glob. Chang. Biol.* **2017**, *23*, 3808–3824. [CrossRef] [PubMed]

49. Zeng, F.P.; Chen, X.; Huang, B.; Chi, G.Y. Distribution changes of phosphorus in soil-plant systems of larch plantations across the chronosequence. *Forests* **2018**, *9*, 563. [CrossRef]
50. Yuan, Z.Y.; Li, L.H.; Han, X.G.; Huang, J.H.; Wan, S.Q. Foliar nitrogen dynamics and nitrogen resorption of a sandy shrub *Salix gordejevii* in northern China. *Plant Soil* **2005**, *278*, 183–193. [CrossRef]
51. Liu, J.T.; Gu, Z.J.; Shao, H.B.; Zhou, F.; Peng, S.Y. N–P stoichiometry in soil and leaves of *Pinus massoniana* forest at different stand ages in the subtropical soil erosion area of China. *Environ. Earth Sci.* **2016**, *75*, 1091. [CrossRef]
52. Escudero, A.; Del Arco, J.M.; Sanz, I.C.; Ayala, J. Effects of leaf longevity and retranslocation efficiency on the retention time of nutrients in the leaf biomass of different woody species. *Oecologia* **1992**, *90*, 80–87. [CrossRef] [PubMed]
53. Norris, M.D.; Reich, P.B. Modest enhancement of nitrogen conservation via retranslocation in response to gradients in N supply and leaf N status. *Plant Soil* **2009**, *316*, 193–204. [CrossRef]
54. Zhao, G.S.; Shi, P.L.; Wu, J.S.; Xiong, D.P.; Zong, N.; Zhang, X.Z. Foliar nutrient resorption patterns of four functional plants along a precipitation gradient on the Tibetan Changtang Plateau. *Ecol. Evol.* **2017**, *7*, 7201–7212. [CrossRef] [PubMed]
55. Hayes, P.; Turner, B.L.; Lambers, H.; Laliberte, E. Foliar nutrient concentrations and resorption efficiency in plants of contrasting nutrient-acquisition strategies along a 2-million-year dune chronosequence. *J. Ecol.* **2014**, *102*, 396–410. [CrossRef]
56. Deng, M.F.; Liu, L.L.; Sun, Z.Z.; Piao, S.L.; Ma, Y.C.; Chen, Y.W.; Wang, J.; Qiao, C.L.; Wang, X.; Li, P. Increased phosphate uptake but not resorption alleviates phosphorus deficiency induced by nitrogen deposition in temperate *Larix principis-rupprechtii* plantations. *New Phytol.* **2016**, *212*, 1019–1029. [CrossRef] [PubMed]
57. Yu, Q.; Chen, Q.; Elser, J.J.; He, N.; Wu, H.; Zhang, G.; Wu, J.; Bai, Y.; Han, X. Linking stoichiometric homoeostasis with ecosystem structure, functioning and stability. *Ecol. Lett.* **2010**, *13*, 1390–1399. [CrossRef] [PubMed]
58. Wang, Z.N.; Lu, J.Y.; Yang, H.M.; Zhang, X.; Luo, C.L.; Zhao, Y.X. Resorption of nitrogen, phosphorus and potassium from leaves of lucerne stands of different ages. *Plant Soil* **2014**, *383*, 301–312. [CrossRef]
59. Kobe, R.K.; Lepczyk, C.A.; Iyer, M. Resorption efficiency decreases with increasing green leaf nutrients in a global data set. *Ecology* **2005**, *86*, 2780–2792. [CrossRef]
60. Tsujii, Y.; Onoda, Y.; Kitayama, K. Phosphorus and nitrogen resorption from different chemical fractions in senescing leaves of tropical tree species on Mount Kinabalu, Borneo. *Oecologia* **2017**, *185*, 171–180. [CrossRef] [PubMed]

Permissions

All chapters in this book were first published in MDPI; hereby published with permission under the Creative Commons Attribution License or equivalent. Every chapter published in this book has been scrutinized by our experts. Their significance has been extensively debated. The topics covered herein carry significant findings which will fuel the growth of the discipline. They may even be implemented as practical applications or may be referred to as a beginning point for another development.

The contributors of this book come from diverse backgrounds, making this book a truly international effort. This book will bring forth new frontiers with its revolutionizing research information and detailed analysis of the nascent developments around the world.

We would like to thank all the contributing authors for lending their expertise to make the book truly unique. They have played a crucial role in the development of this book. Without their invaluable contributions this book wouldn't have been possible. They have made vital efforts to compile up to date information on the varied aspects of this subject to make this book a valuable addition to the collection of many professionals and students.

This book was conceptualized with the vision of imparting up-to-date information and advanced data in this field. To ensure the same, a matchless editorial board was set up. Every individual on the board went through rigorous rounds of assessment to prove their worth. After which they invested a large part of their time researching and compiling the most relevant data for our readers.

The editorial board has been involved in producing this book since its inception. They have spent rigorous hours researching and exploring the diverse topics which have resulted in the successful publishing of this book. They have passed on their knowledge of decades through this book. To expedite this challenging task, the publisher supported the team at every step. A small team of assistant editors was also appointed to further simplify the editing procedure and attain best results for the readers.

Apart from the editorial board, the designing team has also invested a significant amount of their time in understanding the subject and creating the most relevant covers. They scrutinized every image to scout for the most suitable representation of the subject and create an appropriate cover for the book.

The publishing team has been an ardent support to the editorial, designing and production team. Their endless efforts to recruit the best for this project, has resulted in the accomplishment of this book. They are a veteran in the field of academics and their pool of knowledge is as vast as their experience in printing. Their expertise and guidance has proved useful at every step. Their uncompromising quality standards have made this book an exceptional effort. Their encouragement from time to time has been an inspiration for everyone.

The publisher and the editorial board hope that this book will prove to be a valuable piece of knowledge for researchers, students, practitioners and scholars across the globe.

List of Contributors

Nguyen Van Bich
Tasmanian Institute of Agriculture (TIA), University of Tasmania, Private Bag 98, Hobart, Tasmania 7001, Australia
Silviculture Research Institute (SRI), Vietnamese Academy of Forest Sciences (VAFS), 46 Duc Thang, Bac Tu Liem, Hanoi 11910, Vietnam

Alieta Eyles, David Ratkowsky, Katherine J. Evans and Caroline Mohammed
Tasmanian Institute of Agriculture (TIA), University of Tasmania, Private Bag 98, Hobart, Tasmania 7001, Australia

Tran Lam Dong, Vo Dai Hai, Hoang Van Thanh and Nguyen Van Thinh
Silviculture Research Institute (SRI), Vietnamese Academy of Forest Sciences (VAFS), 46 Duc Thang, Bac Tu Liem, Hanoi 11910, Vietnam

Daniel Mendham
CSIRO Land and Water, Private Bag 12, Hobart, Tasmania 7001, Australia

Natalia Startsev
Natalia Startsev. Canadian Forest Service, Northern Forestry Centre, 5320 122 Street, Edmonton, AB T6H 3S5, Canada

Rachhpal S. Jassal
Biometeorology and Soil Physics Group, University of British Columbia, Vancouver, BC V6T 1Z4, Canada

Xiuqing Nie
Key Laboratory of Tree Breeding and Cultivation of the State Forestry Administration, Research Institute of Forestry, Chinese Academy of Forestry, Beijing 100091, China
Key Laboratory of Tibetan Medicine Research, Northwest Institute of Plateau Biology, Chinese Academy of Science, Xining 810008, China
Research Institute of nature protected Area, Chinese Academy of Forestry, Beijing 100091, China

Dong Wang
Key Laboratory of Tree Breeding and Cultivation of the State Forestry Administration, Research Institute of Forestry, Chinese Academy of Forestry, Beijing 100091, China
Research Institute of nature protected Area, Chinese Academy of Forestry, Beijing 100091, China

Lucun Yang and Guoying Zhou
Key Laboratory of Tibetan Medicine Research, Northwest Institute of Plateau Biology, Chinese Academy of Science, Xining 810008, China
Qinghai Key Laboratory of Qing-Tibet Biological Resources, Xining 810008, China

Marcella L.C. Menegale, Robert Harrison Jason N. James and Stephani Michelsen-Correa
School of Environmental and Forest Sciences, University of Washington, Seattle, WA 98195-2100, USA

Jose Henrique T. Rocha, Jose Leonardo de M. Goncalves and Ayeska Hubner
Forest Science Department, "Luiz de Queiroz" College of Agriculture, University of São Paulo, Piracicaba SP 13418-900, Brazil

Rodrigo F. Almeida and Marisa de C. Piccolo
Center of Nuclear Energy in Agriculture, University of São Paulo, Piracicaba, São Paulo 13416-000, Brazil

Jose Carlos Arthur Junior
Federal Rural University of Rio de Janeiro, Seropédica, Rio de Janeiro 23890-000, Brazil

Alexandre de Vicente Ferraz
Forest Science and Research Institute (IPEF), Piracicaba, São Paulo 13400-970 Private Bag 530, Brazil

Yang Zhang
Jiangxi Provincial Key Laboratory of Silviculture, College of Forestry, Jiangxi Agricultural University, Nanchang 330045, China

Wen-Sheng Bu, Han-Jiao Gu, Can-can Zhang, Xiang-Min Fang, Jing Fan and Fu-Sheng Chen
Jiangxi Provincial Key Laboratory of Silviculture, College of Forestry, Jiangxi Agricultural University, Nanchang 330045, China
Jiulianshan National Observation and Research Station of Chinese Forest Ecosystem, 2011 Collaborative Innovation Center of Jiangxi Typical Trees Cultivation and Utilization, Jiangxi Agricultural University, Nanchang 330045, China

Anand Narain Singh
Department of Botany, Panjab University, Chandigarh 160014, India

Hui-Min Wang
Qianyanzhou Ecological Station, Key Laboratory of Ecosystem Network Observation and Modeling, Institute of Geographic Sciences and Natural Resources Research, Chinese Academy of Sciences, Beijing 100101, China

Dennis R. Parent
DRP Forestry, Hayden, ID 83835, USA

Mark D. Coleman
Forest, Rangeland and Fire Sciences, University of Idaho; Moscow, ID 83844, USA

Fangchao Wang and Xiangmin Fang
Jiangxi Provincial Key Laboratory of Silviculture, Jiangxi Agricultural University, Nanchang 330045, China

Fusheng Chen, Rong Mao and Wensheng Bu
Jiangxi Provincial Key Laboratory of Silviculture, Jiangxi Agricultural University, Nanchang 330045, China
Jiulianshan National Observation and Research Station of Chinese Forest Ecosystem, Key Laboratory of National Forestry and Grassland Administration on Forest Ecosystem Protection and Restoration of Poyang Lake Watershed, Jiangxi Agricultural University, Nanchang 330045, China

G. Geoff Wang
Department Forestry and Environmental Conservation, Clemson University, Clemson, SC 29634, USA

Huimin Wang
Qianyanzhou Ecological Station, Key Laboratory of Ecosystem Network Observation and Modeling, Institute of Geographic Sciences and Natural Resources Research, Chinese Academy of Sciences, Beijing 100101, China

Huimei Wang
Key Laboratory of Forest Plant Ecology, Northeast Forestry University, Harbin 150040, China

Wenjie Wang
Key Laboratory of Forest Plant Ecology, Northeast Forestry University, Harbin 150040, China
Northeast Institute of Geography and Agricultural Ecology, Chinese Academy of Sciences, Changchun 130102, China

Scott X. Chang
Department of Renewable Resources, University of Alberta, 442 Earth Science Building, Edmonton, AB T6G 2E3, Canada

Jay E. Raymond, Thomas R. Fox and Brian D. Strahm
Department of Forest Resources and Environmental Conservation, Virginia Polytechnic Institution and State University, Blacksburg, VA 24061, USA

Haixue Bian, Qinghong Geng, Hanran Xiao, Honghua Ruan, Zheng Shi and Xia Xu
College of Biology and the Environment, Nanjing Forestry University, Nanjing, Jiangsu 210037, China
Co-Innovation Centre for Sustainable Forestry in Southern China, Nanjing Forestry University, Nanjing, Jiangsu 210037, China

Caiqin Shen
Dongtai Forest Farm, Yancheng, Jiangsu 224200, China

Qian Li
Advanced Analysis and Testing Center, Nanjing Forestry University, Nanjing, Jiangsu 210037, China

Xiaoli Cheng
Key Laboratory of Aquatic Botany and Watershed Ecology, Wuhan Botanical Garden, Chinese Academy of Sciences, Wuhan 430074, China

Yiqi Luo
Center for Ecosystem Science and Society, Northern Arizona University, Flagstaff, AZ 86011, USA

Juha M. Metsaranta and Jagtar S. Bhatti
Natural Resources Canada, Canadian Forest Service, Northern Forestry Centre, 5320 122 Street, Edmonton, AB T6H 3S5, Canada

Yu Cong, Kai Liu, Yong-Cai Dang and Hu-Dong Han
School of Geographical Sciences, Northeast Normal University, Changchun 130024, China

Mai-He Li
School of Geographical Sciences, Northeast Normal University, Changchun 130024, China
Swiss Federal Institute for Forest, Snow and Landscape Research WSL, Zuercherstrasse 111, CH-8903 Birmensdorf, Switzerland

Hong S. He
School of Geographical Sciences, Northeast Normal University, Changchun 130024, China
School of Natural Resources, University of Missouri, Columbia, MO 65211, USA

Masazumi Kayama
Kyushu Research Center, Forestry and Forest Products Research Institute, Kumamoto 860-0862, Japan

Takashi Yamanaka
Forestry and Forest Products Research Institute, Tsukuba 305-8687, Japan

Yang Liu
College of Forestry, Nanjing Forestry University, Nanjing 210037, China
Department of Forest and Conservation Sciences, University of British Columbia, 3041-2424 Main Mall, Vancouver, BC V6T 1Z4, Canada

Qingliang Liu
College of Forestry, Nanjing Forestry University, Nanjing 210037, China

Tongli Wang
Department of Forest and Conservation Sciences, University of British Columbia, 3041-2424 Main Mall, Vancouver, BC V6T 1Z4, Canada

Shengzuo Fang
College of Forestry, Nanjing Forestry University, Nanjing 210037, China
Co-Innovation Center for Sustainable Forestry in Southern China, Nanjing Forestry University, Nanjing 210037, China

Jing Guo, Yaqiong Wu, Bo Wang, Yan Lu, Fuliang Cao and Guibin Wang
Nanjing Forestry University, Co-Innovation Center for Sustainable Forestry in Southern China, 159 Longpan Road, Nanjing 210037, China

Dalong Jiang
College of Biology and the Environment, Co-Innovation Center for Sustainable Forestry in Southern China, Nanjing Forestry University, Nanjing 210037, China
School of Forest Resources and Conservation, University of Florida, Gainesville, FL 32608, USA

Jason Vogel
School of Forest Resources and Conservation, University of Florida, Gainesville, FL 32608, USA

Index

Printed in the USA
CPSIA information can be obtained
at www.ICGtesting.com
LVHW080041211023
761596LV00010BA/1853

9 781639 897568